R. L. Timings
C.Eng., F.I.P.C., M.I.Prod.E., M.B.I.M.

Workshop processes and materials
Level 1
(Second edition)

Longman London and New York

Longman Group Limited
Longman House, Burnt Mill, Harlow,
Essex CM20 2JE, England
Associated companies throughout the world

*Published in the United States of America
by Longman Inc., New York*

© Longman Group Limited, 1978, 1984

*First published 1978
Reprinted 1980, 1982
Second edition 1984*

British Library Cataloguing in Publication Data

Timings, R. L.
    Workshop processes and materials, Level 1.
    —— 2nd ed
    1. Machine-shop practice
    I. Title
    621.7'5   TJ1125
    ISBN 0—582—41354—0

Printed in Singapore by Selector Printing Co

Longman Technician Series

Mechanical and Production Engineering

General Editors — Mechanical and Production Engineering

H. G. Davies,
Vice Principal, Head of Department of Science, Carmarthen Technical and Agricultural College
G. A. Hicks,
Lecturer in the Department of Engineering, Carmarthen Technical and Agricultural College

Books published in Mechanical and Production Engineering sector:

Manufacturing technology Level 2 Second edition   R. L. Timings
Manufacturing technology Level 3 Second edition   R. L. Timings
Engineering drawing for technician engineers   J. D. Poole
Engineering science for mechanical technicians Level 2   J. O. Bird and A. J. C. May
Engineering science for mechanical technicians Level 3   J. O. Bird and A. J. C. May
Motor vehicle engineering drawing for technicians Level 1   S. J. Zammit
Motor vehicle engineering drawing for technicians Level 2   S. J. Zammit
Engineering science for technicians Level 1   D. R. Browning

Contents

Preface

Workshop Processes and Materials: Level I (Second edition), is the first in a series of books concerned with mechanical engineering workshop and manufacturing processes published in the Longman Technician Series. It has been written to satisfy the requirements of students following the Business & Technician Education Council's (B/TEC) standard unit U80/681. This unit covers the workshop processes and materials syllabus at level I in the mechanical and production engineering programme A.5.

In writing this book the author has had in mind the fact that first-year apprentices — both technician and craft — follow a similar programme of skill training in accordance with the requirements of the Engineering Training Board. Therefore, this book not only covers the requirements of the TEC standard unit on workshop processes and materials but has been deliberately expanded in some areas to cover the theoretical knowledge that should complement the skill training of first-year mechanical engineering apprentices. In this way an attempt has been made to achieve a balance between the technician apprentice's academic requirements and the job-knowledge associated with his skill training.

Subsequent books in this series cover the requirements of B/TEC standard units: Manufacturing Technology I (unit U80/736); Manufacturing Technology II (unit U80/737), at levels II and III respectively. Further books by the same author cover Materials Technology at levels II and III, also Manufacturing Technology at level IV. This latter book being in the Longman Higher Technician Series.

Finally, I wish to thank my friends and colleagues who have assisted me in writing this book and checking the proofs; to the many companies who have provided up-to-date technical data; to the Series editors and the publishers for their help and advice in the preparation of the manuscript; to Mr F. J. M. Smith for allowing me to use some of his drawings in the book, and to Mrs Jean Smith for typing the manuscript and supporting documents.

R. L. Timings
1984

Acknowledgements

We are grateful to the following for permission to reproduce copyright material: Edward Arnold (Publishers) Ltd for Fig. 3.8 from *Metallurgy for Engineers* 4th edn. by E. C. Rollanson; *Electrical Times* for Fig. 1.9; Engineering Industry Training Board for Figs 1.2 and 1.8; International Tin Research Institute for Fig. 5.11; J. E. Baty and Co. Ltd for Fig. 7.21; Colchester Lathe Co. for Figs 12.21, 12.22 and 12.35; Duplex Electrical Machines for Figs 10.30 and 10.31; Jones and Shipman Co. For Fig. 12.40; Moore and Wright Ltd for Figs 7.5, 7.6, 7.8, 8.5 and 8.7; Rabone Chesterman for Figs 7.13 and 7.14; and the British Standards Institution, 2 Park Street, London W1A 2BS for the Figures from British Standards, complete copies of which are obtainable from the above address. Cover photograph by Paul Brierley.

Chapter 1

Workshop hazards

1.1 Introduction

Section 143(2) of the Factories Act states that: 'No person employed in a factory shall wilfully and without reasonable cause do anything likely to endanger himself or others.' Not only must an employer ensure that his premises and plant constitute a safe place to work, but every employee, trainee or experienced worker must ensure that he is a fit and capable person to carry out his or her assigned tasks. By law every employee must:

(a) obey all the safety rules of his or her place of employment;
(b) understand and use, as instructed, the safety practices incorporated in particular activities or tasks;
(c) not proceed with his or her task if any safety requirement is not understood. Guidance must be sought;
(d) maintain his or her working area tidy and maintain his or her tools in good condition;
(e) draw the attention of the safety officer or his or her immediate superior of any potential hazard;
(f) report all accidents or incidents (even if injury does not result from that incident) to the responsible person;
(g) understand the fire drill procedure in the event of an alarm;

to better suit a child with special needs. The special need does not have to be CP.

Research how this issue impacts the child and use this research to modify the equipment. You need to present your new or modified piece of equipment to the local city council to show them how it appeals to all children. The city council also wants to update the mural to include local heritage. Be sure to make sure your modifications meet safety requirements for our region. The budget for the park is $50,000, although you could be approved for a grant of an extra $20,000 which meets the cities "Bring Culture to Life" initiative. The parks modifications will be made to existing parks, so land cost is not an issue unless your plan extends the current footprint of the park.

In this assessment, students created both a drawing and a model to scale as well as using a program to design an accurate model of the playground equipment. This assessment is an authentic assessment because it represents a real-world product. The students will see how people would actually solve this problem. The rubric included specific criteria that helped the students understand what was required of them to complete the models and the drawings.

Authentic assessments are a way that teachers can support connected learning theory in their classrooms. Specifically, authentic assessments, when designed in a way to solve the problem, often tap into students' interests through choice, as well as tapping into the design principles of connected learning. These design principles include learning by doing, everyone can participate, challenge is constant, and everything is interconnected. In the playground example, the students were engaged in learning by doing activities by creating scale drawings. Everyone was able to participate because the students were divided into groups and assigned tasks to do to complete the project. Challenge was constant in that the students first created a drawing, then a design in CAD, then a 3D model. In this way, the tasks built on one another and increased in complexity. It modeled everything is interconnected in that the disciplines to solve the problem were connected. The students needed to use a variety of skills in the project. For example, when creating the scale drawings (ensuring that a specific scale was maintained throughout the project), they used engineering principles in the modeling, and they needed to understand how certain abilities could affect motor function and apply this understanding in their modeling. Authentic assessment also provides opportunity for fostering engagement and self-expression because typically students are able to "add their own spin" on the project, as one student, Max, described. He stated that he was able to demonstrate that he was good at building and art during the 3D modeling.

By fostering engagement and providing a space for self-expression, teachers are tapping into the multiple abilities that we described in Chapter 1. This is important so that students can showcase their strengths. However,

(*b*) understand how to give the alarm in case of fire;

(*i*) cooperate promptly with the senior person in charge in the event of an accident or a fire.

Therefore, for the young worker just entering industry, safety, health and welfare are very personal matters. It is the young worker who is most likely to be involved in an accident because of his inexperience.

This chapter sets out to identify the main hazards and discuss how they can be avoided. Some hazards require more detailed treatment and will be developed, as appropriate, in the later chapters of this book.

Factory life, and particularly engineering, is potentially dangerous, and a positive approach must be taken towards safety, health and welfare.

1.2   Accidents

Accidents do not happen, they are caused. If you are honest with yourself you cannot think of a single accident that could not have been prevented by care and forethought on somebody's part.

Accidents can and must be prevented. They cost millions of lost man-hours of production each year, but even this is of little importance compared with the immeasurable cost in human suffering.

In every eight-hour shift nearly 1000 workers are the victims of industrial accidents. Many of these will be blinded, maimed for life or confined to a

Fig. 1.1   Causes of industrial accidents

hospital bed for months. At least two of them will be killed. Figure 1.1 shows the main causes of these accidents.

1.3 Protective clothing

For general workshop purpose the boiler suit is the most practical and the safest form of body protection. However, to be completely effective certain

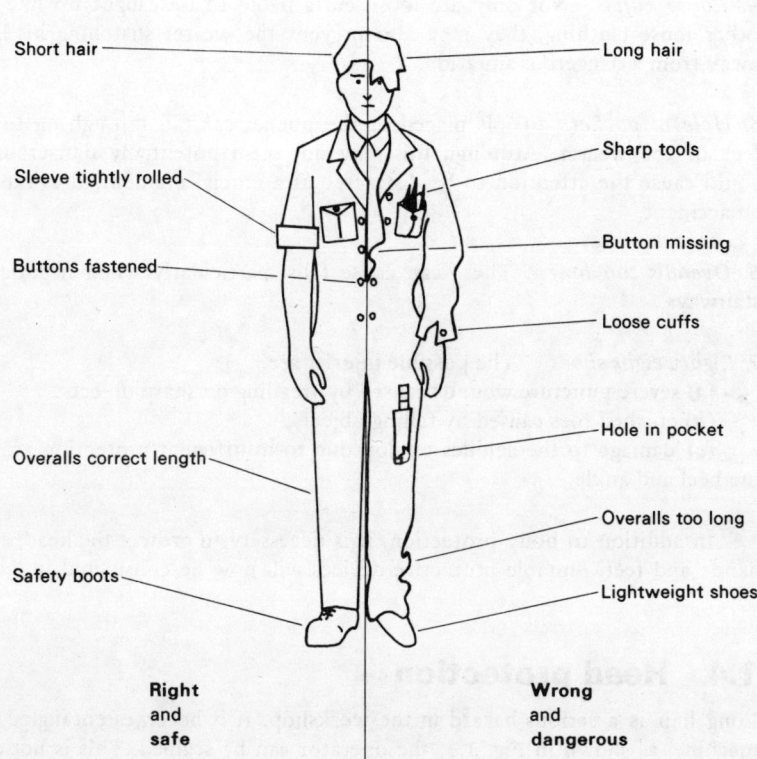

Fig. 1.2   Correct dress

precautions must be taken as shown in Fig. 1.2. The hazards indicated in Fig. 1.2 may be summarised as follows:

1. Long Hair: (*a*) Long hair is liable to be caught in moving machinery, particularly drilling machines and lathes. The resulting wound (i.e. the hair and scalp being torn away) is both extremely painful and dangerous. The brain damage may be permanent.

(*b*) Long hair is also a health hazard, as it is almost impossible to keep it clean and free from infection in the workshop environment. (See section 1.4.)

2. Sharp tools: Sharp tools protruding from the breast pocket can cause severe wounds to the wrist. Since the motor nerves of the fingers are near the surface in the wrist, these wounds can result in paralysis.

3. Button missing: Since the overall cannot be properly fastened, it becomes as dangerous as any other loose clothing and liable to be caught in moving machinery.

4. Loose cuffs: Not only are loose cuffs liable to be caught up like any other loose clothing, they may also prevent the wearer snatching his hand away from a dangerous situation.

5. Hole in pocket: Tools placed in the pocket can fall through on to the feet of the wearer. Although this may not seem potentially dangerous, it could cause the attention to be distracted at a crucial moment, thus causing an accident.

6. Overalls too long: These can cause falls, particularly when negotiating stairways.

7. Lightweight shoes: The possible injuries are:
    (*a*) severe puncture wounds caused by treading on sharp objects,
    (*b*) crushed toes caused by falling objects,
    (*c*) damage to the achilles tendon due to insufficient protection around the heel and ankle.

In addition to body protection, it is necessary to protect the head, eyes, hands and feet. Suitable protective devices will now be considered in detail.

1.4   Head protection

Long hair is a serious hazard in the workshop. If it becomes entangled in a machine, as shown in Fig. 1.3, the operator can be scalped. This is not only incredibly painful, but permanent brain damage and even death can result. If the young engineer persists in retaining a long hair style in the interests of fashion, then it must be contained within a suitable cap. This also helps to keep the hair and scalp clean and healthy.

1.5   Eye protection

Whilst it is possible to walk on a wooden leg, nobody has ever seen out of a glass eye. Therefore eye protection is possibly the most important safety precaution to take in the workshop. Eye injuries fall into three main categories:

(a) pain and inflammation due to abrasive grit and dust getting between the
lid and the eye;

(b) damage due to exposure to ultraviolet radiation and high intensity
visible radiation;

(c) loss of sight due to the eyeball being punctured or the optic nerve being
severed by flying splinters of metal.

In all instances protection is provided by wearing suitable goggles or
visors. Examples of such eye-wear are shown in Fig. 1.4.

1.6 Hand protection

The engineer's hands are in constant use and because of this they are at risk,
handling dirty, oily, greasy, rough, sharp, brittle, hot and maybe toxic and
corrosive materials. Gloves and 'palms' of a variety of styles and types of
materials are available to protect the hands, whatever the nature of the work.

Where gloves cannot be worn a 'barrier cream' should be rubbed into the
hands before commencing work. This is a mildly antiseptic, water-soluble
cream that fills the pores of the skin and prevents the ingress of dirt and
infection. The cream is easily removed by washing and carries away the dirt
and sources of infection.

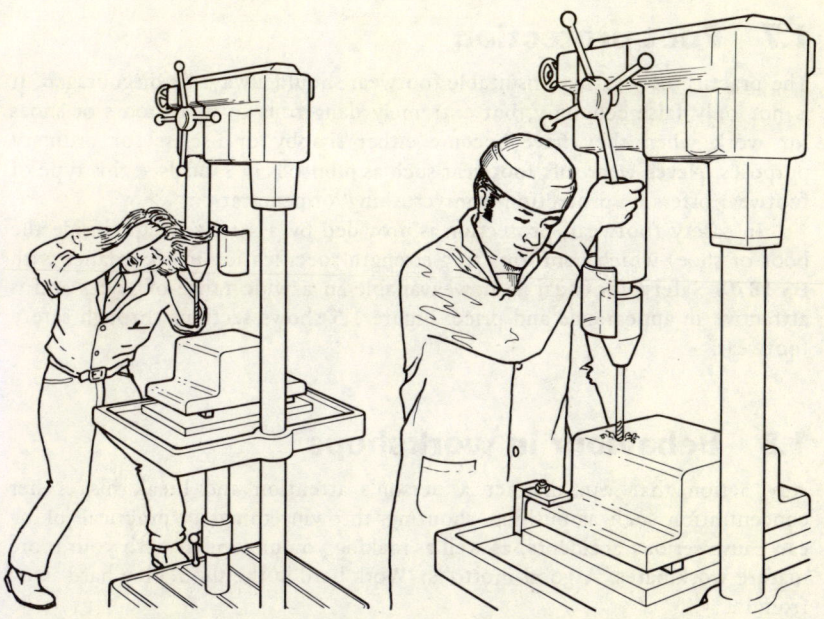

Fig. 1.3   The hazard of long hair

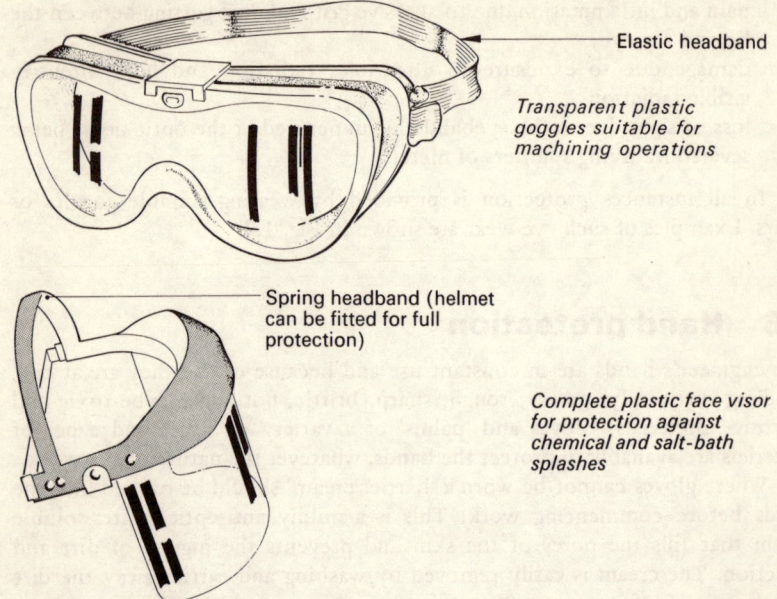

Elastic headband

Transparent plastic goggles suitable for machining operations

Spring headband (helmet can be fitted for full protection)

Complete plastic face visor for protection against chemical and salt-bath splashes

Fig. 1.4 Safety goggles and visors

1.7 Foot protection

The practice of wearing unsuitable footwear should always be discouraged. It is not only false economy, but extremely dangerous to wear boots or shoes for work when they have become either shabby or useless for ordinary purposes. **Never wear soft footwear such as plimsolls or sandals** − this type of footwear offers no protection from 'crushing' or penetration.

In safety footwear, protection is provided by a steel toe-cap (inside the boot or shoe) which conforms to a strength specification in accordance with BS 1870. Safety footwear is now available in a wide range of styles and is attractive in appearance and price. Figure 1.5 shows sections through safety footwear.

1.8 Behaviour in workshops

Any action that can distract a person's attention and break his or her concentration such as pushing, shouting, throwing things or practical joking can cause serious accidents, as well as making you unpopular with your more mature workmates. A good motto is: 'Work hard in the shop; play hard away from it.'

In the industrial environment horseplay infers reckless, foolish and boisterous behaviour by a person or group of persons that leads to accidents.

There is no place for such foolish activity in industry; yet it occurs in some way in every type of firm every day.

Horseplay observes no safety rules. It has no regard for white warning lines, interlocked guards, safety nets. It can defeat safe working procedures and undo the painstaking work of the safety officer by the sheer foolishness and thoughtlessness of the participants.

Types of accidents due to horseplay depend largely on the work of the factory concerned and the circumstances leading to the accident. Generally they are caused when someone is knocked against machinery or factory transport; when they fall against and dislodge heavy, stacked components; when electricity, compressed air and dangerous chemicals are involved.

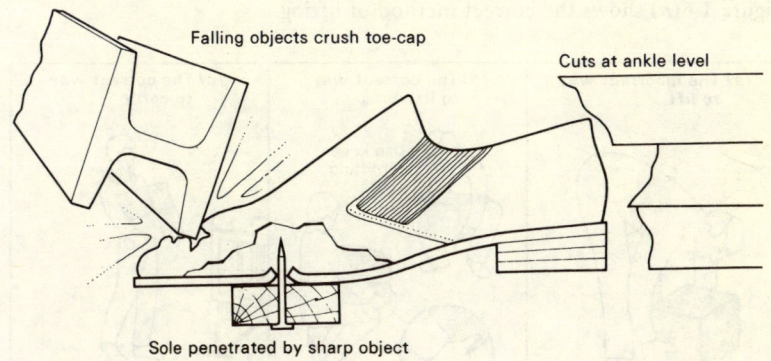

Falling objects crush toe-cap

Cuts at ankle level

Sole penetrated by sharp object

Light-weight shoes offer ***No*** *protection*

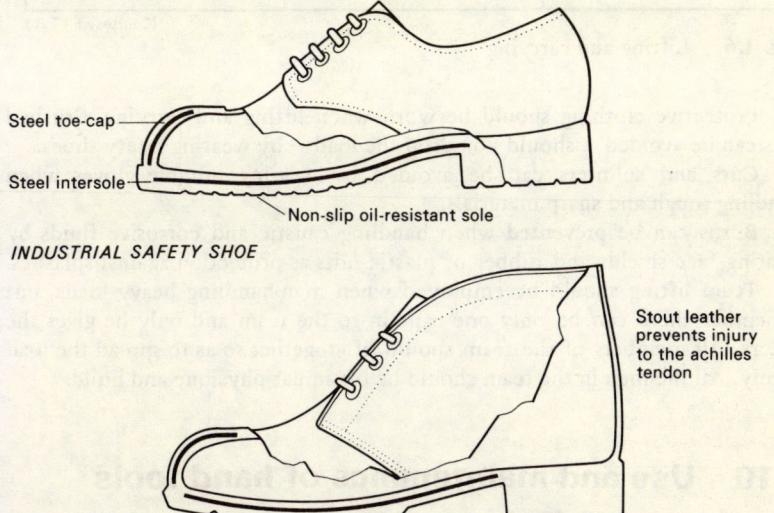

Steel toe-cap

Steel intersole

Non-slip oil-resistant sole

INDUSTRIAL SAFETY SHOE

Stout leather prevents injury to the achilles tendon

INDUSTRIAL SAFETY BOOT

Fig. 1.5   Safety footwear

1.9 Lifting and carrying

As will be seen from Fig. 1.1, the movement of materials is the biggest single cause of factory accidents. Manual handling accidents can be traced to one or more of the following:

(*a*)   incorrect lifting technique;
(*b*)   carrying too heavy a load;
(*c*)   incorrect gripping;
(*d*)   failure to wear protective clothing.

Figure 1.6(*a*) shows the wrong method of lifting that can lead to ruptures, strained backs, sprains, slipped discs and other painful injuries. Figure 1.6(*b*) shows the correct method of lifting.

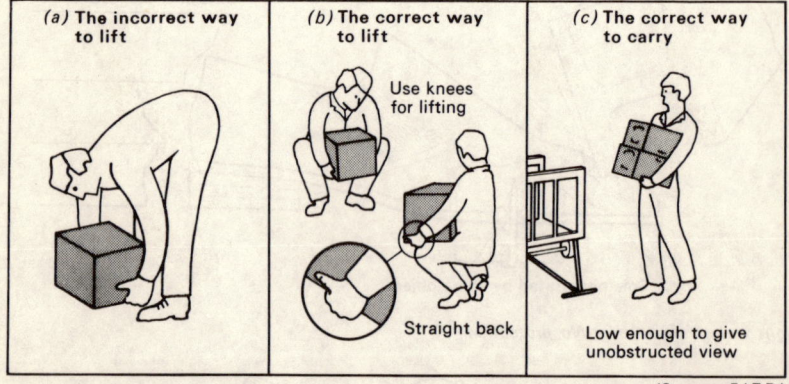

(Courtesy E.I.T.B.)

Fig. 1.6   Lifting and carrying

Protective clothing should be worn when lifting and carrying. Crushed toes can be avoided — should you drop the load — by wearing safety shoes.

Cuts and splinters can be avoided by wearing suitable gloves when handling rough and sharp materials.

Burns can be prevented when handling caustic and corrosive fluids by wearing face shields and rubber or plastic suits as protection against splashes.

Team lifting should be employed when man-handling heavy loads, but remember there can be only **one** captain to the team and only he gives the orders. All members of the team should lift together so as to spread the load evenly. All the men in the team should be of similar physique and build.

1.10   Use and maintenance of hand tools

The newcomer to industry often does not realise the potential danger existing in badly maintained and incorrectly used tools. Unfortunately he is often influenced by older men — who should know better — misusing the more

common hand tools often through laziness. The time and effort taken in 9
fetching the correct tool from the stores or in servicing a worn tool is
considerably less than the time taken in convalescing from an injury. Figure
1.7 shows some of the hazards arising from the incorrect use of hand tools.

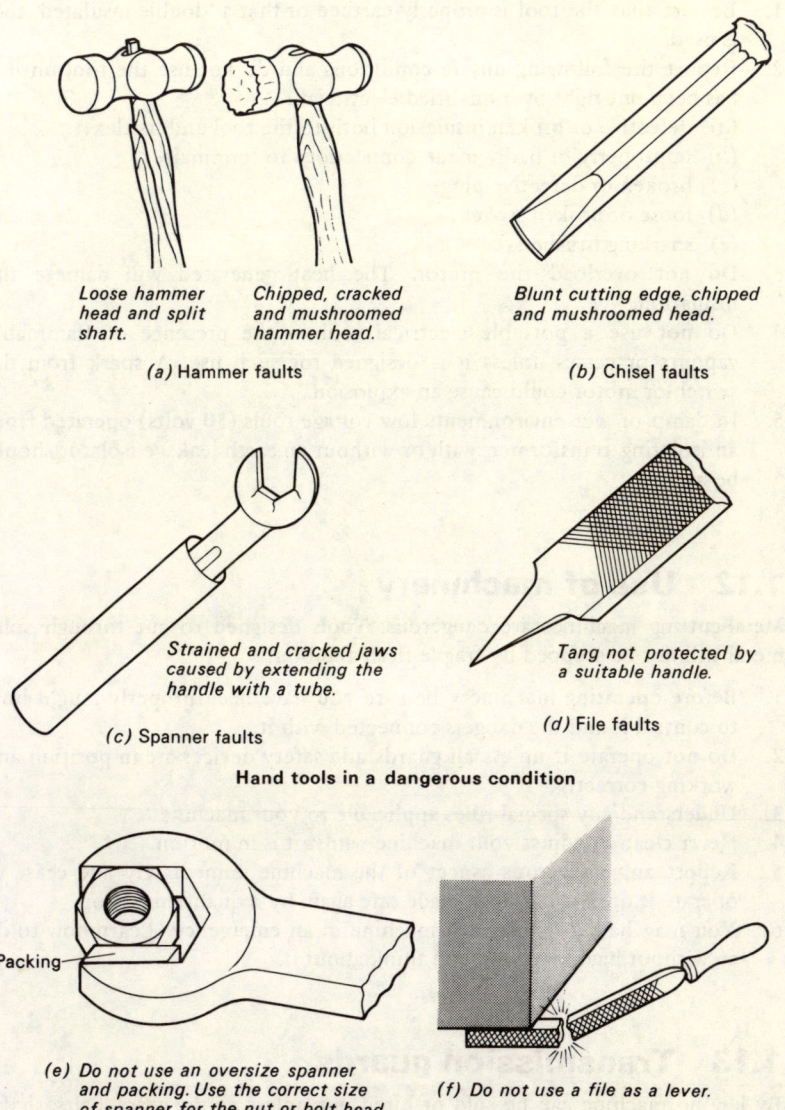

Loose hammer
head and split
shaft.

Chipped, cracked
and mushroomed
hammer head.

(a) Hammer faults

Blunt cutting edge, chipped
and mushroomed head.

(b) Chisel faults

Strained and cracked jaws
caused by extending the
handle with a tube.

(c) Spanner faults

Tang not protected by
a suitable handle.

(d) File faults

Hand tools in a dangerous condition

Packing

(e) Do not use an oversize spanner
and packing. Use the correct size
of spanner for the nut or bolt head.

(f) Do not use a file as a lever.

Misuse of hand tools

Fig. 1.7 Dangers in use of hand tools

1.11   Hand-operated power tools

In addition to maintaining the tools in good mechanical order, it is imperative that the equipment is electrically sound or a fatal electric shock can occur to the user.

1. Be sure that the tool is properly earthed or that a 'double insulated' tool is used.
2. Report the following unsafe conditions and do not use the tool until it has been put right by a qualified electrician:
    (*a*) defective or broken insulation both to the tool and its flex;
    (*b*) improperly or badly made connections to terminals;
    (*c*) broken or defective plug;
    (*d*) loose or broken switch;
    (*e*) sparking brushes.
3. Do not overload the motor. The heat generated will damage the insulation.
4. Do not use a portable electrical tool in the presence of flammable vapours or gasses unless it is designed for such use. A spark from the switch or motor could cause an explosion.
5. In damp or wet environments low voltage tools (50 volts) operated from an isolating transformer, with or without an earth leakage isolator should be used.

1.12   Use of machinery

Metal-cutting machines are dangerous. Tools designed to cut through solid metal will not be stopped by fragile flesh and bone.

1. Before operating machinery be sure you have been properly taught how to control it and the dangers connected with it.
2. Do not operate it unless all guards and safety devices are in position and working correctly.
3. Understand any special rules applicable to your machine.
4. Never clean or adjust your machine whilst it is in motion.
5. Report any dangerous aspect of the machine immediately and cease to operate it until it has been made safe again by a qualified person.
6. You may have to stop your machine in an emergency. Learn how to do so without having to stop and think about it.

1.13   Transmission guards

By law no machine can be sold or hired out unless all the gears, belts, shafts and couplings making up the transmission equipment are guarded so that they cannot be touched whilst in motion.

Sometimes guards have to be removed to service the components they are covering. Before removing the guards or covers:

(*a*)   stop the machine;
(*b*)   isolate the machine from its energy supply;
(*c*)   lock the isolating switch so that no one can turn it on again whilst you are working on the exposed equipment. Keep the key in your pocket;
(*d*)   if it is not possible to lock the isolating switch, then draw the fuses and keep them in your pocket.

If an 'interlocked' guard is removed, an electrical or mechanical trip will stop the machine. This trip is only provided in case you forget to isolate the machine and is not a substitute for full isolation.

1.14 Cutter guards

The machine manufacturer does not normally provide cutter guards, because of the wide range of work a machine may have to do.
1. It is the responsibility of the owner or hirer of the machine to supply his own cutter guards.
2. It is the responsibility of the operator to make sure the guards are fitted and working correctly before operating his machine.
3. The craftsman may have to fit and adjust a guard for an unskilled operator. This is a great responsibility so make sure you understand thoroughly how the guard should function.
4. If ever you are doubtful about the adequacy of a guard or the safety of a process, consult your safety officer immediately.

1.15 Use of grinding wheels

Because of its apparent simplicity the double-ended off-hand grinding machine comes in for more than its fair share of abuse. If it is realised that the grinding wheel does not 'rub off' the metal, but is a precision multi-tooth cutter in which each grain has a definite geometry and produces chips just like a milling cutter, then it must be mounted and used correctly. The reason for this is that a damaged or improperly mounted grinding wheel can burst at speed and cause serious injury. For this reason the mounting of grinding wheels, except by persons trained and registered for the purpose under the Abrasive Wheel Regulations: 1970, is prohibited.

1.16 Electrical hazards

The mechanical engineer is well advised to leave the installation and maintenance of electrical equipment to the specialist. However, he should still have an understanding of electrical installations and protective devices so as to avoid the misuse of equipment provided for his safety.

An electrical fault can cause two kinds of accident:

(a) shock;
(b) fire.

An electric shock from a factory power supply can easily kill a man or woman. Even if the shock is not this severe, the convulsion it causes can throw the victim from a ladder or against a machine and this can result in serious injury or even death. When pulling the victim clear of the fault that caused the shock be very careful, as it is possible to receive a shock from the victim. Always pull the victim clear by his or her clothing, if it is dry, or otherwise use a dry sack or other insulating material. **Never** touch the flesh of the victim, until he or she is clear of the fault, as flesh can act as a conductor. Figure 1.8 shows some causes of shocks.

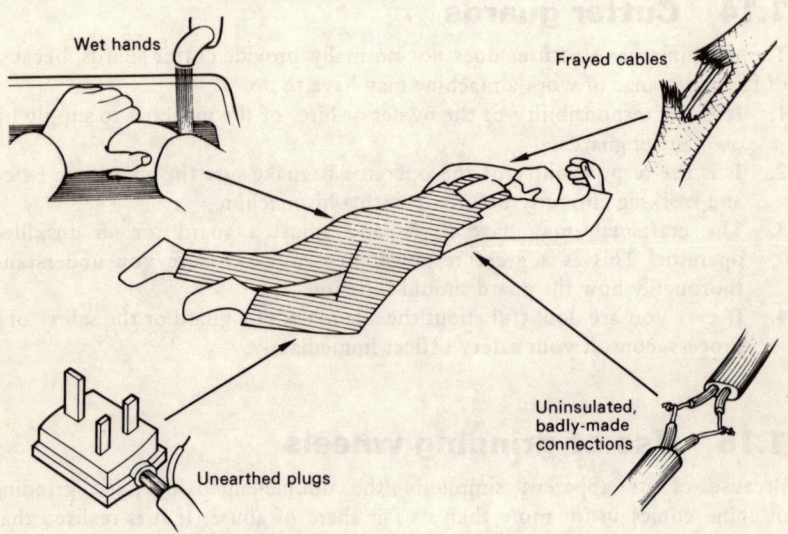

Fig. 1.8   Causes of electric shock

Artificial respiration must be commenced immediately the victim is pulled clear from the live conductor. Figure 1.9 – reproduced by courtesy of *Electrical Times* – gives details of how to treat the victim of a severe shock.

1.17   Earthing electrical equipment

Electrical switchgear and appliances are made in two ways:

(a) fully insulated;
(b) metal clad.

Most domestic and office equipment is fully insulated. That is, all current-carrying parts are enclosed in an insulated (usually plastic) casing so that in normal use it is not possible to come into contact with the current. This type of apparatus cannot be earthed neither does it need earthing.

Most industrial equipment is metal-clad. That is, it is encased in a metal container that may be part of the structure or may merely be giving mechanical protection (sheathing) to the equipment. This metal casing must be carefully earthed.

1.18   The control of electric motors

Nowadays, virtually all workshop machinery is driven by electric motors. Under section 12 of the Factories Act, paragraph (3): 'every part of an electric motor shall be securely fenced unless in such a position or of such construction as to be safe to every person employed or working on the premises'.

In addition, under section 13 of the Act: 'Devices or appliances for promptly cutting off the power from the transmission machinery must be provided in every room or place where work is carried on.'

In addition, where the power source is an electric motor, Regulation 12 of the memorandum on the Electricity Regulations of the Factories Act also applies. This states:

(a)   every electrical motor shall be controlled by an efficient switch or switches for starting and stopping, so placed as to be easily worked by the person in charge of the motor;

(b)   in every place in which machines are being driven by any electric motor there shall be means at hand for either switching off or stopping the machines if necessary to prevent danger (emergency stop).

From the regulations quoted above it will be apparent that three stages of control are required.

1.   The press button starter used for routine starting and stopping of the motor.
2.   Emergency stop buttons provided at strategic points around the machine and workshop.
3.   An isolating switch for each motor in order that the supply can be disconnected whilst the machine is being serviced, cleaned or adjusted.

Starting and stopping

Figure 1.10 shows a direct on-line (DOL) starter suitable for an alternating current induction motor.

In addition to providing push-button control of the motor it also provides two protective devices.

'No-volt' release: Since the contacts are closed by a solenoid, any failure in the supply causes the magnetic field of the solenoid to collapse and the contacts are released. The supply cannot be reconnected to the motor until the start button is again operated, re-energising the solenoid.

Order of action

1 Switch off current
Do this immediately. If not possible do not waste time searching for the switch.

2 Secure release from contact
Safeguard yourself when removing casualty from contact. Stand on non-conducting material (rubber mat, DRY wood, DRY linoleum). Use rubber gloves, DRY clothing, a length of DRY rope or a length of DRY wood to pull or push the casualty away from the contact.

3 Start artificial respiration
If the casualty is not breathing artificial respiration is of extreme urgency. A few seconds delay can mean the difference between success or failure. Continue until the casualty is breathing satisfactorily or until a doctor tells you to stop.

4 Send for doctor and ambulance
Tell someone to send for a doctor and ambulance immediately and say what has happened. Do not allow the casualty to exert himself by walking until he has been seen by a doctor. If burns are present, ask someone to cover them with a dry sterile dressing.

If you have difficulty in blowing your breath into the casualty's lungs, press his head further back and pull chin further up. If you still have difficulty, check that his lips are slightly open and that the mouth is not blocked, for example, by dentures. If you still have difficulty, try the alternative method, mouth-to-mouth, or mouth-to-nose, as the case may be.

Method - mouth-to-mouth

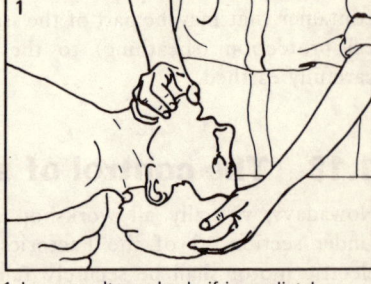

1. Lay casualty on back; if immediately possible, on a bench or table with a folded coat under shoulders to let head fall back. Kneel or stand by casualty's head. Press his head fully back with one hand and pull chin up with the other.

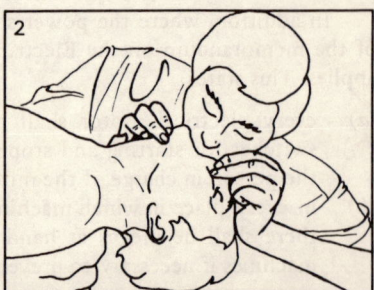

2. Breathe in deeply. Bend down, lips apart and cover casualty's mouth with your well open mouth. Pinch his nostrils with one hand. Breathe out steadily into casualty's lungs. Watch his chest rise.

3. Turn your own head away. Breathe in again.

Repeat 10 to 12 times per minute.

Fig. 1.9   Treatment for electric shock

If the patient does not respond proceed as follows:

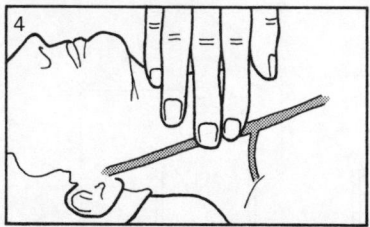

 NORMAL PUPILS  DILATED PUPILS

4. Check carotid pulse, pupils of eyes and colour of skin (see Fig.4)

5. Pulse present, pupils normal - continue inflations until recovery of normal breathing (Figs 1, 2 and 3)

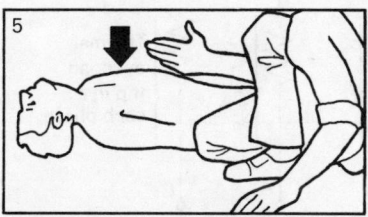

6. Pulse absent, pupils dilated, skin grey - strike smartly to the left part of breast bone with edge of hand (see Fig.5)

7. Response of continued pulse, pupils contract - continue inflations until recovery of normal breathing

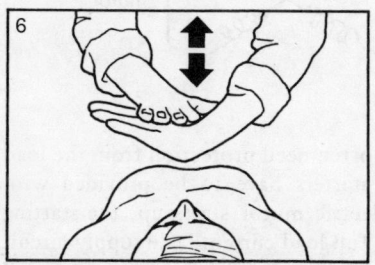

8. No response of continued pulse, pupils unaltered, skin grey - commence external heart compression (see Fig.6) *

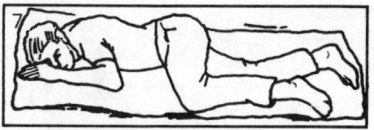

9 When normal breathing commences, keep warm, place casualty in the recovery position (see Fig. 7)

* Method - external heart compression (Fig.6)

1. Place yourself at the side of the casualty

2. Feel for the lower half of the breastbone

3. Place the heel of your hand on this part of the bone, keeping the palm and fingers off the chest

4. Cover this hand with the heel of the other hand

5. With arms straight, rock forwards pressing down on the lower half of the breastbone (in an unconscious adult it can be pressed towards the spine for about one and a half inches (4 cm))

6. The action should be repeated about once a second

Continue as above until a continued pulse is felt and pupils contract

Continue inflations until recovery of normal breathing

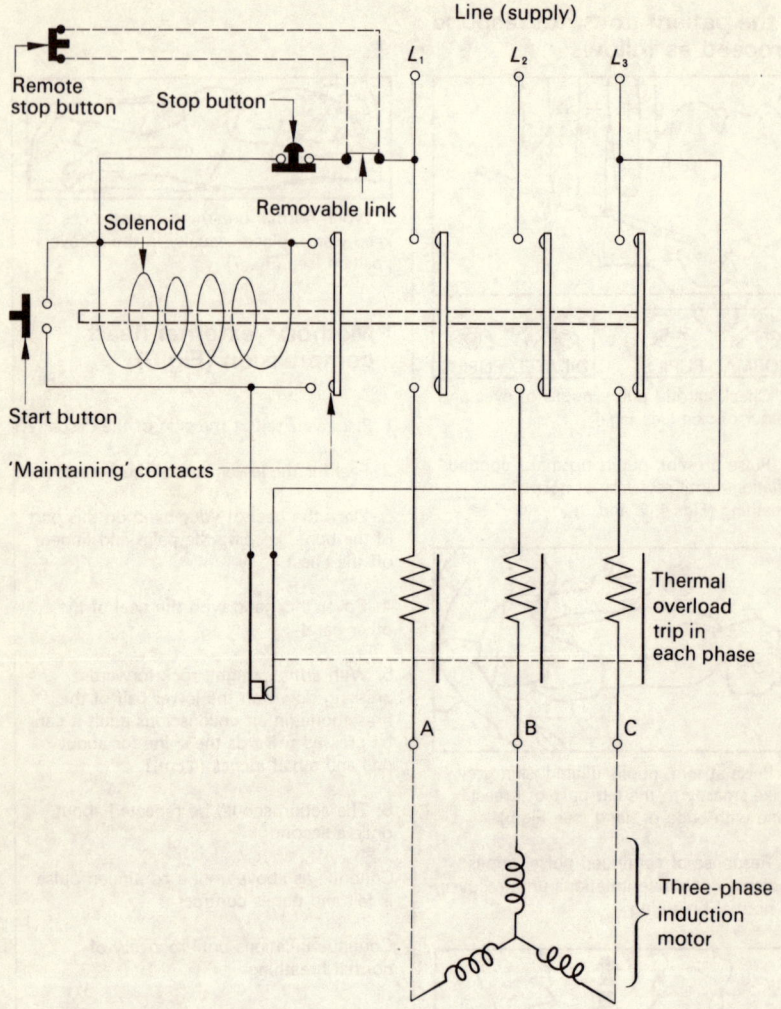

Fig. 1.10(a) Direct on-line motor starter

Thermal overload trip: Electric motors often need protection from the load they are driving. Therefore all motor starters have to be provided with **overload protection devices.** When an electric motor starts up, the starting current may surge up to many times the full load current. The supply circuit and the fuses will be designed to withstand the starting surge. The motor will also withstand the surge, but for a short time only. The fuses, having too high a rating to protect the motor from a slight but continuous overload, may allow it to overheat and burn out. Hence the need for an overload protection device. Figure 1.10(b) shows a typical arrangement based on a bi-metal strip. It is so designed that it takes an overload lasting several seconds to make it

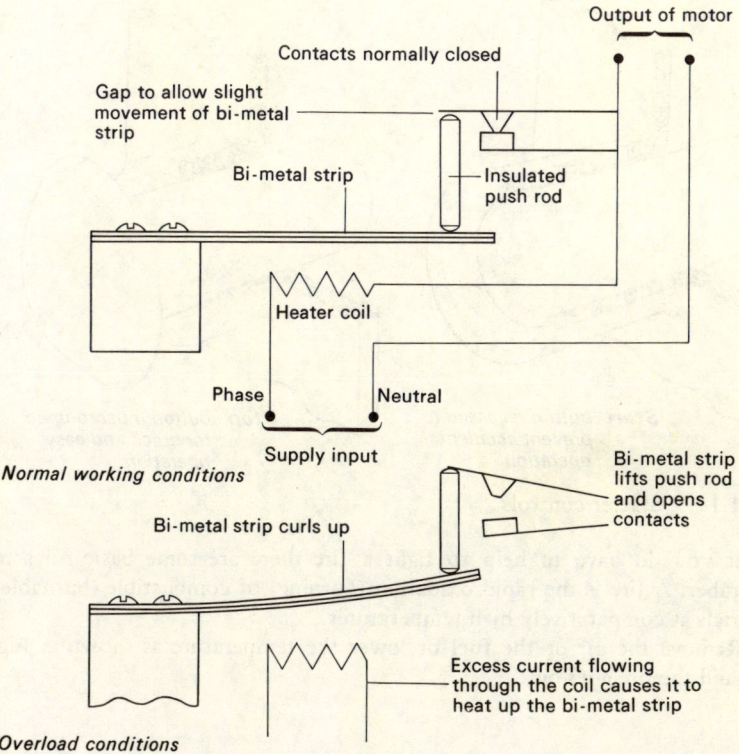

Output of motor

Contacts normally closed

Gap to allow slight
movement of bi-metal
strip

Bi-metal strip

Insulated
push rod

Heater coil

Phase   Neutral

Supply input

Normal working conditions

Bi-metal strip
lifts push rod
and opens
contacts

Bi-metal strip curls up

Excess current flowing
through the coil causes it to
heat up the bi-metal strip

Overload conditions

Fig. 1.10(b)   Thermal overload release

operate and is too sluggish to respond to the brief starting surge.

Even the shape and positioning of the control buttons can affect the
safety of operation and some examples are shown in Fig. 1.11.

1.19   Fire fighting

Fire fighting is a highly skilled operation and most medium and large size
firms have properly trained teams who can fight the fire locally or at least
contain it until the professional service arrives.

The best way you can help is to learn the correct fire drill; both how to
give the alarm and how to leave the building. It only requires one person to
panic and run in the wrong direction to cause a disaster.

In an emergency never lose your head.

Smoke is the main cause of panic. It spreads quickly through a building,
reducing visibility and increasing the risk of falls down staircases. It causes
choking and even death by asphyxiation. Always keep fire doors in corridors
and staircases closed but never locked.

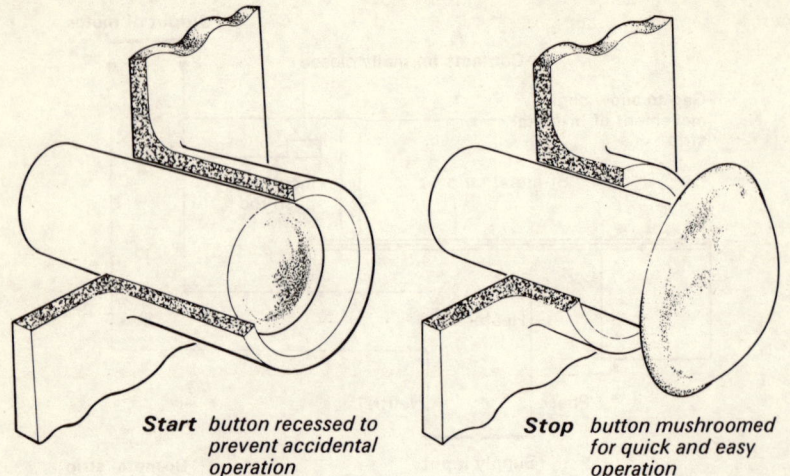

Start button recessed to prevent accidental operation

Stop button mushroomed for quick and easy operation

Fig. 1.11   Starter controls

If you do have to help to fight a fire there are some basic rules to remember. A fire is the rapid oxidation (burning) of combustible (burnable) materials at comparatively high temperatures.

Remove the air or the fuel or lower the temperature as shown in Fig. 1.12 and the fire goes out.

Table 1.1   Fires and extinguishers

Water
Used in large quantities this reduces the temperature and puts out the fire. The steam generated also helps to smother the flames. However, for various technical reasons it should only be used on burning solids such as wood, paper, some plastics, etc.

Foam extinguishers
These are used for fighting oil and chemical fires and act by smothering the flames and preventing the oxygen in the air from feeding the fire.

Note: Since both water and foam are conductive, do not use on electrical fires or the person wielding the hose or the extinguisher will be electrocuted.

Carbon dioxide extinguishers
Used on burning gases and vapours. It displaces the air and stops the fire. To be effective it must be used in a confined space so that it isn't dispersed by drafts.

Note: If the fire can't breath neither can you, so care must be taken to evacuate living creatures from the vicinity before operating the extinguisher.

Vapourising liquid extinguishers

These cover CTC, CBM and BCF extinguishers. The heat from the fire causes rapid vaporisation of the spray from the extinguisher and this smothers the fire. It will also smother you unless precautions are taken. This type of extinguisher is used on gas and vapour fires. Like carbon dioxide they are safe to use on electrical fires.

Dry powder extinguishers

These are suitable for small fires involving flammable liquids and some solids such as paper. They are suitable for offices and canteens as they leave little residual mess to clean up, neither do they contaminate anything with which they come into contact.

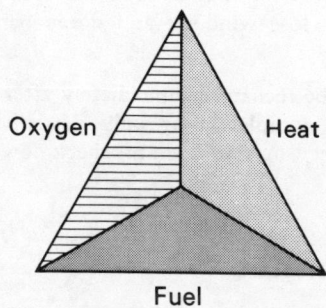

These are the three 3 essentials to start a fire

Note : *Once the fire has started it produces sufficient heat to maintain its own combustion reactions and sufficient surplus heat to spread the fire.*

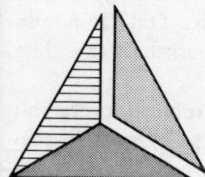

When solids are on fire remove heat by applying water.

(a) **Remove heat**

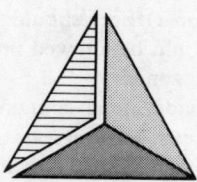

Liquids, such as petrol etc. on fire can be extinguished by removing oxygen with a foam or dry powder extinguisher.

(b) **Remove oxygen (air)**

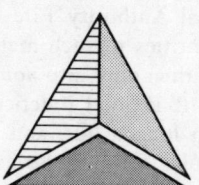

Electrical or gas fires can usually be extinguished by turning off the supply of energy.

(c) **Remove energy source (fuel, gas, electricity etc.)**

Fig. 1.12   Fire fighting

It will be seen from Fig. 1.12 that different fires require to be dealt with in different ways. Table 1.1 correlates the main classes of fires to the normally available portable extinguishing equipment.

Fire spreads quickly; a speedy attack is essential. If anyone is near, tell him to report the outbreak and return to give you assistance.

Extinguishers are only to deal with small fires.

Take position between the fire and the exit, so that your escape cannot be cut off.

Do not continue to fight a fire:

(a)   if it is dangerous to do so;
(b)   if there is a possibility that the escape route may be cut off by fire or smoke;
(c)   if the fire continues to grow in spite of your efforts;
(d)   if there are gas cylinders or explosives in the vicinity of the fire. If you have to withdraw, close windows and doors behind you whenever possible.

Extinguishers should be recharged immediately after use, irrespective of whether they have been completely or only partially discharged. Some extinguishers must be turned over to operate; check now, and see how yours works.

1.20   Fire prevention

Prevention is always better than cure, and fire prevention is better than fire fighting. Tidiness is of paramount importance in reducing outbreaks of fire.

Fires have a small beginning and it is amongst rubbish left lying about that many fires originate. So make a practice of constantly removing rubbish, shavings and off-cuts; remove cans and any other unwanted materials at regular intervals to a safe place.

Highly flammable materials should be stored in specially designed and equipped compounds away from the main working areas. The advice of the Local Authority Fire Prevention Officer should be sought. Only minimum quantities of such materials should be allowed into the workshop at a time, and then only into *non-smoking* zones.

It is good practice to provide metal containers for rubbish, preferably with hinged lids, and with proper markings as to the type of rubbish they should contain. Some types of rubbish, when mixed, will ignite spontaneously. Take care.

1.21   Workshop layout

Much can be done to prevent accidents in the layout of the machines and equipment in a workshop.

Figure 1.13 shows a well laid out workshop. It will be seen that ample gangways are left and that these are clearly marked. The machines are arranged so that bar stock does not protrude into the gangway. They are also arranged so that the operators' attention is not distracted by other workers constantly passing close to them.

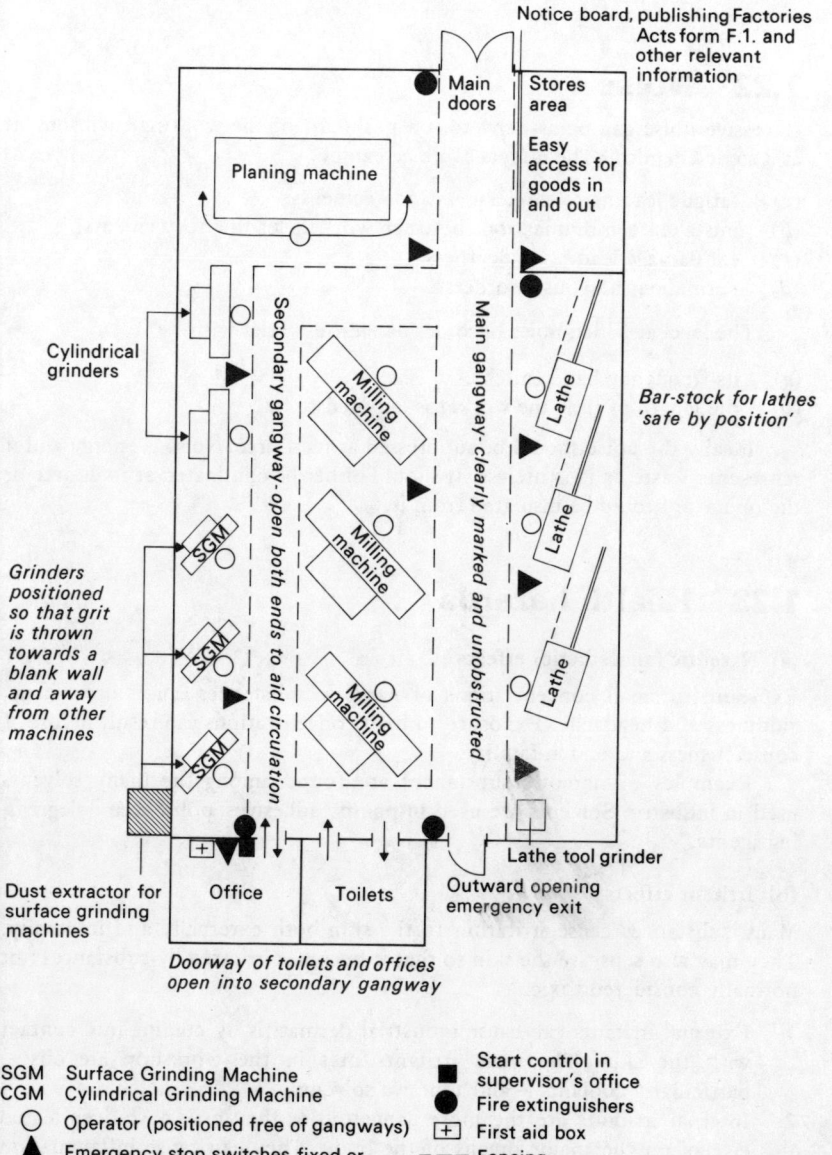

Fig. 1.13   Machine shop layout

Scale models of the more common types and makes of machine tools are available, and these can be used in conjunction with scale plans of the workshop floor for experimenting with different layouts in order to achieve an efficient and safe workshop.

1.22   Noise

Excessive noise can be as dangerous a pollutant of the working environment as a toxic chemical. The effects of noice can be:

(*a*)   fatigue leading to carelessness and accidents;
(*b*)   mistaken communications between workers leading to accidents;
(*c*)   ear damage leading to deafness;
(*d*)   permanent nervous disorders.

The level at which noise becomes dangerous varies with:

(*a*)   its frequency band (pitch);
(*b*)   the length of time the worker is exposed to it.

Ideally the noise should be suppressed at its source. Noise is energy and it represents waste as it is useless. It should either be eliminated at its source or the operator should be insulated from it.

1.23   Health hazards

(a) Narcotic (anaesthetic) effects

Exposure to small concentrations of narcotic substances causes drowsiness, giddiness and headache. Exposure to high concentrations can result in loss of consciousness and end in fatality.

Examples of narcotic substances are found among the many solvents used in industry. Solvents are used in paints, adhesives, polishes and degreasing agents.

(b) Irritant effects

Many substances cause irritation to the skin both externally and internally. They may also sensitise the skin so that it becomes irritated by substances not normally considered toxic.

1.   External irritants can cause industrial dermatitis by coming into contact with the skin. The main irritants met in the workshop are oils — particularly coolants — and adhesive solvents.
2.   Internal irritants are the more dangerous as they may have deep-seated effects on the major organs of the body. They can cause inflammation and ulceration, poisoning and the formation of cancerous tumours. Internal irritants are usually air pollutants in the form of dusts, fumes and vapours.

Substances known as systemics affect the fundamental organs and bodily functions. They affect the heart, the brain, the liver, the kidneys, the lungs, the central nervous system and the bone marrow. Their effect cannot be reversed and thus lead to chronic ill-health and ultimately an early death.

The toxic substance may enter the body in various ways.

1. Dust, vapours and gases can be breathed in through the nose.
2. Liquids and powders which contaminate the hands can be transferred to the digestive system by handling food or cigarettes with dirty hands.
3. Liquids, powders, dust, vapours may all enter the body through the skin:
   (a) directly through the pores;
   (b) by destroying the outer horny layers of the skin and attacking the sensitive layers underneath;
   (c) by entering through undressed puncture wounds.

It is important, therefore, to recognise the main hazards to be met with in engineering workshops so that adequate preventive measures can be taken.

1.24   Personal hygiene

Personal hygiene is most important. There is nothing 'cissy' about rubbing a barrier cream into your hands before work; about washing thoroughly with soap and hot water after work; about changing your overalls regularly so that they can be cleaned.

Personal hygiene can go a long way towards preventing skin diseases; both irritations and infections. In some processes, where gloves would hinder the finger dexterity required, a barrier cream is the only protection available.

Skin disease due to continual contact with oil forms the main health hazard in the engineering industry. The effects range from skin irritations and dermatitis to skin cancers.

The effects will depend on the nature of the oil, its temperature and the degree and length of time of exposure. They will also depend upon the condition of the skin; cuts and abrasions from swarf, irritation from additives and infection.

The first effect is usually simple irritation accompanied by redness and pimples. If treatment is not sought the condition deteriorates until cracking, scaling and skin growths may appear. Even in mild cases sensitisation of the skin may occur in which case the operator may have to change his job for one where oil and other irritants are not present.

Soluble oils (suds) are particularly difficult as the water content causes the skin to macerate (become soggy); this reduces its natural resistance.

If excessive contact with the oil cannot be reduced by modification to the plant or processes then additional water- and oil-proof protective clothing should be worn.

Problems

Section A

1. Goggles are worn when working with machine tools to: (a) improve vision; (b) prevent glare; (c) protect your eyes; (d) reduce eye fatigue.

2. A guard is provided on a drilling machine to: (a) protect the drill from damage; (b) prevent coolant being thrown out by the drill; (c) prevent swarf being thrown out by the drill; (d) prevent the operator from coming into contact with the drill.

3. The most suitable type of starter switch for a machine tool is operated by a: (a) mushroom headed push button; (b) shrouded (recessed) push button; (c) rotary knob; (d) tumbler knob.

4. The manufacturer of a machine tool is responsible for providing: (a) transmission guards only; (b) cutter guards only; (c) both transmission and cutter guards; (d) neither transmission nor cutter guards.

5. When lifting heavy loads you should: (a) bend your back and your knees; (b) keep your back and your knees straight; (c) keep your knees straight and bend your back; (d) keep your back straight and bend your knees.

Section B

6. Describe in detail six checks that should be made to ensure that a portable power tool is electrically safe to use.

7. Describe what precautions should be taken to ensure a reasonable level of personal hygiene when working in an engineering workshop. State **three** harmful effects that may result from not taking reasonable precautions to maintain personal cleanliness.

8. Explain why the following basic protective clothing should be worn when working in a machine shop: (a) boiler suit type overall; (b) reinforced protective shoes; (c) safety goggles.

9. (a) What type of fire extinguisher should be used when fighting: (i) burning oil; (ii) burning paper and rag; (iii) an electrical fire? (b) What action should you take if you discover a fire has started in a building where you are working?

10. A workmate has been rendered unconscious by a severe electric shock. Describe in reasonable detail how you would: (a) separate the victim from the source of the shock; (b) render artificial respiration.

Chapter 2

Engineering materials

2.1 Engineering materials

Almost every substance known to man has found its way into the engineering workshop at some time or other. Neither this chapter nor in fact the entire book could hold all the facts about all these materials. Therefore, it is intended to show how engineering materials can be grouped into families; then some of the more common materials will be taken from these groups and examined in greater detail.

Table 2.1 shows the main groupings of engineering materials. The specialist in each group could break them down further and, in fact, no engineer or scientist can be expert in all materials.

2.2 Properties

In order that the various materials can be compared properly, it is important to thoroughly understand the meaning of their more common properties. These properties are set out below. First, properties concerned with strength.

Tensile strength

The ability of a material to withstand a stretching load without breaking.

Table 2.1   Engineering materials

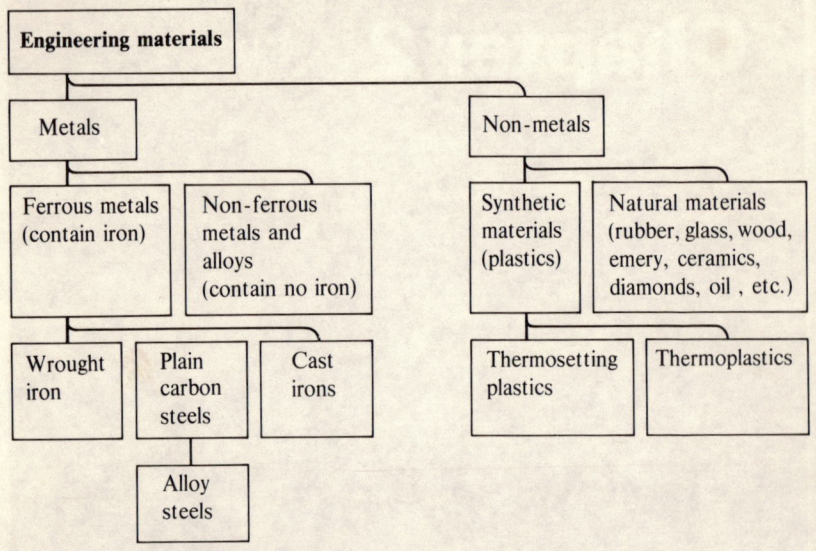

Compressive strength
The ability of a material to withstand a squeezing load without breaking.

Shear strength
The ability of a material to withstand off-set loads, or transverse cutting actions.

Toughness
The ability of a material to withstand an impact or hammering load.

These properties are illustrated in Fig. 2.1.

Second, the properties concerned with flow in the solid state will be considered. Such properties are highly important during manufacturing processes.

Elasticity
The ability of a material to deform under load and return to its original shape when the load is removed, as in a spring.

Plasticity
The ability of a material to deform under load and retain its new shape when the load is removed. When soft steel is bent it shows this property.

Ductility
This is the term used when the deforming load described in plasticity is a tensile load. As when wire-drawing.

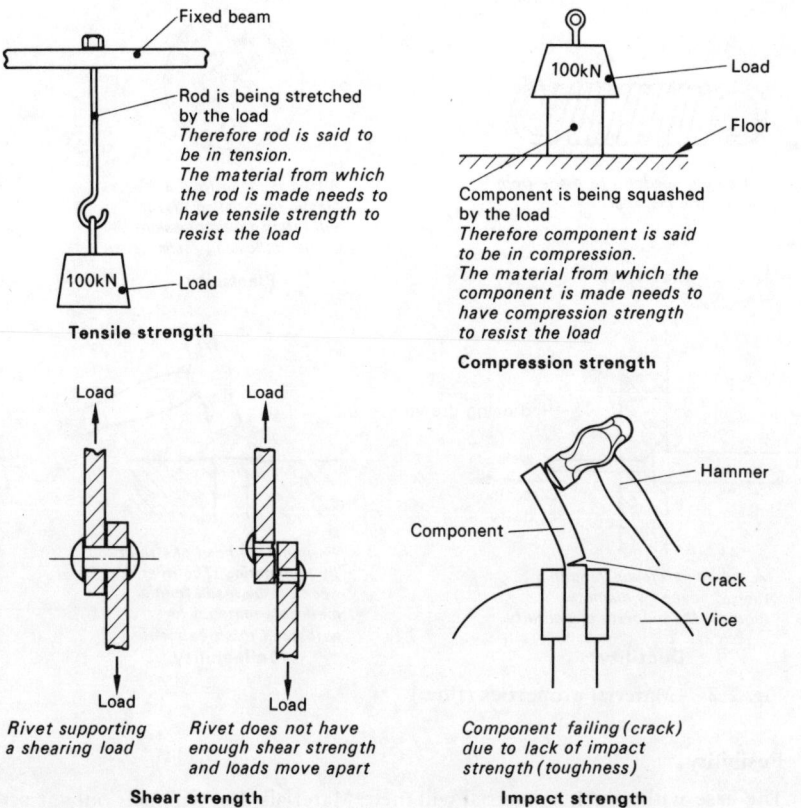

Fig. 2.1 Material properties (strength)

Malleability

This is the term used when the deforming load described in plasticity is a compressive load. As when forging.

These properties are shown in Fig. 2.2.

Third, the properties concerned with durability will be considered together with some miscellaneous properties.

Hardness

The ability to withstand scratching or indentation by another hard body.

Corrosion resistance

The ability to withstand chemical and electro-chemical attack.

Conductivity

(Electrical or thermal) refers to the ability of a material to allow the passage of electricity or heat.

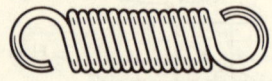

A spring needs to be made from an elastic material

Elasticity

A coin is made from a plastic material so that it will retain the impression of the embossing tools

Plasticity

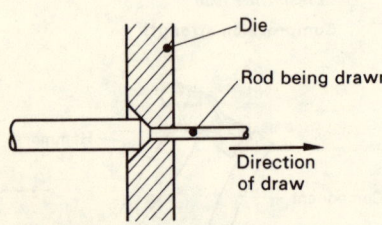

A rod being drawn through a die to reduce its diameter requires the property of ductility

Ductility

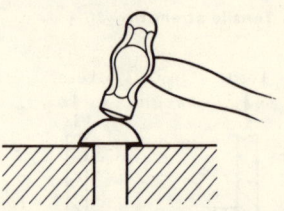

Forming the head of a rivet by hammering. The rivet needs to be made from a malleable material to withstand this treatment

Malleability

Fig. 2.2   Material properties (flow)

Fusibility

The ease with which a material will melt. Materials that will melt only at very high temperatures are called **refractories.**

These properties are shown in Fig. 2.3

2.3   Heat treatment

From time to time in this chapter it is necessary to refer to some basic heat treatment terms in relation to the properties of materials. These terms will be fully explained in Chapter 3, but in the meantime their meanings are briefly summarised in Table 2.2.

2.4   Hot- and cold-working

The terms hot-working and cold-working are used from time to time in this chapter and elsewhere in this book. The properties of materials change when they are subjected to hot- or cold-working so, although these processes are considered more fully in Chapter 3, it is appropriate to introduce these processes at this point.

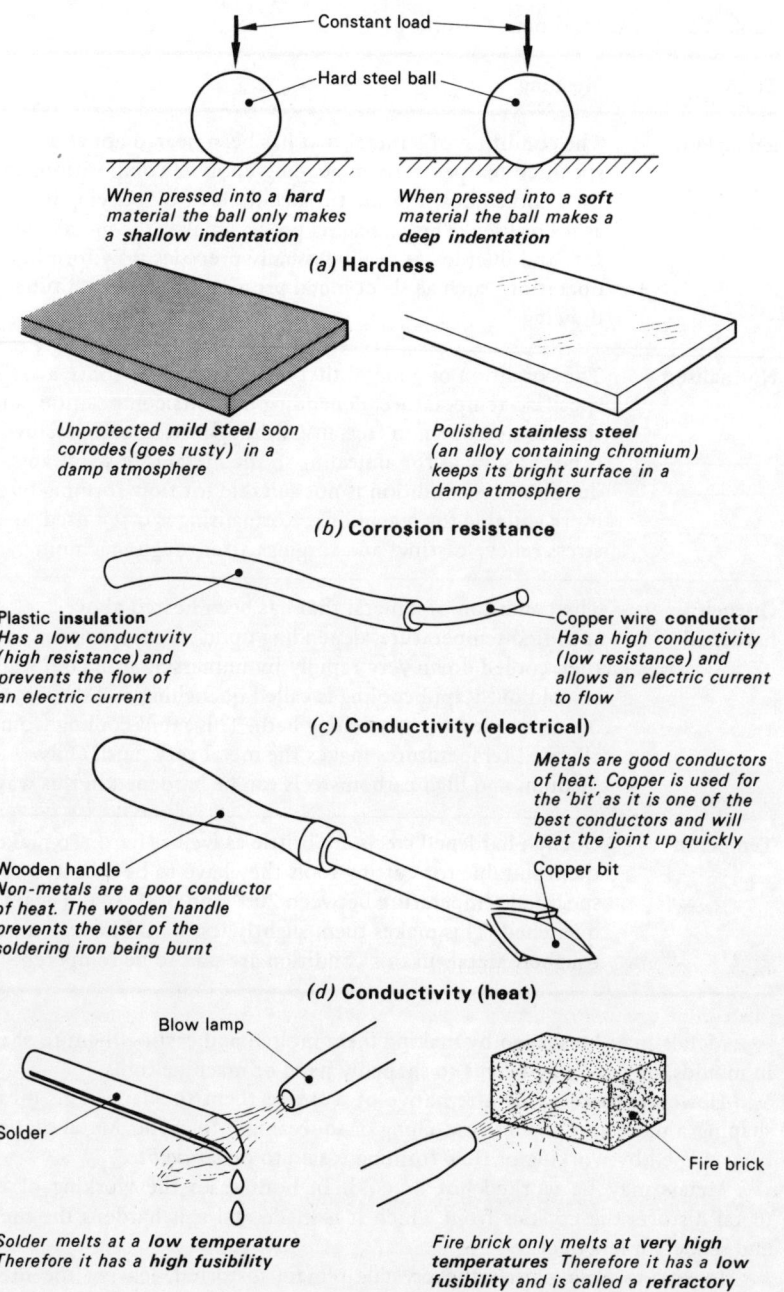

When pressed into a **hard**
material the ball only makes
a **shallow indentation**

When pressed into a **soft**
material the ball makes a
deep indentation

(a) **Hardness**

Unprotected **mild steel** soon
corrodes (goes rusty) in a
damp atmosphere

Polished **stainless steel**
(an alloy containing chromium)
keeps its bright surface in a
damp atmosphere

(b) **Corrosion resistance**

Plastic **insulation**
*Has a low conductivity
(high resistance) and
prevents the flow of
an electric current*

Copper wire **conductor**
*Has a high conductivity
(low resistance) and
allows an electric current
to flow*

(c) **Conductivity (electrical)**

*Metals are good conductors
of heat. Copper is used for
the 'bit' as it is one of the
best conductors and will
heat the joint up quickly*

Wooden handle
*Non-metals are a poor conductor
of heat. The wooden handle
prevents the user of the
soldering iron being burnt*

Copper bit

(d) **Conductivity (heat)**

Blow lamp

Solder

*Solder melts at a **low temperature**
Therefore it has a **high fusibility***

Fire brick

*Fire brick only melts at **very high
temperatures** Therefore it has a **low
fusibility** and is called a **refractory***

(e) **Fusibility**

Fig. 2.3   Material properties (miscellaneous)

Table 2.2   Heat treatment definitions

Term	Meaning
Annealed	The condition of a metal that has been heated above a specified temperature, depending upon its composition, and then cooled down in the furnace itself or by burying it in ashes or lime. This annealing process makes the metal very soft and ductile. Annealing usually precedes flow forming operations such as sheet metal pressing and wire and tube drawing.
Normalised	The condition of a metal that has been heated above a specified temperature, depending upon its composition, and then cooled down in free air. Although the cooling is slow, it is not as slow as for annealing so the metal is less soft and ductile. This condition is not suitable for flow forming but more suitable for machining. Normalising is often used to stress relieve castings and forgings after rough machining.
Quench hardened	The condition of a metal that has been heated above a specified temperature, depending upon its composition, and then cooled down very rapidly by immersing it in cold water or cold oil. Rapid cooling is called **quenching** and the water or oil is called the **quenching bath**. This rapid cooling from elevated temperatures makes the metal very hard. Only medium and high carbon steels can be hardened in this way.
Tempered	Quench-hardened steels are brittle as well as hard. To make them suitable for cutting tools they have to be reheated to a specified temperature between 200° and 300° C and again quenched. This makes them slightly less hard but very much tougher. Metals in this condition are said to be tempered.

Metals may be shaped by making them molten and casting them to shape in moulds, or they may be cut to shape by hand or machine tools.

However, there is the alternative of *working* them to shape. This means shaping a piece of metal by stretching or squeezing it to shape. Metal that has been shaped by working or flow forming is said to be *wrought*.

Metals may be worked hot or cold. In both cases the working of the metal distorts the crystals from which it is made and this hardens the metal and renders it less ductile.

In *cold-worked* metals the crystals remain distorted, leaving the metal hard and lacking in ductility after processing. The metal is said to be *work hardened*.

In *hot-working* the metal is sufficiently hot for the crystals to reform after being distorted by the working process. Thus hot-worked metal remains soft and ductile no matter how severely it is worked. The temperature at which the crystals reform depend upon the composition of the metal and is called the temperature of *recrystallisation*. This will be discussed further in Chapter 3. Figure 2.4 shows some examples of hot- and cold-working.

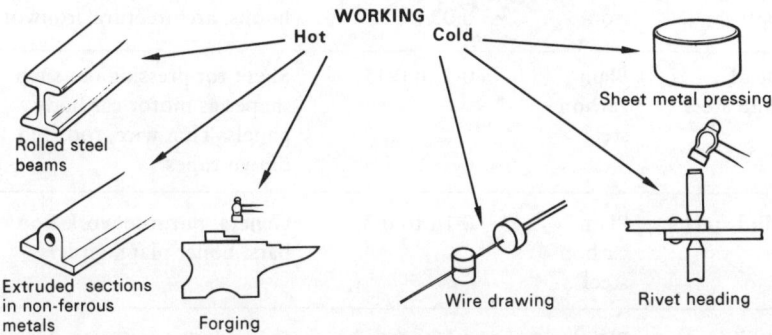

Fig. 2.4   Example of hot- and cold-working

2.5   Ferrous metals (basic composition)

Ferrous metals and alloys are based on the metal iron which is their main constituent. It is a soft, grey metal and is rarely found in the pure state outside the laboratory. The engineer usually finds it alloyed, or associated, with the non-metal **carbon**. Coke is a form of carbon and it is from the coke used in the blast furnace that the iron/carbon compounds are initially formed. This association with carbon greatly modifies the behaviour of the iron and makes it harder, stronger and of greater use to the engineer. Slight variations in the amount of carbon can make very great differences in the properties of the metal.

Table 2.3 shows how the addition of varying amounts of carbon to the metal iron can produce a wide range of ferrous metals. In all the examples given the metals are in the annealed (soft) state; see section 3.7.

The reason why the presence of carbon in varying amounts can produce such widely ranging variations in properties will now be considered. At room temperature ferrous metals, in the annealed condition, contain three main constituents. These are:

Ferrite: This is a very weak, solid solution of carbon in iron (0·006%) and for all practical purposes can be considered pure iron. This constituent is very soft and ductile.

Cementite: This is a compound of iron and carbon. It is called *iron carbide* by the chemist and *cementite* by the engineer and the metallurgist. Cementite is very hard and brittle.

Table 2.3 Ferrous metals

Name	Group	Carbon content %	Some uses
Wrought iron	Wrought iron	Less than 0·05	Chain for lifting tackle, crane hooks, architectural ironwork.
Dead mild steel	Plain carbon steel	0·1 to 0·15	Sheet for pressing out such shapes as motor car body panels. Thin wire, rod, and drawn tubes
Mild steel	Plain carbon steel	0·15 to 0·3	General purpose workshop bars, boiler plate, girders
Medium carbon steel	Plain carbon steel	0·3 to 0·5	Crankshaft forgings, axles
		0·5 to 0·8	Leaf springs, cold chisels
High carbon steel	Plain carbon steel	0·8 to 1·0	Coil springs, wood chisels
		1·0 to 1·2	Files, drills, taps and dies
		1·2 to 1·4	Fine-edge tools (knives, etc.)
Grey cast iron	Cast iron	3·2 to 3·5	Machine castings

Pearlite: This is a crystal structure containing alternate layers of ferrite and cementite. This laminated structure (like plywood) is very strong and the crystals are referred to as *lamnellar* crystals. Pearlite is very strong and tough.

Figure 2.5 shows the effect of the carbon content upon the properties of plain carbon steels. It will be seen that the upper limit of carbon content is 1·7 per cent. This is the maximum amount of carbon that can combine with iron at room temperature. In actual practice there is little advantage in raising the carbon content above 1·2 per cent, and a very real danger that above this figure some carbon may precipitate out destroying the properties of the steel. For this reason the maximum carbon content shown in Fig. 2.5 has been limited to 1·2 per cent.

It will be seen from Fig. 2.5 that low carbon steels are relatively soft, weak and ductile. This is because they consist mainly of crystals of ferrite, and ferrite is relatively weak, soft and ductile.

The increased amount of carbon in medium carbon steels promotes the formation of cementite. This results in an increased presence of pearlite

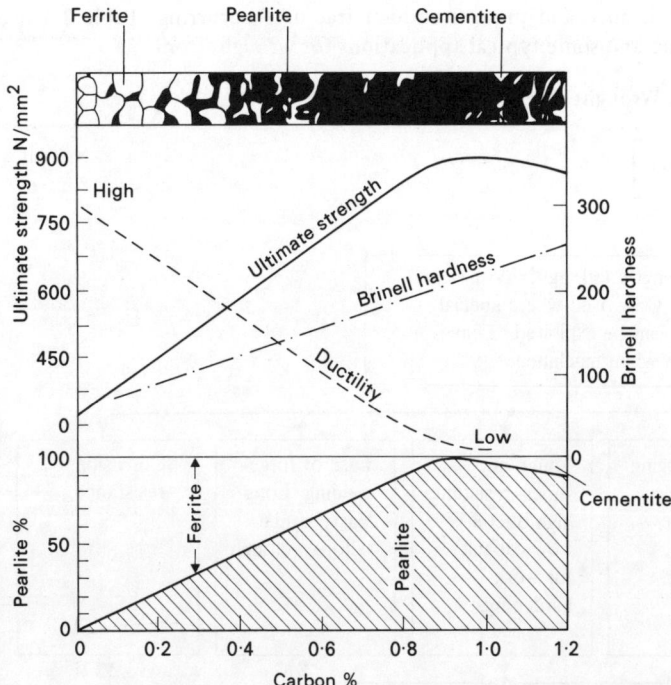

Fig. 2.5 The effect of carbon content on the properties of plain carbon steels (annealed)

making such steels strong and tough but not so ductile as the low carbon steels.

When the carbon content reaches approximately 0·85 per cent the steel consists entirely of pearlite. This is said to be the *eutectoid* composition. Steels with less that 0·85 per cent carbon are called *hypo-eutectoid* steels, and those with more than 0·85 per cent carbon are called *hyper-eutectoid* steels. The eutectoid composition produces plain carbon steel of maximum toughness and strength.

Increasing the carbon content still further increases the amount of iron carbide or cementite in the steel. Since the maximum amount of combined cementite occurred at 0·85 per cent carbon, the formation of further cementite results in it appearing around the crystal boundaries. This further reduces the toughness and ductility of the steel, but increases its hardness.

2.6 Ferrous metals (wrought iron)

Wrought irons are a group of ferrous metals with such a low carbon content that the iron/carbon compounds essential to a steel cannot be formed. In addition they contain fibres of slag trapped in the metal that promote corrosion resistance and prevent sudden fractures occurring. Table 2.4 lists the properties and some typical applications for wrought iron.

Table 2.4   Wrought iron

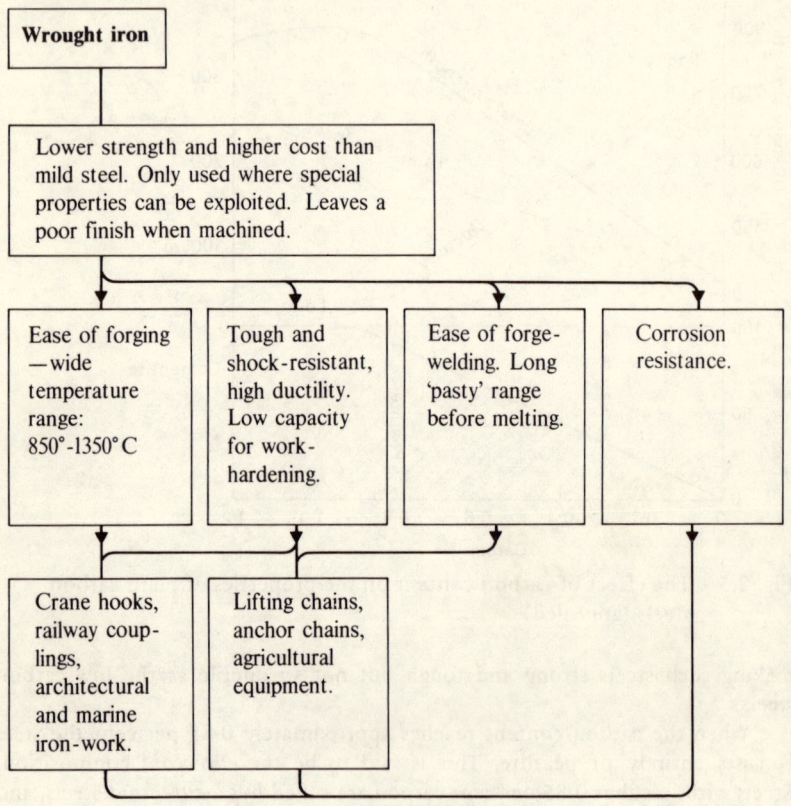

2.7   Ferrous metals (plain carbon steels)

Plain carbon steels are defined as alloys of iron and carbon in which the iron and carbon are chemically combined at all times. Only the range of ferrous metals with carbon contents lying between 0·1 and 1·7 per cent satisfy the definition. In practice an upper limit of 1·4 per cent is rarely exceeded.

In addition to carbon, all plain carbon steels contain the following elements:

Manganese: up to 1·0%
Silicon: up to 0·3%
Sulphur: up to 0·05%
Phosphorus: up to 0·05%

Manganese is an essential constituent element since it ensures a sound ingot free from blow holes. Further, it combines with any sulphur present which would otherwise weaken the steel. In general, manganese raises the yield point, together with the strength and toughness of the steel. However it increases the tendency of the steel to crack and distort when quench hardened and for this reason the content should be kept below 0·5 per cent in medium and high carbon steels.

Silicon

Silicon is an impurity carried over from the ore. It should be limited to 0·1 or 0·2 per cent in steels otherwise it can cause breakdown of the cementite which would result in weakness. It has little direct effect upon the mechanical properties providing the amount present is limited to the percentage quoted above.

It is often added to cast irons to prevent chill hardening (see section 2.9).

Sulphur

Sulphur is an impurity carried over from the coke used in the blast furnace. It tends to combine with the iron to form ferrous sulphide which greatly weakens the steel. Fortunately it has a greater affinity for manganese, and manganese sulphide does not weaken the steel. For this reason the amount of sulphur present should be kept below 0·05 per cent and there should always be at least five times as much manganese present as there is sulphur. Some *free-cutting* steels contain up to 0·2 per cent of sulphur to improve their machinability at the expense of strength for lightly stressed turned parts.

Phosphorus

Phosphorus is an impurity carried over from the ore. It forms compounds that make the steel brittle and, therefore, should be refined out as far as possible. It should not exceed 0·5 per cent.

Dead mild steel

The carbon content is deliberately left low so that the steel will have a high ductility. This enables it to be pressed into complicated shapes even while it is cold. It is slightly weaker than mild steel and is not usually machined since its softness would cause it to tear and leave a poor finish. It is used extensively for motor car body panels.

Mild steel

This is relatively soft and ductile, it can be forged and drawn in the hot or cold conditions, and is easily machined using high-speed steel cutting tools (see Table 2.5).

Medium carbon steel

This is harder, tougher and less ductile than mild steel, and cannot be bent or

Table 2.5 Mild steel

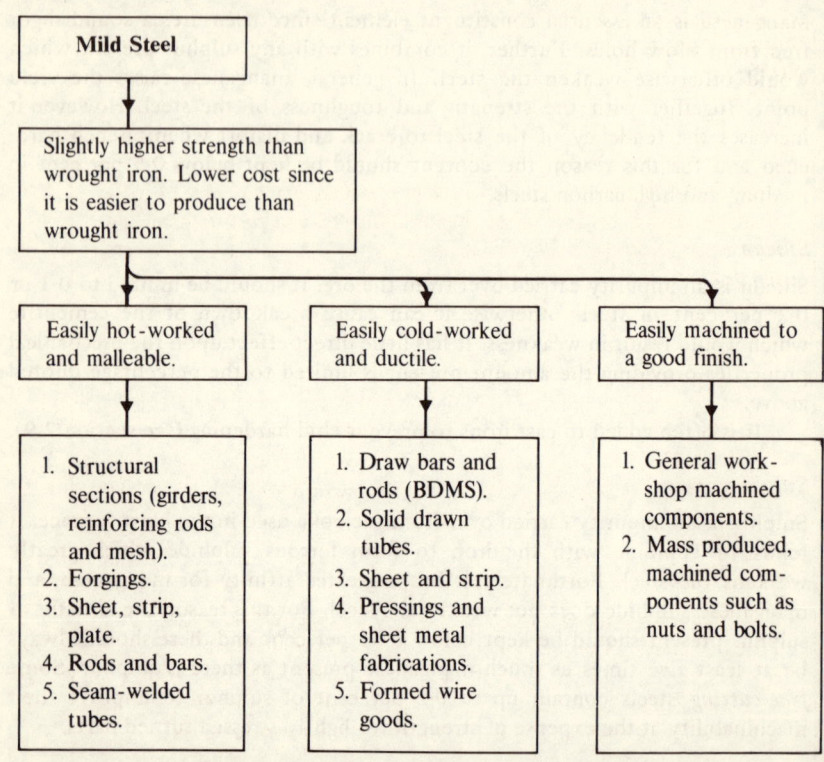

formed in the cold condition to any great extent without cracking. It hot forges well, but close temperature control is required to prevent:

(*a*) 'burning' at high temperatures (over 115° C), which leads to embrittlement;

(*b*) cracking below 700° C, due to work hardening (see Table 2.6).

High carbon steel

This is harder, less ductile and slightly less tough than medium carbon steel. Cold forming is not recommended, but it hot forges well, providing the temperature is even more closely controlled, with an upper limit of 900° C and a low limit of 700° C (see Table 2.7).

2.8 Ferrous metals (alloy steels)

Alloy steels are plain carbon steels to which other metals (alloying elements) have been added in sufficient quantities to materially alter the properties of the steel.

Table 2.6   Medium carbon steel   37

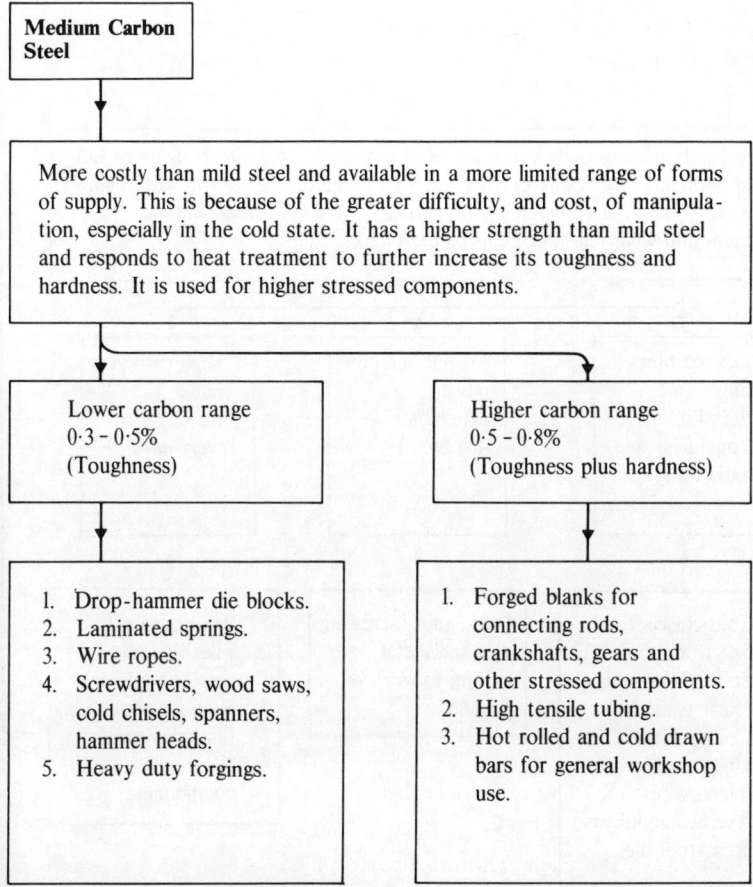

The most common alloying metals added to steels are:

1. *Nickel*, to refine the grain and strengthen the steel.
2. *Chromium*, to improve the response of the steel to heat treatment; to improve the corrosion resistance of the steel.
3. *Molybdenum*, reduces temper brittleness, and enables an alloy steel to operate continuously at high temperatures without becoming brittle.
4. *Manganese*, improves wear resistance. Steels containing a high percentage (14 per cent) of manganese are used for bulldozer and plough blades.
5. *Tungsten and cobalt*, improve the ability of a steel to remain hard at high temperatures and are used extensively in cutting tool materials.

High-speed steel, which is an alloy steel containing tungsten and cobalt, has largely superseded high carbon steel for power operated cutting tools. It

Table 2.7 High carbon steel

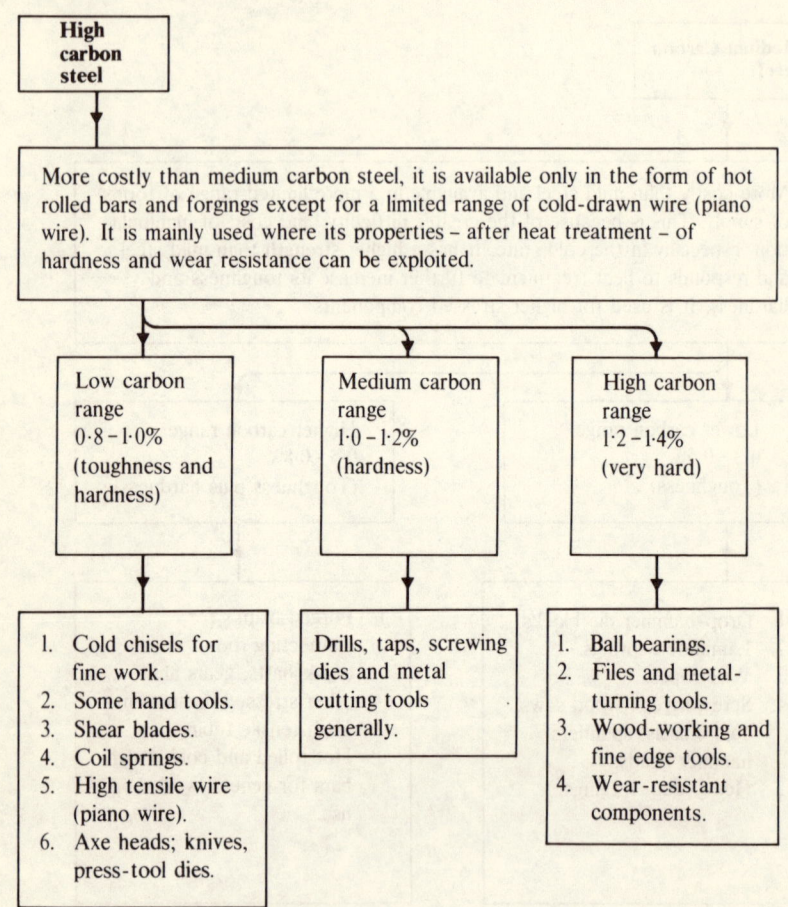

High carbon steel		

More costly than medium carbon steel, it is available only in the form of hot rolled bars and forgings except for a limited range of cold-drawn wire (piano wire). It is mainly used where its properties – after heat treatment – of hardness and wear resistance can be exploited.

Low carbon range 0·8 – 1·0% (toughness and hardness)	Medium carbon range 1·0 – 1·2% (hardness)	High carbon range 1·2 – 1·4% (very hard)
1. Cold chisels for fine work. 2. Some hand tools. 3. Shear blades. 4. Coil springs. 5. High tensile wire (piano wire). 6. Axe heads; knives, press-tool dies.	Drills, taps, screwing dies and metal cutting tools generally.	1. Ball bearings. 2. Files and metal-turning tools. 3. Wood-working and fine edge tools. 4. Wear-resistant components.

has the advantage of retaining its hardness at operating temperatures as high as 700° C, whereas high carbon steel starts to lose its hardness at 220° C and is useless for cutting purposes at 300° C (see Table 2.8).

2.9 Ferrous metals (grey cast iron)

Grey cast iron is very similar in composition and properties to the crude pig iron produced by the blast furnace. It does not require the complex and expensive refinement processes of steels and, therefore, provides a useful low-cost engineering material (see Table 2.9).

Cast iron contains substantially more than the 1·7 per cent carbon that forms the upper limit for steel. In fact, the distinguishing characteristic of cast irons is their uncombined or **'free' carbon** content. In grey cast iron this

Table 2.8   High-speed steel   39

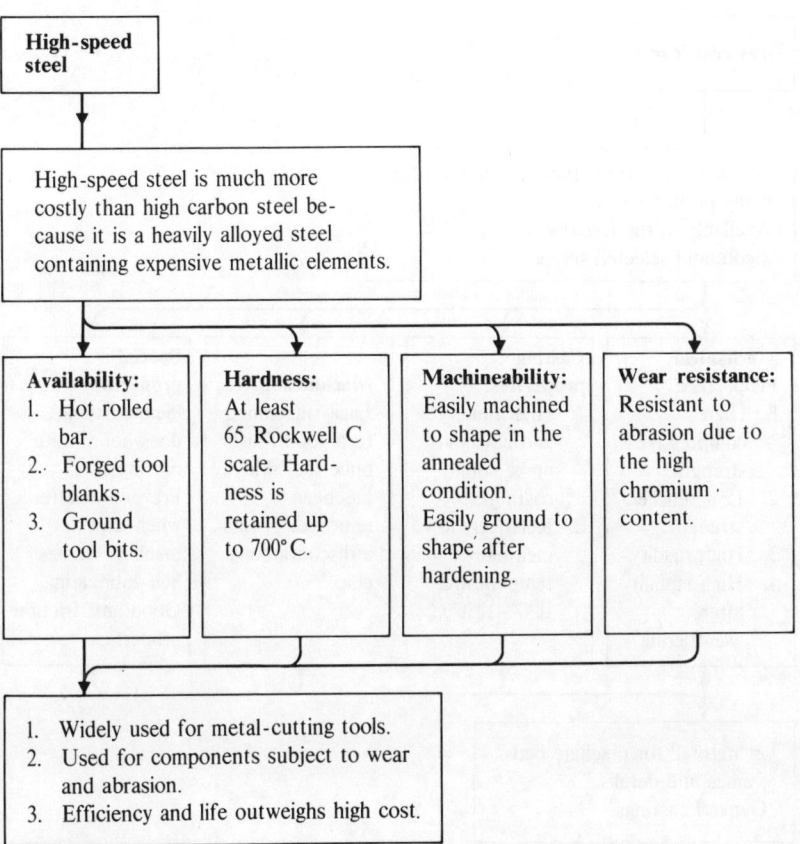

free carbon appears as flakes of graphite as shown in Fig. 2.6(*a*). It is these flakes of graphite that give cast irons their characteristic grey colour when fractured, its 'dirtiness' when machined or filed, and its weakness when subjected to a tensile load. The graphite also promotes good machining characteristics by acting as an internal lubricant and produces a discontinuous chip. The cavities containing flake graphite have a damping effect upon vibrations — cast iron is *anti-resonant* — and this property makes it particularly suitable for machine tool frames and beds.

Silicon is used to *soften* cast iron by promoting the formation of 'free' graphite. Its content is increased in irons used for light or thin components that might chill harden by cooling too quickly in the mould and become unmachineable.

Manganese is important in all ferrous metals (see section 2.7) as it reduces the weakening effect of any residual sulphur. Unfortunately it promotes the formation of cementite and hardens the iron. Therefore a balance has to be achieved between the manganese and silicon content.

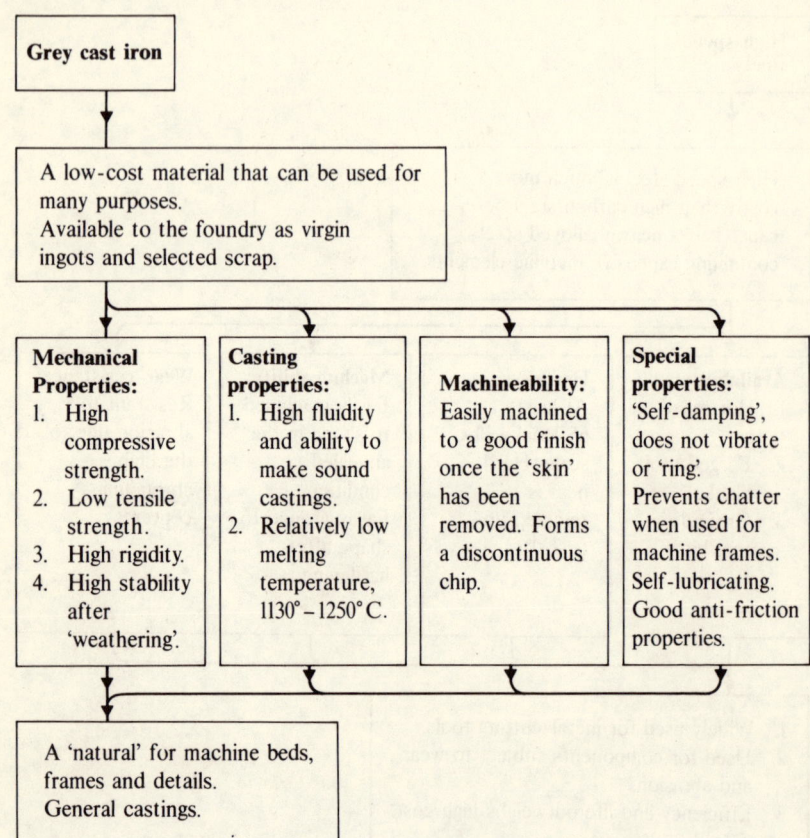

Grey cast iron

A low-cost material that can be used for many purposes.
Available to the foundry as virgin ingots and selected scrap.

Mechanical Properties:
1. High compressive strength.
2. Low tensile strength.
3. High rigidity.
4. High stability after 'weathering'.

Casting properties:
1. High fluidity and ability to make sound castings.
2. Relatively low melting temperature, $1130°–1250°C$.

Machineability: Easily machined to a good finish once the 'skin' has been removed. Forms a discontinuous chip.

Special properties: 'Self-damping', does not vibrate or 'ring'. Prevents chatter when used for machine frames. Self-lubricating. Good anti-friction properties.

A 'natural' for machine beds, frames and details.
General castings.

Phosphorus is another impurity that embrittles and weakens the iron without improving any other property. It is neutralised by the manganese and should be kept to a minimum.

A typical composition for grey cast iron could be:

Carbon: 3·3%
Silicon: 1·5%
Manganese: 0·75%
Sulphur: 0·05%
Phosphorus: 0·5%

2.10   Ferrous metals (malleable cast iron)

This is a 'white' cast iron of low carbon content (2·5–3·0%) that has been cast to shape in the ordinary way. It is then subjected to a heat treatment

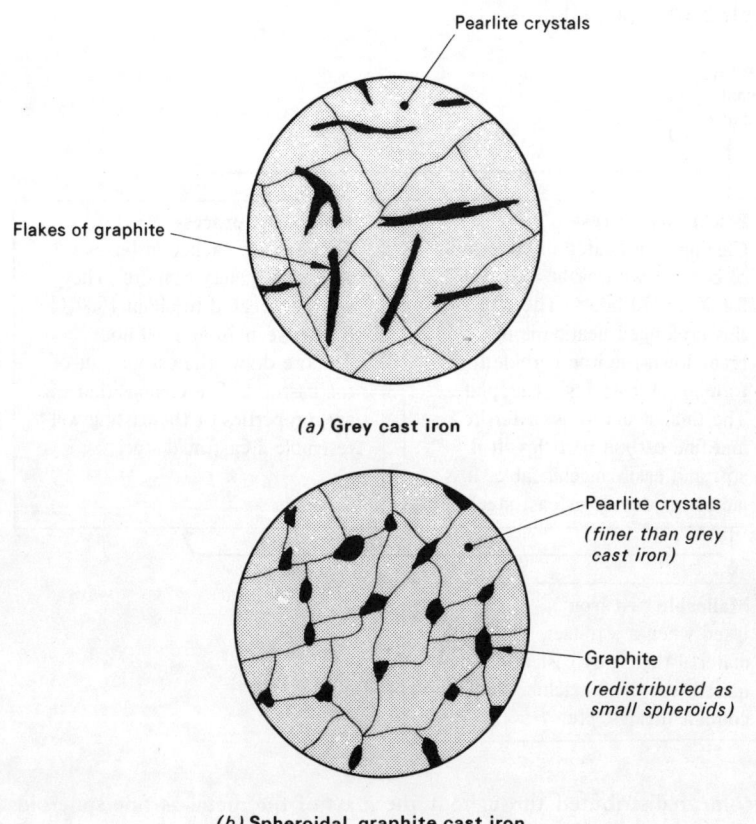

(a) **Grey cast iron**

(b) **Spheroidal graphite cast iron**

Magnification x 100

Fig. 2.6 Cast iron micro-structures

process that increases its strength and ductility very considerably.

There are two heat treatment processes:

(a) blackheart;
(b) whiteheart.

These names refer to the colour of a broken casting after heat treatment (see Table 2.10).

2.11 Ferrous metals (spheroidal graphite (S.G.) cast iron)

This is also known as 'high duty' cast iron. When traces of the metals *magnesium* or *cerium* are added to ordinary grey cast iron, the graphite flakes

Table 2.10   Malleable cast iron

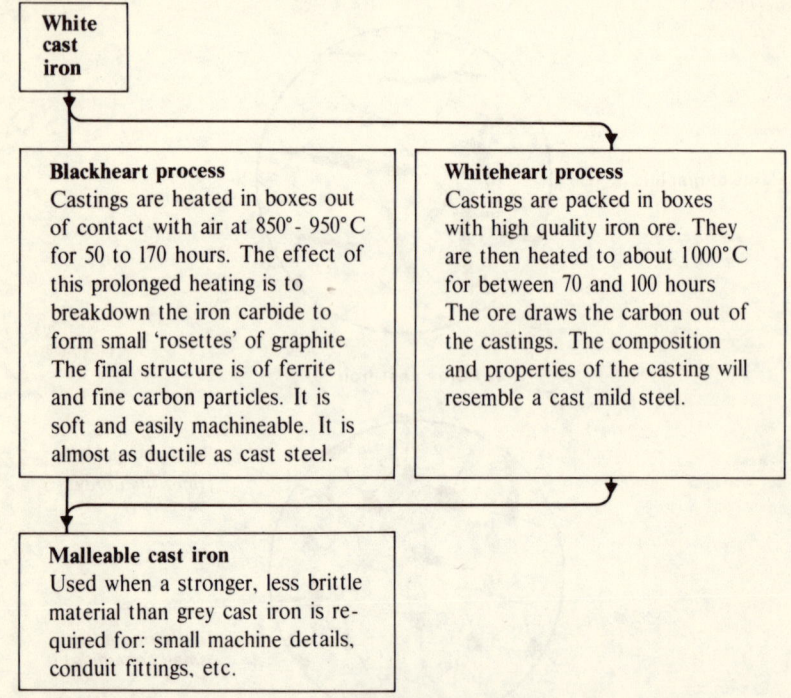

become redistributed throughout the mass of the metal as fine spheroids as shown in Fig. 2.6(*b*). This greatly improves the mechanical properties of the metal so that it can be used for more highly stressed applications than is possible for grey cast irons and malleable cast irons (see Table 2.11).

2.12   Non-ferrous metals

The term 'non-ferrous metals' refers to the thirty-eight metals other than iron that are known to man. The non-ferrous metals that are most commonly used by engineers are listed in Table 2.12. As well as being used as alloying elements, nickel and chromium are also electroplated on to a variety of metals both as a decorative finish and to give protection against corrosion (see section 4.8).

Two of the most important non-ferrous metals are aluminium and copper. They not only form the bases of many alloys, but they are widely used in their own right as pure metals (see Tables 2.13 and 2.14).

A list of non-ferrous metals is not complete in these days without mention of the **new metals**. Although known for many years, these metals are new to engineering. It is only since the Second World War that it has been

Table 2.11   Spheroidal graphite (S.G.) cast iron   43

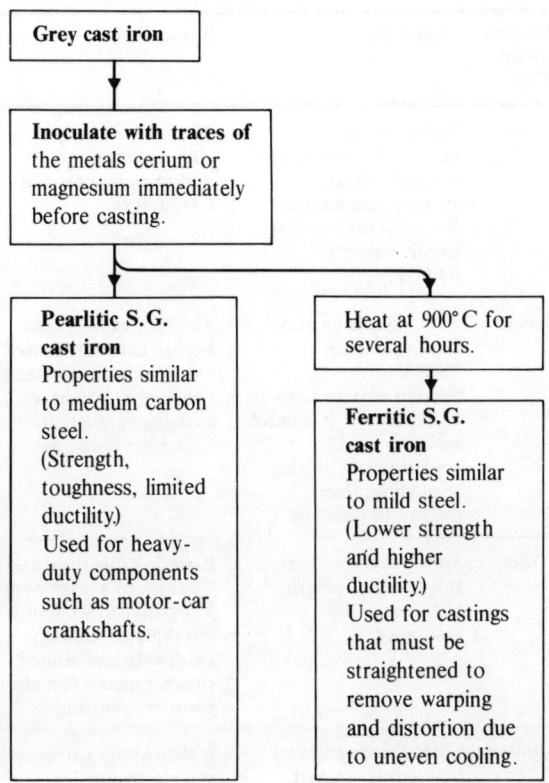

possible to produce these metals in bulk, and it is only recently that there has been a commercial demand for these metals.

The new metals

Niobium, Tantalum, Zirconium

Used for atomic reactor components.

Tellurium

Used instead of lead in free-cutting alloys.

Titanium

Used in supersonic aircraft and rockets as it has a higher strength/weight ratio than aluminium, and retains its strength at high temperatures.

Beryllium

Used as an alloying element with copper to make instrument springs. Beryl-

Table 2.12 Common non-ferrous metals

Metal	Density (kg/m³)	Melting point (°C)	Properties	Typical uses
Aluminium	2700	660	Lightest of the commonly used metals. High electrical and thermal conductivity. Soft, ductile and low tensile strength 93 MN/m²	The base of many engineering alloys. Lightweight electrical conductors
Copper	8900	1083	Soft, ductile and low tensile strength 232 MN/m². Second only to silver in conductivity, it is much easier to joint by soldering and brazing than aluminium. Corrosion resistant	The base of brass and bronze alloys. It is used extensively for electrical conductors and heat exchangers, such as motor car radiators
Lead	11 300	328	Soft, ductile and very low tensile strength. High corrosion resistance	Electric cable sheaths. The base of 'solder' alloys. The grids for 'accumulator' plates. Lining chemical plant. Added to other metals to make them 'free-cutting'
Silver	10 500	960	Soft, ductile and very low tensile strength. Highest conductivity of any metal	Widely used in electrical and electronic engineering for switch and relay contacts
Tin	7300	232	Resists corrosion	Coats sheet mild steel to give 'tin plate'. Used in soft solders. One of the bases of 'white metal' bearings. An alloying element in bronzes
Zinc	7100	420	Soft, ductile and low tensile strength. Corrosion resistant	Used extensively to coat sheet steel to give 'galvanized iron'. The base of die-casting alloys. An alloying element in brass
Chromium	7500	1890	Resists corrosion. Raises strength but lowers ductility of steels. Improves heat-treatment properties	Used as an alloying element in high-strength and corrosion-resistant steels. Used for electro-plating

Metal	Density (kg/m^3)	Melting point (°C)	Properties	Typical uses
Cobalt	8900	1495	Improves wear resistance and 'hot hardness' of high-speed steels	Used as an alloying element in 'super' high-speed steels and in permanent magnet alloys
Manganese	7200	1260	High affinity for oxygen and sulphur soft and ductile	Used to de-oxidise steels and to offset the ill-effects of the impurity sulphur. Larger amounts improve wear resistance
Molybdenum	9550	2620	A heavy, heat-resistant metal that alloys readily with other metals	Used as an alloying element in high-strength nickel-chrome steels to improve mechanical and heat-treatment properties. It reduces mass effect and temper-brittleness
Nickel	8900	1458	A strong, tough, corrosion-resistant metal widely used as an alloying element	Used as an alloying element to improve the strength and mechanical properties of steel. Tends to unstabilise the carbon during heat-treatment, and chromium has to be added to counteract this effect in medium and high carbon steels. Used for electro-plating

lium copper alloys can be hardened to provide 'non-sparking' tools for use in oilfields and on gas rigs.

These new metals are very expensive compared with the more common engineering materials and are only used where their special properties can be fully exploited.

The pure non-ferrous metals are used mainly where their properties of corrosion resistance and electrical and thermal conductivity can be exploited. They are not widely used for mechanical engineering applications as their mechanical strength is too low. Their mechanical properties are greatly improved by alloying them together.

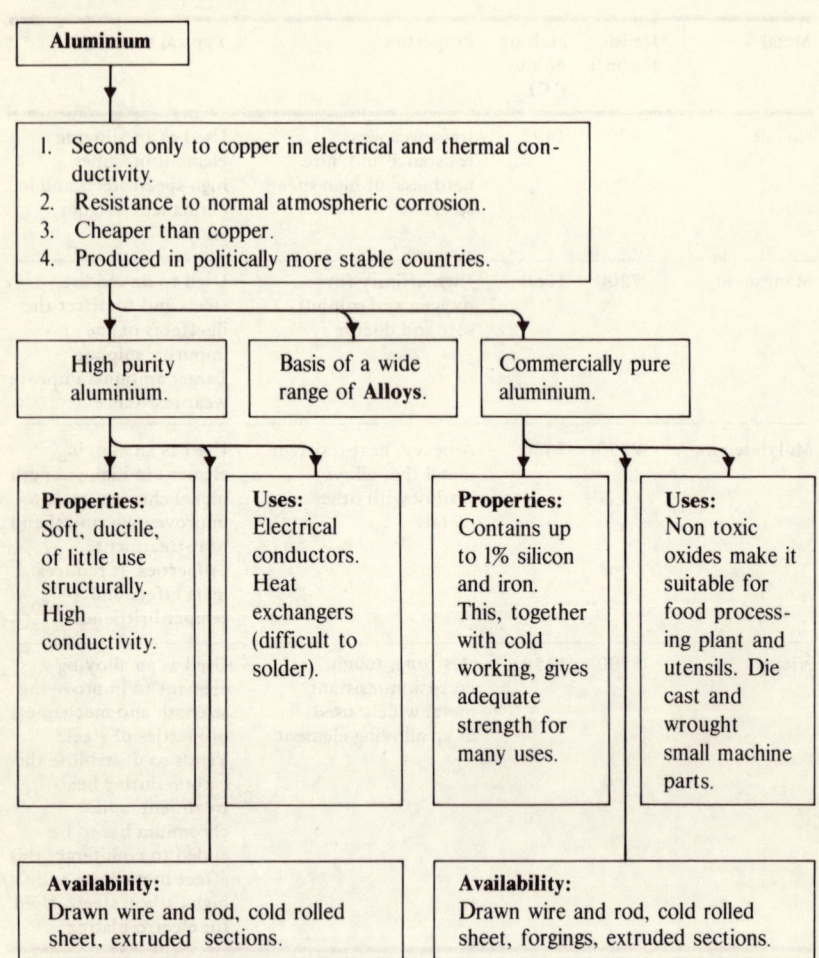

2.13   Non-ferrous alloys (high copper content)

The first group of copper alloys to be considered are the high copper content alloys. That is, the additional alloying element represents only a very small percentage of the total; yet these small additions make a significant change to the properties of the alloy compared with pure copper.

Silver copper

The addition of only 0.1 per cent of silver to high conductivity copper raises its annealing temperature by up to 150° C. This is very important where electrical conductors have to be soldered to hard-drawn copper contacts. If

Table 2.14   Copper   47

Copper

Properties:
1. Relatively high strength.
2. Very ductile (easily cold-worked).
3. Corrosion resistant.
4. Second only to silver as a conductor of heat and electricity.
5. Easily joined by soldering, brazing and welding.

One of the few metals of use to the engineer as a structural material in the pure state, although commercial grades contain some trace elements.
Availability: Cold drawn rods, bars, wire and tubes. Cold rolled sheet and strip. Extruded sections. Castings. Powder for sintered components.

Cathode copper
Used in the production of copper alloys.

Phosphorus-de-oxidised non-arsenical copper
Welding quality copper. Removal of the dissolved oxygen content prevents gassing and porosity. Used for fabrication, casting, cold impact extrusion and severe presswork.

High conductivity copper
Better than 99·9% pure. Used for electrical conductors and heat exchangers.

Refined tough pitch copper
General purpose copper. Used for roofing, chemical plant, general presswork, decorative metalwork and applications where special properties are not required.

Arsenical tough pitch and phosphorus de-oxidised copper
The addition of arsenic improves the strength at high temperatures. Used for boiler and firebox plates, stays, flue tubes and domestic plumbing.

pure copper is used the heat required to make the soldered joint would soften the copper and render its useless as a contact material.

Cadmium copper

Like silver, cadmium has little effect upon the conductivity of copper. Again like silver, it also raised the softening temperature. In addition, however, it also strengthens and toughens the copper, increasing its resistance to fatigue. As cadmium copper is substantially free from oxygen it is not susceptible to 'gassing' when it is braze welded.

Cadmium copper is used for low and medium voltage overhead transmission lines where its high conductivity and high tensile strength enable it to be used over relatively long spans.

Cadmium copper is also used for traction purposes, e.g. the overhead conductors on the electrified lines of railways. It is also recommended, in the soft condition, for aircraft wiring where its flexibility is combined with its resistance to the effects of vibration.

Chromium copper

This is one of the few non-ferrous materials that can be heat treated to improve its mechanical properties rather than relying upon cold-work hardening. A typical alloy contains 0·5 per cent chromium and is quenched from $1000°$ C. This leaves the alloy in a soft and ductile condition with rather a low electrical conductivity. However, if the metal is reheated to $500°$ C for approximately two hours, the mechanical and electrical properties are improved.

Since the properties of chromium copper depend upon heat treatment rather than cold-working, it can be used in cast as well as wrought forms. Similarly, components can be formed from annealed sheets or extruded rod and then hardened after manipulation and machining.

Tellurium copper

Tellurium forms stable compounds with copper and the addition of 0·5 per cent makes the copper as machineable as a free cutting brass whilst retaining a high electrical conductivity. It has a very high corrosion resistance and is used extensively for heavy duty contacts and commutators on machines and switch gear situated in hostile environments such as mines and chemical plants. The addition of traces of nickel and silicon allows it to be heat treated similarly to chromium copper.

Beryllium copper

Beryllium copper is used where mechanical rather than electrical properties are required. Beryllium copper is softened by heating to $800°$ C and quenching. In this condition the material can be extensively cold-worked. Subsequent heat treatment consists of heating the metal to $300°$ C to $320°$ C for upwards of two hours. The subsequent mechanical properties will depend to some extent upon the degree of cold-working that took place between the first and second treatments.

Beryllium copper is used widely for instrument springs, flexible bellows, corrugated diaphragms (aneroid altimeters and barometers) and the bourdon tubes for pressure gauges.

Because hand tools made from beryllium copper are almost as strong as those made from steel, but will not strike sparks from other metals or from flint stones, such tools are widely used in hazardous locations where there is a high risk of explosion such as mines, oil refineries, oil rigs, chemical plant, etc. Its high cost precludes its use for more conventional situations.

2.14 Non-ferrous alloys (copper-based)

Unlike the alloys discussed in section 2.13, the alloys to be considered in this section contain elements present in much greater amounts. For example, in some brasses zinc may make up to 40 per cent of the alloy. Three groups of alloys will be considered:

(a)   the brass alloys;
(b)   the tin-bronze alloys;
(c)   the aluminium-bronze alloys.

Brass alloys

These are alloys of copper and zinc. They tend to give rather poor quality, porous castings and depend upon hot- or cold-working to consolidate the metal and improve its mechanical properties. The more common brasses are considered in Table 2.15.

Tin bronze alloys

These are alloys of copper and tin together with a de-oxidiser. The de-oxidiser is essential to prevent the tin content oxidising during casting and hot-working. Oxidation of the tin would result in a weak, brittle, 'scratchy' bronze. Two de-oxidisers are commonly used:

(a)   phosphorus in the '*phosphor bronze*' alloys;
(b)   zinc in the '*gun metal*' alloys.

Some typical tin bronze alloys are listed in Table 2.16.

Unlike the brasses which are largely used in the wrought condition (rod, sheet, etc.), only low tin content bronzes can be worked and most bronze components are in the form of castings. Tin bronzes are more expensive than the brasses, but are stronger and give sound, pressure-tight castings that are widely used for steam and hydraulic valve bodies and mechanisms. They are highly resistant to wear and corrosion.

Aluminium bronze alloys

These are more expensive than 'tin bronzes' but are more corrosion resistant at high temperatures. They are also more ductile and can be cold-worked into tubes for boilers and condensers in steam plant.

Table 2.15 Brass alloys

Name	Composition (%)			Applications
	Copper	Zinc	Other elements	
Cartridge brass	70	30	—	Most ductile of the copper–zinc alloys. Widely used in sheet metal pressing for severe deep drawing operations. Originally developed for making cartridge cases, hence its name
Standard brass	65	35	—	Cheaper than cartridge brass and rather less ductile. Suitable for most engineering processes
Basis brass	63	37	—	The cheapest of the cold working brasses. It lacks ductility and is only capable of withstanding simple forming operations
Muntz metal	60	40	—	Not suitable for cold-working, but hot-works well. Relatively cheap due to its high zinc content, it is widely used for extrusion and hot-stamping processes
Free-cutting brass	58	39	3% lead	Not suitable for cold-working, but excellent for hot-working and high-speed machining of low strength components
Admiralty brass	70	29	1% tin	This is virtually cartridge brass plus a little tin to prevent corrosion in the presence of salt water
Naval brass	62	37	1% tin	This is virtually Muntz metal plus a little tin to prevent corrosion in the presence of salt water

Table 2.16 Tin-bronze alloys

Name	Composition (%)					Application
	Copper	Zinc	Phosphorus	Tin		
Low-tin bronze	96	—	0·1 to 0·25	3·9 to 3·75		This alloy can be severely cold-worked to harden it so that it can be used for springs where good elastic properties must be combined with corrosion resistance, fatigues resistance and electrical conductivity, e.g. contact blades
Drawn phosphor-bronze	94	—	0·1 to 0·5	5·9 to 5·5		This alloy is used in the work-hardened condition for turned components requiring strength and corrosion resistance, such as valve spindles
Cast phosphor-bronze	Rem.	—	0·03 to 0·25	10		Usually cast into rods and tubes for making bearing bushes and worm wheels. It has excellent anti-friction properties
Admiralty gunmetal	88	2	—	10		This alloy is suitable for sand casting where fine-grained, pressure-tight components such as pump and valve bodies are required
Leaded-gunmetal (free-cutting)	85	5 (5% lead)	—	5		Also known as 'red brass', this alloy is used for the same purposes as standard, Admiralty gunmetal. It is rather less strong but has improved pressure tightness and machining properties
Leaded (plastic) bronze	74	(24% lead)	—	2		This alloy is used for lightly loaded bearings where alignment is difficult. Due to its softness, bearings made from this alloy 'bed in' easily

Table 2.17 Some typical aluminium alloys

Composition (%) (Only elements other than aluminium are shown)						Category	Applications
Copper	Silicon	Iron	Manganese	Magnesium	Other elements		
0·1 max	0·5 max	0·7 max	0·1 max	—	—	Wrought Not heat-treatable	Fabricated assemblies. Electrical conductors. Food and brewing processing plant. Architectural decoration
0·15 max	0·6 max	0·75 max	1·0 max	4·5 to 5·5	0·5 Chromium	Wrought Not heat-treatable	High-strength shipbuilding and engineering products. Good corrosion resistance
1·6	10·0	—	—	—	—	Cast Not heat-treatable	General purpose alloy for moderately stressed pressure die-castings
—	10·0 to 13·0	—	—	—	—	Cast Not heat-treatable	One of the most widely used alloys. Suitable for sand, gravity and pressure die-castings. Excellent foundry characteristics for large marine, automotive and general engineering castings

Table 2.11 (continued)

4.2	0.7	0.7	0.7	0.3 Titanium (optional)	Wrought Heat-treatable	Traditional 'Duralumin' general machining alloy. Widely used for stressed components in aircraft and elsewhere
—	0.5	—	0.6	—	Wrought Heat-treatable	Corrosion-resistant alloy for lightly stressed components such as glazing bars, window sections and automotive body components
1.8	2.5	1.0	0.2	0.15 Titanium 1.2 nickel	Cast Heat-treatable	Suitable for sand and gravity die-casting. High rigidity with moderate strength and shock resistance. A general purpose alloy
—	—	—	10.5	0.2 Titanium	Cast Heat-treatable	A strong, ductile and highly corrosion-resistant alloy used for aircraft and marine castings both large and small

These are expensive copper-based alloys that have special properties where extremes of strength and corrosion resistance are required. They are used for such applications as high duty boiler and condenser tubes, bullet envelopes and resistance wires. Monel metal has exceptionally high corrosion resistance and is widely used in chemical plant. Copper nickel alloys are increasingly used for marine applications because of their superior corrosion resistance.

2.15   Aluminium alloys

Aluminium alloys cover a vast range of engineering materials sold under a bewildering number of proprietary trade names. However, for general engineering purposes, the alloys are listed under BS 1470—77 for wrought metals, and BS 1490 for cast metals and ingots. Air frame and aircraft engine materials are covered by the 'L' series of British Standards and the DTD specifications. DTD stands for the Directorate of Technical Development of the Ministry of Supply.

Aluminium alloys can be classified into four groups:

(*a*)   wrought alloys (not heat-treatable);
(*b*)   cast alloys (not heat-treatable);
(*c*)   wrought alloys (heat-treatable);
(*d*)   cast alloys (heat-treatable).

(See Table 2.17.)

The heat treatment of aluminium alloys is in two parts: *solution treatment* and *precipitation treatment*, and will be considered in detail in Chapter 3.

2.16   Tin-lead alloys

Tin and lead form a continuous range of alloys from pure lead and no tin to pure tin and no lead. The most useful alloys are those ranging from 33 per cent tin and 67 per cent lead (plumber's solder) to those containing 62 per cent tin and 38 per cent lead (tinman's solder). In addition, tin—lead alloys form the basis of many bearing metals and these will be considered in section 2.22.

Table 2.18 lists some typical tin—lead solders together with their solidification range and applications.

2.17   Common forms of supply

There is an almost unlimited range to the forms in which metal can be supplied. Figure 2.7 shows some of these forms. The process by which the metal is produced will have a profound effect upon its properties. Tables

Table 2.18 Soft solders

BS solder	Composition (%)			Melting range (°C)	Remarks
	Tin	Lead	Antimony		
A	65	34·4	0·6	183–185	Free running solder ideal for soldering electronic and instrument assemblies. Commonly referred to as **electrician's solder**
K	60	39·5	0·5	183–188	Used for high-class tinsmith's work, and is known as **tinman's solder**
F	50	49·5	0·5	183–212	Used for general soldering work in coppersmithing and sheet metal work
G	40	59·6	0·4	183–234	**Blow-pipe solder.** This is supplied in strip form with a D cross-section 0·3 mm wide
J	30	69·7	0·3	183–255	**Plumber's solder.** Because of its wide melting range this solder becomes 'pasty' and can be moulded and wiped

2.19, 2.20 and 2.21 compare the advantages and limitations of hot-working, cold-working and casting processes respectively.

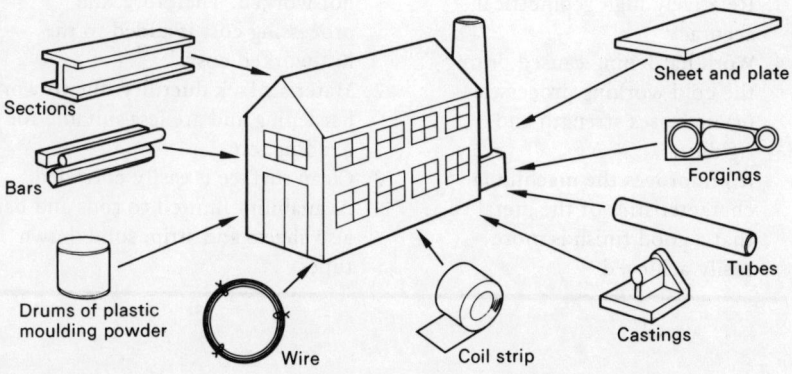

Fig. 2.7 Forms of supply

Table 2.19   Hot-working processes

Advantages	Limitations
1. Low cost	1. Poor surface finish — rough and scaly
2. Grain refinement from cast structure	2. Due to shrinkage on cooling the dimensional accuracy of hot-worked components is of a low order
3. Materials are left in the fully annealed condition and are suitable for cold-working (heading, benching, etc)	3. Due to distortion on cooling and to the processes involved, hot-working generally leads to geometrical inaccuracy
4. Scale gives some protection against corrosion during storage	4. Fully annealed condition of the material coupled with a relatively coarse grain leads to a poor finish when machined
5. Availability as sections (girders) and forgings as well as the more usual bars, rods, sheets and strip and butt welded tube	5. Low strength and rigidity for metal considered
	6. Damage to tooling from abrasive scale on metal surface

Table 2.20   Cold working processes

Advantages	Limitations
1. Good surface finish	1. Higher cost than for hot-worked materials. It is only a finishing process for material previously hot-worked. Therefore, the processing cost is added to the hot-worked cost.
2. Relatively high dimensional accuracy	
3. Relatively high geometrical accuracy	2. Materials lack ductility due to work hardening and are less suitable for bending, etc
4. Work hardening caused during the cold-working processes: (*a*) increases strength and rigidity; (*b*) improves the machining characteristics of the metal so that a good finish is more easily achieved	3. Clean surface is easily corroded
	4. Availability limited to rods and bars also sheets and strip, solid drawn tubes

Table 2.21   Casting processes (gravity, sand only)   57

Advantages	Limitations
1. Virtually no limit to the shape and complication of the component to be cast 2. Virtually no limit to the size of the casting 3. Low cost, as no expensive machines and tools are required as in forging 4. Scrap metal can be reclaimed in the melting furnace. (Wrought and machined components have to be made from relatively expensive pre-processed materials)	1. Strength and ductility low, as structure is un-refined 2. Quality is uncertain, as local differences of structure and mechanical defects such as blow-holes cannot be controlled or corrected 3. Low accuracy due to shrinkage 4. Poor surface finish 5. Component must be designed without sudden changes of section, so that molten metal flows easily and cooling cracks and warping will not occur 6. Not all metals are suitable for casting. The best metals have a low shrinkage, a short freezing range and high fusibility (melt at relatively low temperatures), and have a high fluidity when molten

2.18   Identification of metals

The similarity in appearance between many materials of different physical properties makes it essential that some form of permanent identification should be marked on them, e.g. colour coding.

However, bar 'ends' are often used up for 'one-off' jobs and Table 2.22 gives some simple workshop tests of identification.

2.19   Non-metallic materials (natural)

Rubber: Anti-vibration mountings; Coolant and compressed-air hoses; Transmission belts.

Glass: Spirit level vials (the tube containing the bubble); Optical measuring instrument lenses.

Wood: Casting patterns.

Emery: Abrasive wheels, and grinding pastes (nowadays usually produced artificially to control quality).

Ceramics: Cutting-tool tips; Electrical insulators.

Table 2.22 Workshop identification tests (*These are not foolproof and require some experience*)

Metal	Appearance	Hammer cold	Type of chip	'Spark test' on grinding wheel
Mild steel ('black')	Smooth scale with blue/black sheen	Flattens easily	Smooth, curly ribbon-like	Stream of yellow white sparks, varying in length: slightly 'fiery'
Mild steel ('bright')	Smooth, scale-free, silver grey surface			
Medium carbon steel	Smooth scale, black sheen	Fairly difficult to flatten	Chip curls more tightly and discolours light brown	Yellow sparks, shorter than m/s, and finer and more feathery
High carbon steel	Rougher scale, black	Difficult to flatten	Chip curls even more tightly and discolours dark blue	Sparks less bright, starting near grinding wheel, and more feathery with secondary branching (distinctive acrid smell)
High-speed steel	Rougher scale, black with reddish tint	Very difficult to flatten. Tends to crack easily	Long ribbon-like chip. Distinctive smell. Over-heats tool easily	Faint red streak ending in fork (distinctive acrid smell)
Cast iron	Grey and sandy	Crumbles	Granular, grey in colour	Faint red spark, ending in bushy yellow sparks (distinctive acrid smell)
Copper	Distinctive 'red' colour	Flattens very easily	Ribbon-like, with razor edge	Should not be ground

Diamonds: Turning tools for fine finishing light alloys; Lapping wheels.

This is not intended as an exhaustive list, but to give an indication of some of the materials available and their uses.

2.20   Non-metallic materials (synthetic)

There is an ever increasing number of synthetic materials available under the name of *plastic*. These materials rarely show plastic properties (section 2.2) in their finished condition, in fact many of them are elastic, but during the moulding operation by which they are formed they are reduced to a plastic condition by heating to just above the temperature of boiling water.

There are two main groups of plastic materials:

Thermo-setting plastics

These undergo a chemical change during moulding and can never be softened by re-heating. These materials are hard, rigid and rather brittle.

Thermoplastics

These can be softened as often as they are re-heated. They are not so rigid as the first group and tend to be tougher.

Some typical examples of plastic materials that are met with in the workshop are given in Tables 2.23 and 2.24.

The strength of thermosetting plastics can be greatly increased by reinforcing them with fibrous materials.

Laminated plastic (Tufnol)

Fibrous material such as paper, woven cotton cloth or woven asbestos cloth is impregnated with phenolic resin (this is the basic material of bakelite). The sheets of fabric are then laid up in hydraulic presses and squeezed and heated so that they become solid sheets, rods, tubes, etc.

This material can be machined dry with ordinary engineering machine tools, using low rake tools and fairly high cutting speeds. Care must be taken because of the 'grain' of the material which causes it to behave rather like plywood.

It is widely used for making bearings, gears and other engineering components.

Table 2.23   Some typical thermo-setting plastic materials

Material	Characteristics
Phenolic resins and powders	These are used for dark-coloured parts because the basic resin tends to become discoloured. These are heat-curing materials
Amino (containing nitrogen) resins and powders	These are colourless and can be coloured if required; they can be strengthened by using paper-pulp fillers, and used in thin sections
Polyester resins	Polyester chains can be cross-linked by using a monomer such as styrene; these resins are used in the production of glass-fibre laminates
Epoxy resins	These are also used in the production of glass-fibre laminates

Table 2.24 Some typical thermoplastic materials

Type	Material	Characteristics
Cellulose plastics	Nitrocellulose	Materials of the 'celluloid' type are tough and water resistant. They are available in all forms except moulding powders. They cannot be moulded because of their inflammability
	Cellulose acetate	This is much less inflammable than the above. It is used for tool handles and electrical goods
Vinyl plastics	Polythene	This is a simple material that is weak, easy to mould, and has good electrical properties. It is used for insulation and for packaging
	Polypropylene	This is rather more complicated than polythene and has better strength
	Polystyrene	Polystyrene is cheap, and can be easily moulded. It has a good strength but it is rigid and brittle and crazes and yellows with age
	Polyvinyl chloride (PVC)	This is tough, rubbery, and practically non-inflammable. It is cheap and can be easily manipulated: it has good electrical properties
Acrylics (made from an acrylic acid)	Polymethyl methacylate	Materials of the 'perspex' type have excellent light transmission, are tough and non-splintering, and can be easily bent and shaped
Polyamides (short carbon chains that are connected by amide groups-NHCO)	Nylon	This is used as a fibre or as a wax-like moulding material. It is fluid at moulding temperature, tough, and has a low coefficient of friction
Fluorine plastics	Polytetrafluoro-ethylene (ptfe)	Is a wax-like moulding material; it has an extremely low coefficient of friction. It is very expensive
Polyesters (when an alcohol combines with an acid, an 'ester' is produced)	Polyethylene terephthalate	This is available as a film or as 'Terylene'. The film is an excellent electrical insulator

Woven glass fibre can be bonded together by epoxy and polyester resins to form large and complex mouldings (from crash helmets to 18-metre yachts).

The resin used is a thermo-setting plastic and it is set by chemical action at room temperature; a press is not required.

The impregnated glass fibre is laid up over wooden or plaster patterns. When set it is lifted off and the pattern can be used again.

This material is used to produce large casting patterns, copymilling models, machine castings and guards.

2.21 Properties of plastics

The properties of plastics can vary widely, but all plastic materials have the following properties in common.

Electrical insulation

All plastic materials exhibit good electrical insulation properties. However, their usefulness in this field is limited by their low heat-resistance and softness. Thus they are useless as formers for winding electric radiator elements on, and as insulators for use out of doors where their surface would soon be roughened by the weather. Dirt collecting on this roughened surface would then provide a conductive path, causing a short circuit.

Strength/weight ratio

Plastic materials vary in strength considerably. Some of the stronger, such as nylon, compare favourably with the weaker metals. All of them are much lighter than most metals. Therefore, properly chosen and proportioned their strength/weight ratio compares favourably with many light alloys. They are steadily taking over engineering duties which, until recently, were considered the prerogative of metal.

Corrosion resistance

All plastic materials are inert to most inorganic chemicals. Thus they can be used in environments that are hostile to the most corrosion-resistant metals.

They are superior to rubber in that they are resistant to attack by oils and greases.

Note: Because of the ease with which plastics can be manufactured and used, they are finding increasing use as pipes and conduits. Care must be taken when inserting a length of plastic pipe or conduit into any system that the earth continuity of the system is not being destroyed. This is because cold water pipes and conduits are often used as convenient earthing points for electrical equipment.

2.22 Bearing materials (anti-friction)

There are essentially two types of bearings. Those where the moving parts slide over each other as in plain bearings and machine tool slideways, and those where the moving parts roll over each other as in ball and roller bearings.

Metals and non-metals developed as bearing materials for the former type (sliding contact) require the following characteristics:

(a) Coefficient of friction (μ): This should be kept as low as possible to avoid wasting energy. A sledge will run down a hill much more easily on snow than on the grass that is exposed when the snow melts. This is because snow (ice) has a low coefficient of friction. Energy wasted in overcoming friction causes the bearing to heat up. Eventually the bearing may overheat and be destroyed.

(b) Strength: The bearing must have sufficient compression strength to support the shaft and any load that may be applied to the shaft in service.

(c) Wear resistance: The bearing material must resist wear, but at the same time it is invariably better for the bearing to eventually wear out rather than the journal on the shaft. It is easier and cheaper to replace the bearing shell.

(d) Plasticity: It is almost impossible to obtain perfect alignment in a bearing and its associated shaft. Therefore a bearing material should be capable of slightly distorting and bedding in to ensure alignment. White metals and leaded bronzes are better at aligning themselves than the harder phosphor bronzes.

(e) Surface texture: A perfectly smooth surface would be a poor bearing surface as there would be no provision for the retention of pockets of lubricant. An ideal bearing material consists of facets of a relatively hard, anti-friction material dispersed through a soft matrix. The matrix wears away between the facets to form pockets for the retention of the lubricant. The soft matrix also has sufficient 'give' to assist alignment.

(f) Corrosion resistance: The bearing material should resist corrosion by impurities in the lubricant or impurities that may form in service. It should also be resistant to the attack of any additives in the lubricant intended to give it greater lubricity.

(g) Thermal conductivity: Since the best of bearing materials offer some degree of friction, there will always be some energy loss and a corresponding rise in temperature when the shaft is rotating. This heat energy can only be dissipated through the lubricant or by conduction through the walls of the bearing. If the heat were not conducted away fast enough the temperature could rise to the melting point of the bearing alloy.

The most commonly used bearing materials are cast iron, phosphor bronze, white metal and plastic. Table 2.25 lists some typical bearing materials, their composition, properties and typical applications.

Table 2.25 Some typical bearing materials

Category	Composition (%)					Properties and applications
	Sn	Sb	Cu	Pb	P	
White metal	93	3.5	3.5	—	—	Big-end bearings for light and medium duty, high-speed internal combustion engines
	86	10.5	3.5	—	—	Main bearings for light and medium duty, high-speed internal combustion engines
	80	11.0	3.0	6.0	—	General purpose, heavy duty bearings. Lead improves plasticity where alignment is a problem
	60	10.0	28.5	1.5	—	Heavy duty marine reciprocating engines, electrical machines
	40	10.0	1.5	48.5	—	Low cost, general purpose, medium duty, bearing alloy
Bronze	10.5	—	89	—	0.5	Good anti-friction properties, suitable for heavy loads, rigid
	10.0	—	79.9	10	0.1	Good anti-friction properties, lubrication not critical, lead content reduces rigidity and helps alignment
	3	—	74	23	—	Leaded (plastic) bronze, excellent self-alignment properties due to high lead content. For duty intermediate between white metal and phosphor bronze
	Fe	C	Si	Mn	S:/p	
Cast Iron	94	3.3	1.3	1.0	0.1;/0.3	The flakes of graphite (carbon) in grey cast iron gives it self lubricating properties. Suitable for heavy duty, low-speed applications where lubrication is difficult, e.g. machine tool slideways
Plastic	Polytetrafluoroethylene					**Teflon** .: Can withstand much higher temperatures than most plastics. Very expensive anti-friction coating – very low coefficient of friction. Does not require lubrication
	Polyamide					**Nylon** .: Can be moulded into bushes and gears. Does not require lubrication. Use for office and food processing machinery
	High density polyethylene					Low cost bearings. Does not require lubrication. Cannot support such high loads as Nylon or Teflon

Sn = Tin, Sb = Antimony, Cu = Copper, Pb = Lead, P = Phosphorus, Fe = Iron, C = Carbon, Si = Silicon, Mn = Manganese, S = Sulphur

Problems

Section A

1. Non-metallic components are used in electrical equipment because of their: (a) light weight;   (b) good conductivity;   (c) ease of manufacture;   (d) insulating properties.
2. Low carbon steel that has been shaped by hot-working tends to: (a) become work hardened;   (b) remain soft and ductile;   (c) have an improved surface finish;   (d) have increased elasticity.
3. High carbon (hyper-eutectoid) steels can be used for cutting tools because they contain: (a) cementite and pearlite;   (b) only pearlite; (c) ferrite and pearlite;   (d) only cementite.
4. High duty cast irons are stronger than grey cast iron because the free graphite (uncombined carbon) is: (a) removed altogether;   (b) left in flake form;   (c) spherodised by adding traces of magnesium or cerium;   (d) combined with the iron to form cementite.
5. In 'gun metal' bronze alloys the de-oxidising agent is: (a) copper;   (b) phosphorus;   (c) zinc;   (d) tin.

Section B

6. With reference to the properties of the materials concerned, explain why: (a) the overhead electrical transmission lines of the 'grid system' have a high tensile steel wire core and an outer sheath of pure aluminium wires;   (b) soldering irons have copper 'bits';   (c) beryllium copper tools are frequently used on oil drilling rigs and down coal mines.
7. Explain why laminated plastic (Tufnol) is highly suitable for sea water pump bearings. Your answer should consider such factors as: (a) electrolytic corrosion;   (b) mechanical strength;   (c) coefficient of friction;   (d) lubrication.
8. List the essential properties of a metallic-bearing alloy and explain how 'Babbitt' metal fulfils these requirements.
9. (a) State the difference between thermo-plastic and thermosetting plastic materials;
   (b) Select a suitable plastic material for each of the following applications, giving reasons for your choice: (i) the insulation of flexible cables;   (ii) the rigid casing for an ammeter;   (iii) a safety helmet;   (iv) a heavy duty gear wheel;   (v) a light duty bearing bush for an office machine;   (vi) a transparent moulded cover for an instrument.
10. State a typical composition for each of the following non-ferrous alloys and give a typical application, with reasons for your choice, in each case: (a) cartridge brass;   (b) free-cutting brass;   (c) admiralty brass; (d) phosphor bronze;   (e) gun-metal bronze;   (f) silver copper; (g) duralumin;   (h) tinman's solder.

Chapter 3

Heat treatment

3.1 Critical temperatures

It has already been explained in Chapter 2 that plain carbon steels are alloys of *iron* and *carbon*, and that at ambient (room) temperatures the iron and carbon are combined together to form the compound *iron carbide* (referred to by engineers and metallurgists as *cementite*). There is no 'free' carbon in steel ('free' carbon only appears in the cast irons), all the carbon being combined with some of the iron.

The structures and properties considered in Chapter 2 were produced under equilibrium conditions. That is, the steels were heated to certain critical temperatures and allowed to cool sufficiently slowly that all the changes in the crystal structure of the metal had time to take place, and a state of chemical 'balance' or equilibrium was achieved.

These critical temperatures, and the metallic structures associated with them, are show in Fig. 3.1. This diagram will be fully explained in the next book in this series: *Manufacturing Technology I* (level 2). For the time being, it is sufficient to appreciate that there are critical temperatures at which changes take place in the crystal structure of the metal, and that these temperatures vary according to the carbon content of the steel being considered.

Figure 3.2 summarises the processes by which the structure and therefore the properties of plain carbon steels can be changed by heat treatment.

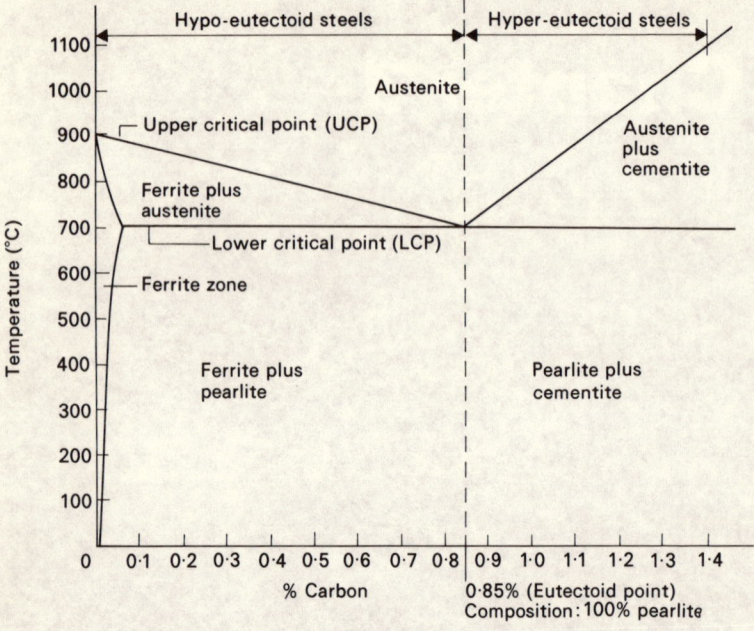

Fig. 3.1   Iron/carbon equilibrium diagram (steel section)

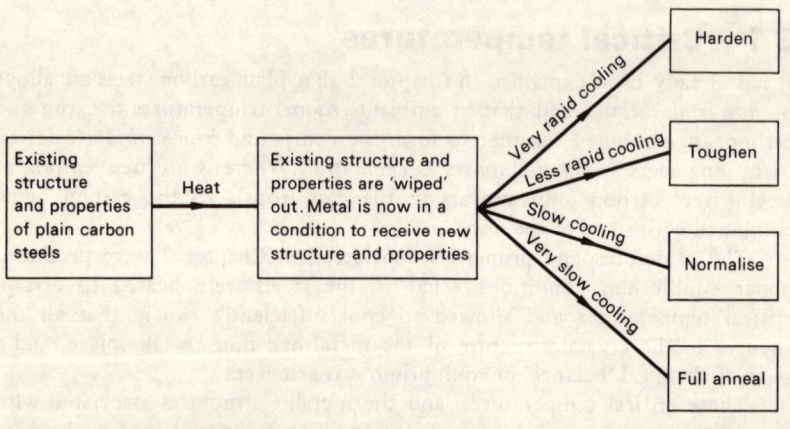

Fig. 3.2   Heat treatment of plain carbon steels

3.2   Quench hardening plain carbon steels

The temperature to which the steel must be heated to wipe out the initial structure and properties will depend upon the carbon content of the steel. Figure 3.3 shows the temperatures from which plain carbon steels should be

quench hardened. It will be seen that these temperatures lie within a band
ranging from 20° to 50° C above the line connecting the upper critical points
for hyper-eutectoid steels (steels with a carbon content below 0·85 per cent).
For hyper-eutectoid stills (carbon content above 0·85 per cent) the temperature is constant and lies within a band ranging from 20° to 50° C above the
line connecting the lower critical points.

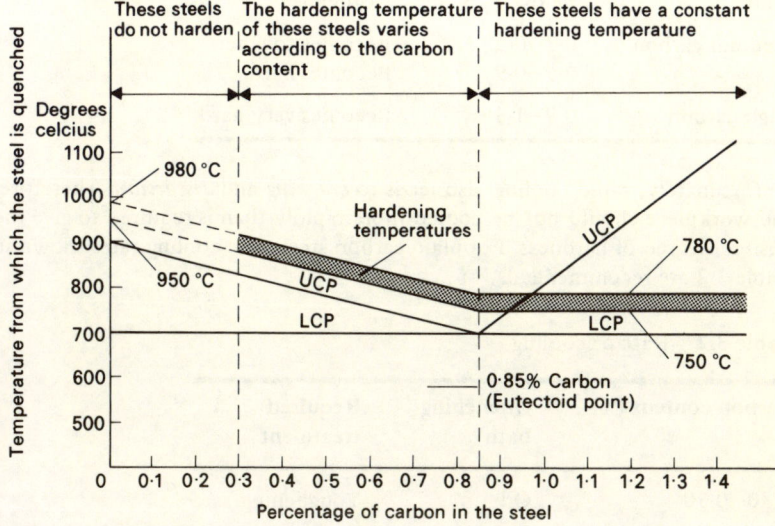

Fig. 3.3   Hardening of plain carbon steels

To quench harden the steel, it is cooled rapidly (quenched) from the
temperatures shown in Fig. 3.3. The degree of hardness the steel achieves is
solely dependent upon:

(*a*)   the carbon content;
(*b*)   the rate of cooling.

(*a*) *Carbon content:*   There must be sufficient carbon present to form the
hard crystal structures in the steel when it is heated and quenched. The effect
of the carbon content on the hardness of the steel after heating and quenching is shown in Table 3.1.

(*b*) *Rate of cooling:*   The rapid cooling necessary to harden steel is known as
quenching. The liquid into which the steel is dipped to cause this rapid
cooling is called the **quenching bath**.

In the workshop, the quenching bath will contain either:

(*a*)   water;
(*b*)   quenching oil (on no account use lubricating oil).

The more rapidly a plain carbon steel is cooled the harder it becomes.

Table 3.1 Effect of carbon content

Type of steel	Carbon content (%)	Effect of heating and quenching (rapid cooling)
Mild	Below 0·25	Negligible
Medium carbon	0·3—0·5	Becomes tougher
	0·5—0·9	Becomes hard
High carbon	0·9—1·3	Becomes very hard

Unfortunately, rapid cooling also leads to *cracking* and *distortion.* Therefore, the workpiece should not be cooled more rapidly than is required to give the desired degree of hardness. For plain carbon steels, the cooling rates shown in Table 3.2 are recommended.

Table 3.2 Rate of cooling

Carbon content (%)	Quenching bath	Required treatment
0·30—0·50	Oil	Toughening
0·50—0·90	Oil	Toughening
0·50—0·90	Water	Hardening
0·90—1·30	Oil	Hardening

Note:
1. Below 0·5 per cent carbon content, steels are not hardened as cutting tools, so water hardening has not been included.
2. Above 0·9 per cent carbon content, any attempt to harden the steel in water could lead to cracking.

3.3 Tempering

Hardened plain carbon steel is very brittle and unsuitable for immediate use. A further process known as **tempering** must be carried out to greatly increase the toughness of the steel at the expense of some hardness.

Tempering consists of reheating the steel to a suitable temperature and quenching it in oil or water. The temperature to which the steel is heated depends upon the use to which the component is going to be put. Table 3.3 gives some suitable temperatures for tempering components made from plain carbon steel.

Table 3.3 Tempering temperatures 69

Component	Temper colour	Temperature ($^\circ$C)
Edge tools	Pale straw	220
Turning tools	Medium straw	230
Twist drills	Dark straw	240
Taps	Brown	250
Press tools	Brownish-purple	260
Cold chisels	Purple	280
Springs	Blue	300
Toughening (crankshafts)	–	450–600

In the workshop, the tempering temperature is usually judged by the colour of the oxide film that appears on a freshly polished surface of the steel when it is heated. Some tools, such as chisels, only need the cutting edge hardened, the shank being left tough to withstand hammer blows.

3.4 Overheating carbon steels

It is a common mistake to overheat a steel in the hope that it will become harder. As already stated, the hardness only depends upon the carbon content of the steel and the rate of cooling. Once the correct hardening temperature has been reached, any further increase in temperature only slows up the time taken to cool the workpiece and this tends to reduce the final hardness. Further, overheating also causes *crystal growth,* resulting in a weak and defective component. If the overheating is excessive then 'burning' occurs. This is oxidation of the crystal boundaries of the metal resulting in great weakness and, unlike overheating, the condition cannot be corrected so the workpiece is useless and can only be melted down as scrap.

On the other hand, failure to reach the hardening temperature results in the component not becoming hard no matter how quickly it is quenched.

3.5 Hardening faults

Figure 3.4 shows some of the causes of components cracking during hardening, whilst Fig. 3.5 shows some causes of distortion.

It will be seen that the careful design of the workpiece and the selection of the appropriate material makes for fewer problems in the hardening shop. However, no matter how much care is taken in the design stage and during

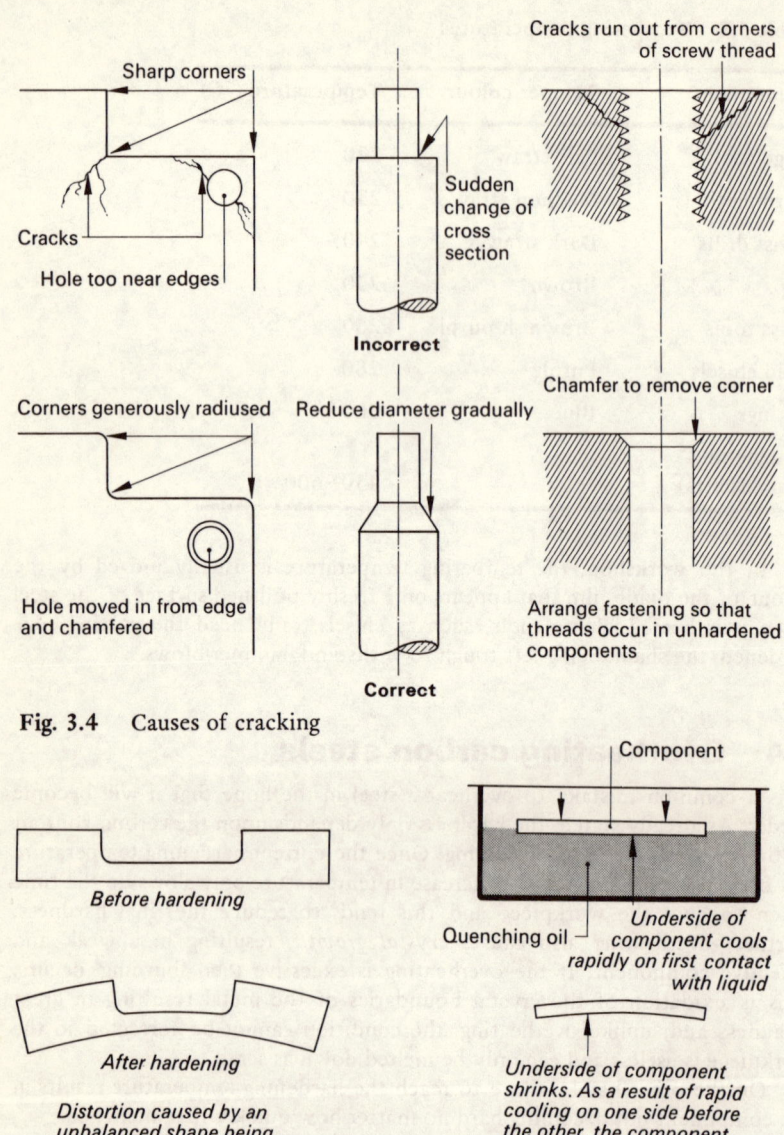

Fig. 3.4 Causes of cracking

Fig. 3.5 Causes of distortion

hardening, some movement will always take place. Further, there is often slight changes in composition of the metal at the surface of the component during heat treatment. Thus for precision components an adequate grinding allowance should be left on during initial manufacture so that surface contamination and distortion can be corrected by grinding after hardening and tempering.

3.6 Recrystallisation

The terms hot- and cold-working have already been introduced in Chapter 2. The term cold-working is relative. For instance, the metal lead hot-works at room temperature, whilst mild steel cold-works at temperatures up to 550° C. In fact, the temperature at which a metal ceases to be *cold-worked* and becomes *hot-worked* is the temperature at which the crystals distorted by the forming process start to change back to normal crystals; this temperature depends upon two factors:

(*a*)   the metal under consideration;
(*b*)   the degree of cold-working to which that metal has been subjected.

The way this happens is simply explained in Fig. 3.6. The cold-worked and distorted crystal contains stress points locked into it. At a critical temperature depending upon the metal and the degree of cold-working it has received, the nucleus of a new crystal starts to grow at each stress point. This phenomenon is called *nucleation* and the temperature at which it occurs the *temperature of re-crystallisation.*

Therefore, *cold-working* is the flow forming of metal *below* the temperature of re-crystallisation for a given metal. Similarly, *hot-working* is the flow forming of metal *above* its temperature of re-crystallisation.

3.7 Annealing plain carbon steels

The annealing processes are used to soften steels that are already hard. This hardness may be imparted in two ways:

(*a*)   *Quench-hardening (section 3.2).*
(*b*)   *Work-hardening.* This occurs when a metal is cold-worked. It becomes hard and brittle at the point where the cold-working is, causing the cystal structure of the metal to become distorted. For example, if a strip of metal is bent backwards and forwards in a vice it starts to harden at the point of bending and breaks off.

Full annealing

This process can be used for either quench-hardened or for work-hardened steels. For economic reasons it is used mainly for quench-hardened steels (see 'Sub-Critical Annealing'). Full annealing consists of heating the metal to the temperatures shown in Fig. 3.7 for its particular carbon content, and cooling

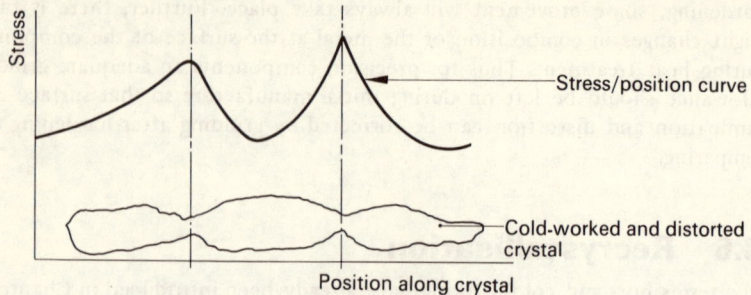

A cold-worked crystal is not of uniform cross-section. Therefore it will be subject to stress 'peaks' at the points of minimum cross-section

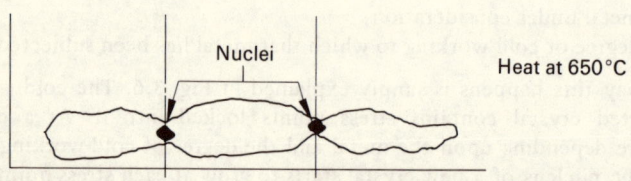

*At 650°C the nucleus of an equi-axed crystal will form at each stress 'peak'. Since 650°C is below the lower critical point this process is referred to as **sub-critical annealing***

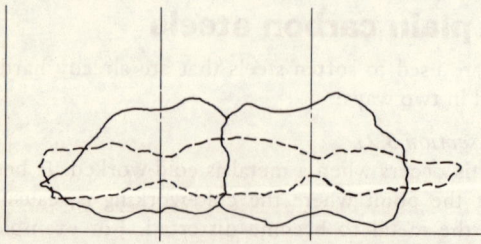

If heating is continued the nuclei grow – feeding on the material of the original cold worked crystal until it is consumed

The structure of the component will now consist of equi-axed crystals and it will be annealed

Note : Since two or more new crystals replace each original crystal, they will be smaller. Grain refinement will have occurred

Fig. 3.6 Nucleation

the metal down very, very slowly indeed — usually by turning off the furnace, closing the dampers and allowing furnace and work to cool down together. Some grain growth will occur and the metal will have maximum ductility but minimum strength for its particular composition. In this state the metal will be in the ideal condition for cold forming, but it will tend to tear and leave a poor finish when machined. It will be seen from Fig. 3.7 that the full annealing temperatures are identical with the quench-hardening temperatures for plain carbon steels.

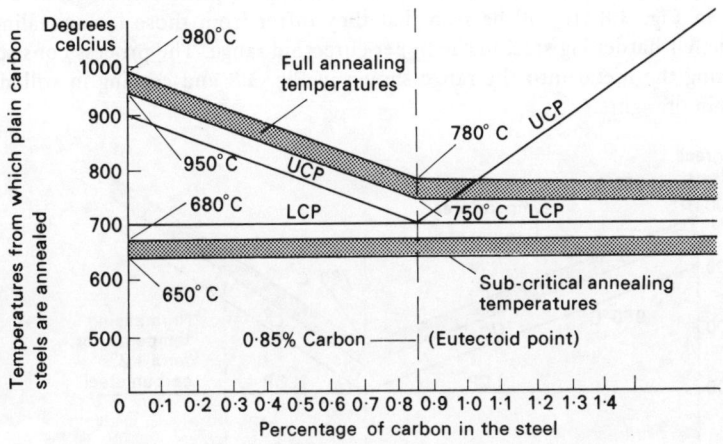

Fig. 3.7 Annealing temperatures for plain carbon steels

Sub-critical annealing

When a low or medium carbon steel is cold-worked so that it becomes work-hardened, it is often necessary to anneal the components before they are put into service or before further processing. For this latter reason sub-critical annealing is often referred to as *process annealing* or *interstage annealing*.

The re-crystallisation temperature for mild steel lies in the band $550°$ to $650°$ C, depending upon the degree of cold-working; and it is within these temperatures that work-hardened steel components are usually annealed. Since these temperatures lie *below* the lower critical temperature this process is referred to as *sub-critical* annealing. The sub-critical annealing temperatures for plain carbon steels are indicated in Fig. 3.7.

Sub-critical annealing consists of heating the work-hardened steel components until their temperatures lie within the $550°$ to $650°$ C band as shown in Fig. 3.7. The components are soaked at this temperature until the required degree of grain growth has been achieved. The components are then slowly cooled as for full annealing. This process cannot be used to anneal quench-hardened components, as such components will not contain the distorted, stressed crystals that trigger off the re-crystallisation process.

Obviously full annealing can be used to soften cold-worked materials, but sub-critical annealing is much more economical. Since the temperature is much lower, furnace maintenance and fuel consumption (and costs) are reduced. Further, little oxidation of the work occurs and there is no need to use a controlled atmosphere furnace.

3.8 Normalising

The temperatures associated with the *normalising* of plain carbon steels are shown in Fig. 3.8. It will be seen that they differ from those for annealing and quench hardening steel in the hyper-eutectoid range. The process consists of heating the metal into the range shown in Fig. 3.8 and cooling in still air free from draughts.

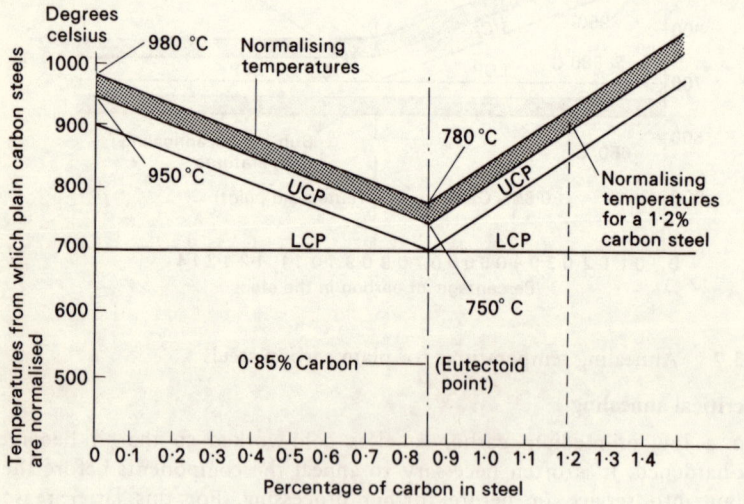

Fig. 3.8 Normalising temperatures for plain carbon steels

This slightly more rapid cooling results in the metal having a finer grain structure than for annealing. Thus the metal is rather less ductile, but the more refined grain structure gives a superior finish when the metal is machined. The finer grain structure also results in superior strength compared with annealed metals. Normalising is also used to remove process stresses and prevent warping.

3.9 Heat treatment of plain carbon steels (examples)

Example 3.1. *To harden and temper a cold chisel made from octagonal chisel steel having a carbon content of 0·6 to 0·7 per cent.*

From Fig. 3.3 it will be seen that the correct hardening temperature for a 0·6–0·7 per cent plain carbon steel lies between 820° and 850° C (bright red heat). As the carbon content is fairly low, and a chisel is a simple shape, it can be safely quenched in water to achieve maximum hardness.

To temper the chisel, the cutting end is first polished so that the temper colours caused by the oxide film can be observed. The shank is then heated and the polished end is watched as the temper colours travel towards it. If the chisel is to cut metal, the cutting edge should be a brownish-purple when it is quenched. If the chisel is to cut brick or concrete the cutting edge should be purple to give it greater impact resistance. Letting the heat travel down the shank ensures that the shank is left in a tough, rather than a hard, condition so that it will not shatter when hit by a hammer.

Example 3.2. *To harden and temper a press-tool die made from 1·2 per cent plain carbon steel.*

From Fig. 3.3 it will be seen that the correct hardening temperature for a 1·2 per cent plain carbon steel lies between 750° and 780° C (dull red heat). As the carbon content is high and the die may be relatively intricate in shape, it should be quenched in oil to prevent distortion and cracking.

Unlike Example 1, the press tool die must be tempered to a uniform hardness throughout. This can only be achieved by uniform heating. When the die is the correct tempering colour — dark brown for brass; brownish purple for steel — it is again quenched in oil.

Example 3.3. *To anneal a component made from 0·6 per cent carbon steel.*

From Fig. 3.7 it will be seen that the full annealing temperature for a 0·6 per cent steel is approximately 850° C (the same as for hardening). The steel is raised to this temperature and soaked until the temperature is uniform throughout. The furnace is turned off; the dampers are closed; the furnace and component cool down slowly together.

Example 3.4. *To anneal a component made from 1·2 per cent plain carbon steel.*

From Fig. 3.7 it will be seen that the full annealing temperature for a 1·2 per cent carbon steel lies between 750° and 780° C; large components will then be cooled in the furnace as described in Example 3. Small components may be cooled slowly by burying them in ashes or in powdered limestone.

Example 3.5. *To anneal a batch of mild steel pressings that have been severely cold-worked.*

Since the components have been work-hardened, as a result of pressing them to shape below the temperature of re-crystallisation, they will be sub-critically annealed. They would be loaded into metal pans or baskets and placed in a furnace at the sub-critical annealing temperature which for mild steel, will lie in the $650°$ to $680°$ C band (see Fig. 3.7). As for the previous examples, after raising the components to their annealing temperature they will be cooled very slowly.

Example 3.6. *To normalise a 1·2 per cent carbon steel forging after rough machining to relieve residual stresses.*

Reference to Fig. 3.8 shows that the normalising temperature for a 1·2 per cent carbon steel lies within the $900°$ to $930°$ C band. It will be seen that this is considerably higher than for annealing or hardening a 1·2 per cent carbon steel. Despite the higher temperature, grain growth is avoided by the faster rate of cooling. After the steel has reached $900-930°$ C it is removed from the furnace and allowed to cool down in still air. Care must be taken to avoid:

(a)   placing the component in a draught;
(b)   placing the component on a surface that will 'chill' it;
(c)   restricting the natural circulation of air around the component.

3.10   Heat treatment of non-ferrous metals

None of the non-ferrous metals and only few of the non-ferrous alloys can be quench hardened like plain carbon steel. The majority of non-ferrous components are hardened by manufacturing them by cold-working processes or manufacturing them from cold-rolled ('spring temper') sheet or cold drawn wire. Some alloys, notably those based upon aluminium, age harden naturally after annealing, or they can be hardened by a precipitation heat treatment process. Since, in the main, most hardened non-ferrous metals are work hardened, it follows that they must be annealed by a re-crystallisation process similar to the sub-critical annealing process for steel. For example:

Aluminium	$500°-550°$ C	(pure metal)
Copper	$650°-750°$ C	(pure metal)
Cold-working brass	$600°-650°$ C	(simple alloy)

The majority of non-ferrous alloys that can be heat treated require a slightly different approach. They are annealed by *solution treatment* and hardened by *precipitation treatment*. For example: duralumin is the generic name for a group of heat-treated, wrought alloys containing traces of copper, magnesium, manganese and zinc. The basis of the duralumin alloys is aluminium as the name implies. Duralumin can be heat treated as follows.

Solution treatment: The alloy is raised to a temperature at which the alloying elements can form a solid solution in the aluminium (e.g. approximately $500°$ C). The alloy is quenched from this temperature to preserve the solution at room temperature.

Precipitation treatment: If the alloy is kept at room temperature after solution treatment, hardening commences spontaneously. That is, particles of copper/aluminium compounds start to precipitate out of the solid solution, hardening and strengthening the alloy.

When hardening occurs naturally it is referred to as '*age-hardening*' and maximum hardness occurs in about four days. Natural ageing can be delayed by refrigerating the solution treated components at $-6°$ to $-10°$ C.

Precipitation hardening can be accelerated by reheating the solution treated components to about $150°$ to $170°$ C for a few hours. This is referred to as artificial-ageing or precipitation-hardening.

The times and temperatures vary for each alloy and the specified treatment must be carefully observed.

3.11 Quenching media

The most widely used quenching media, commencing with the most severe (quickest), are brine solutions, water, oil and air. Water quenching is essential if the maximum hardness is required from plain carbon steels, but its use can lead to cracking with high carbon steel. Water has the disadvantage of forming a steam blanket around the component and this considerably reduces the cooling rate. The steam blanket can be dispersed by agitating the work in the quenching bath. Further, as the temperature of the water in the bath rises, the rate of cooling slows down and the efficiency of the bath is reduced. Therefore the water should be circulated through a cooling and filtering system so that the bath is supplied with clean, cool water. The mass of water in the bath should be large compared with the mass of the components being quenched.

Brine solutions (sodium chloride) in water give a more severe quench for two reasons. First, the solution has a higher boiling point than pure water, and, second, the salt strips the scale from the steel thus improving the metal-to-quenching media contact. This results in a more rapid heat conduction and more rapid cooling of the work.

Quenching oil should have a low viscosity to avoid blanketing effects and a high flash point to lessen the fire hazard. On no account should ordinary mineral lubricating oils be used. Special oils have been developed that give a rapid and uniform cooling with a minimum of fuming and a low fire risk. However, even when using a specially blended quenching oil certain precautions must be taken.

1. The quenching tank *must* be fitted with an airtight metal lid that can be remotely lowered over the tank in the event of fire to blanket the flames.
2. Adequate and suitable fire fighting equipment must be readily available and the personnel in the heat treatment department must be trained in its use.
3. A fume hood and extractor fan should be fitted over the quenching tank.
4. Provision should be available for circulating the oil through a cooling and filtering plant to ensure that it is kept well below its flash point and that it is kept clear of foreign matter.

5. The mass of oil in the tank should be great compared with the mass of the work being quenched in it.

Oil is most widely used in conjunction with alloy steels where the critical cooling rate is slower than for plain carbon steels.

Some alloy steels have such a low critical cooling rate that small components made from such steels may be quenched in a cold air blast.

Problems

Section A

1. The correct hardening temperature for a 0·6 per cent carbon chisel steel is: (a) slightly above its lower critical temperature;   (b) slightly above its upper critical temperature;   (c) slightly below its lower critical temperature;   (d) as high as possible without melting the steel.

2. The purpose of tempering a hardened steel component is to: (a) increase its toughness;   (b) reduce its hardness;   (c) increase its hardness; (d) increase its ductility.

3. The solution treatment of aluminium alloys consists of: (a) raising the temperature of the alloy so that the elements can form a solid solution; (b) heating the alloy in a solution of chemicals;   (c) melting the alloy and adding other solids to form a solution;   (d) a corrosion proofing process.

4. The precipitation treatment of aluminium alloys is used to: (a) soften the alloys;   (b) harden the alloys;   (c) increase the density of castings; (d) remove impurities from the alloys.

5. To achieve the maximum hardness when quench-hardening high carbon (hyper eutectoid) steels it is essential to: (a) raise the temperature of the steel as high as possible;   (b) soak the steel at a high temperature for as long as possible;   (c) cool the steel slowly in oil;   (d) cool the steel quickly in water.

Section B

6. A batch of work-hardened mild steel pressings are to be sub-critically annealed. Describe: (a) what is meant by the term *sub-critical annealing*; (b) how the process is carried out;   (c) the advantages of this process over full annealing in this instance.

7. Describe the essential differences between *normalising* and full *annealing* and give an example of where each process would be used.

8. Describe the difference between hot- and cold-working when flow-forming metal components and compare the advantages and limitations of each process.

9. Explain, in detail, how a chisel made from a silver steel (high carbon) steel rod should be hardened and tempered for cutting cast iron.

10. (a) Explain how duralumin rivets should be softened.   (b) Explain how the process of 'age-hardening' may be delayed after the rivets have

been softened.   (c) Explain, briefly, the cause of age-hardening in
duralumin alloys.

Chapter 4

Corrosion prevention

4.1 The corrosion of metals

Corrosion is the slow but continuous eating away of metallic components by chemical or electro-chemical attack. That this is costly and destructive will be vouched for by any motorist who has seen his valuable 'pride and joy' eaten away before his eyes as body rot sets in. Three factors govern the rate of corrosion.

1. The metal from which the component is made.
2. The treatment the surface of the component receives.
3. The environment in which the component is kept.

The metals used by engineers do not, with the exception of copper and gold, occur in nature in the metallic state. They all occur compounded with other chemicals to form minerals. Extractive metallurgy is concerned with the separation of metals from the other element with which they are naturally combined, and the extractive processes use vast quantities of costly energy. For example, it takes approximately 15 000 kWh (kilowatt hours) to produce one tonne of aluminium. The development of the extractive industry can be related to the development and availability of energy sources.

The expenditure of large amounts of energy in extracting a metal from its ore can be compared with a mechanical system in which the state of the metal is equivalent to an object that has been pushed to the top of a steep incline. The steeper the slope, the greater the amount of energy required, and the greater the tendency for the object to slip back down the slope. That is, the less stable the object is. Similarly the greater the amount of energy that is required to separate a metal from its ore, the greater will be its tendency to slip back into a more stable compound where it will, once more, be combined with other chemical elements. This re-combination process is *corrosion* — the separation process is reverse — resulting in degradation of the metal so that it no longer exhibits the properties that makes it useful to the engineer.

All metals corrode to a greater or lesser degree; even precious metals like gold and silver tarnish in time and this is a form of corrosion. Corrosion-prevention processes are not able to prevent the inevitable failure of a component by corrosion, but slow down the process to a point where it will have worn out and been discarded for other reasons before failing due to corrosion.

Fortunately the reactivity of a metal and the rate at which it corrodes is not related. For example, aluminium is chemically more reactive than iron, hence more energy is required to extract it from its ore. However, immediately it is exposed to the atmosphere it forms an oxide film that seals the surface and prevents further corrosion taking place. On the other hand, iron corrodes more slowly but its products of corrosion are porous and the process continues unabated until the metal is destroyed. Table 4.1 indicates the rates of corrosion of unprotected steelwork.

Table 4.1 Rate of corrosion

Type of environment	Typical rate of rusting for mild steel in temperate climates (mm/year)
Rural	0·025−0·050
Urban	0·050−0·100
Industrial	0·100−0·200
Chemical	0·200−0·375
Marine	0·025−0·150

4.2 Atmospheric corrosion

The cost of corrosion and its prevention is largely related to atmospheric corrosion, and it is this form of corrosion that will be considered in this chapter. The corrosion problems associated with chemical engineering, marine engineering and food processing are more highly specialised and the remedies resorted to after a careful study of the environmental conditions are, correspondingly, more specific.

Any metal exposed to normal atmospheric conditions becomes covered with an invisible, thin film of moisture. This moisture film is invariably contaminated with dissolved solids and gases which increase the rate of corrosion. The most common example of corrosion due to moisture is the rapid surface formation of 'red rust' resulting from the exposure of iron and steel to the atmosphere. This 'red rust' is an oxide of iron, but a different oxide from the blue black 'mill scale' oxide formed under the action of heating in dry air, and is the result of the corrosive action of oxygen and moisture. It is shown in Fig. 4.1 that iron and steel does not rust in dry air, or in pure water from which the dissolved air has been removed by boiling.

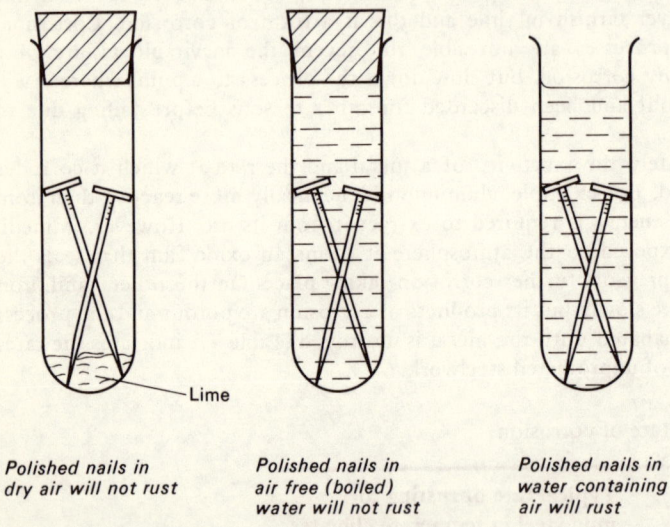

Lime

Polished nails in
dry air will not rust

Polished nails in
air free (boiled)
water will not rust

Polished nails in
water containing
air will rust

Fig. 4.1   Corrosion of steel

Once 'rusting' commences the action is self-generating. That is, it will continue even after the initial supply of moisture and air has been removed. This is why all traces of rust must be removed before painting. The rate of rusting slows down as the rust deepens or thickens, but as rain washes off the surface rust the rate increases again. This cycle is continuous and, once started, is very difficult to control.

Atmospheric pollution rapidly increases the rate of rusting of iron and steel. It also attacks, but much more slowly, copper and zinc. Lead is virtually unaffected and so is aluminium if its surface has been correctly pre-treated and it is regularly washed clean. How the gases given off by burning fuels and by industrial processes are converted into corrosive substances is shown in Fig. 4.2. Gases that become acids when dissolved in water are called *acid anhydrides*. In coastal areas the problem is increased by the presence of salt spray in the atmosphere.

The most important gaseous pollutant is sulphur dioxide. Many 'clean fuels' are guilty of producing large quantities of this gas. Unfortunately the 'clear air' act only controls the visible products of combustion and ignores the

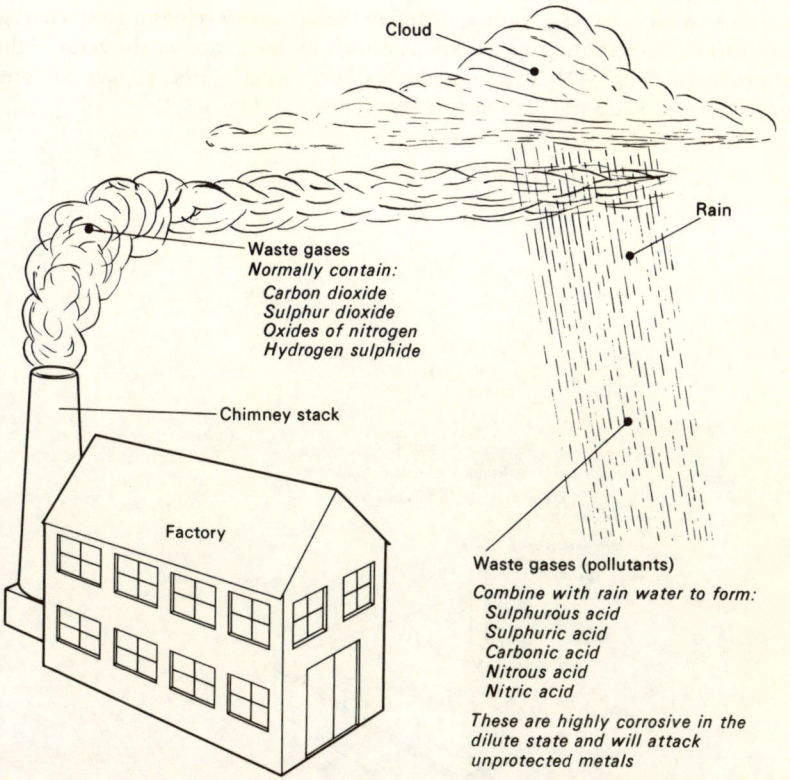

Cloud

Rain

Waste gases
Normally contain:
Carbon dioxide
Sulphur dioxide
Oxides of nitrogen
Hydrogen sulphide

Chimney stack

Factory

Waste gases (pollutants)

Combine with rain water to form:
Sulphurous acid
Sulphuric acid
Carbonic acid
Nitrous acid
Nitric acid

*These are highly corrosive in the
dilute state and will attack
unprotected metals*

Fig. 4.2 Corrosive pollution

fact that many of the most damaging pollutants are colourless gases that are invisible to the naked eye. It has been estimated that, in 1953, 10 million tonnes of sulphurous and sulphuric acid were produced from the combustion of coal alone in the United Kingdom. This is about four times the amount of such acids produced intentionally.

4.3   Metals that resist corrosion

It has already been shown (section 4.1) that metals are chemically unstable substances that combine with atmospheric oxygen or atmospheric pollutants to a greater or lesser extent. The following metals, which resist corrosion, react to form impervious, homogeneous coatings on their surfaces which prevent further corrosion from taking place, providing these coatings of the products of initial corrosion remain undisturbed.

Copper

When copper is exposed to atmospheric pollution for a long time, as for example when used as a roofing material, the surface develops a green coating or 'patina'. This coating or 'patina' is caused by the action of the acids in the atmosphere (Fig. 4.2) attacking the oxide coating of the copper to form protective sulphate and carbonate salts as shown in Fig. 4.3.

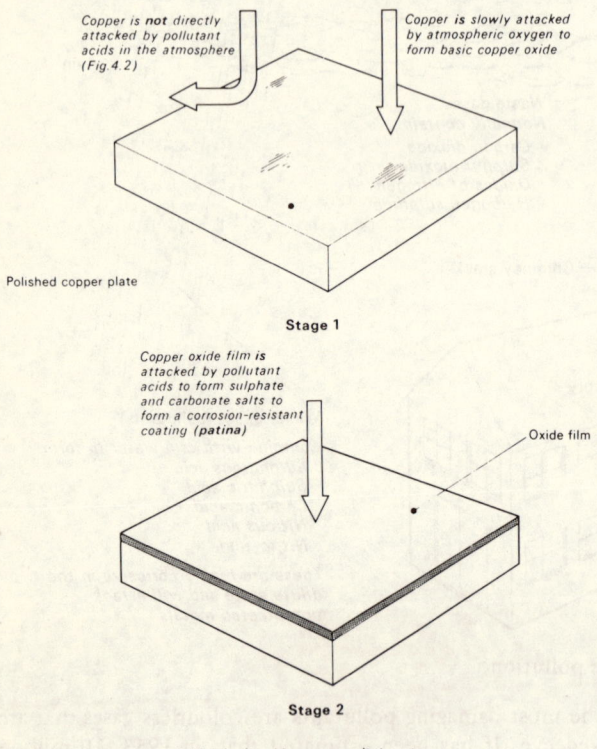

Copper is **not** directly attacked by pollutant acids in the atmosphere (Fig. 4.2)

Copper is slowly attacked by atmospheric oxygen to form basic copper oxide

Polished copper plate

Stage 1

Copper oxide film **is** attacked by pollutant acids to form sulphate and carbonate salts to form a corrosion-resistant coating (**patina**)

Oxide film

Stage 2

Acid + basic oxide ⟶ salt + water

Fig. 4.3 Formation of 'Patina' on copper

The 'patina' which forms on copper giving it a protective skin must not be confused with the green compound *'verdigris'* which forms on the surface of copper by the action of *organic matter*. Verdigris corrosion will, in time, completely destroy copper. Thus copper vessels used for industrial processing, food processing and cooking are heavily coated with tin to prevent verdigris forming.

Zinc

The results of atmospheric attack upon zinc is very similar to copper, expecially when used for outdoor purposes. In this instance a carbonate coating forms on the surface after a period of exposure, and this forms a protective film that gradually strengthens with time. This coating is grey is colour, not unlike the colour of the parent metal itself, and does not crack or peel off with any expansion or contraction of metal due to temperature changes. For this reason zinc is an excellent exterior building material. Similarly it gives excellent protection when coated on to steel (see: galvanising, section 4.6). Its useful life tends to be foreshortened in areas of high industrial pollution where there is a high concentration of sulphur compounds in the air. Galvanised hardware has a noticeably shorter life in an urban environment than it has in a rural environment.

Aluminium

The reaction of aluminium and its alloys with the atmosphere is somewhat different to that of copper and zinc, but the effect of the initial reaction is, as in the previous cases, to retard further corrosion.

Aluminium has a greater affinity for oxygen and even highly polished aluminium surfaces quickly develop a thin, transparent film of aluminium oxide or 'alumina' which, if the metal is kept indoors, prevents further oxidation and retains the bright appearance. However, exterior use of aluminium results in the oxide film thickening. When this happens the film becomes grey in colour but, when sufficiently developed, protects the parent metal from further attack. The oxide film on aluminium and its alloys can be artificially thickened by a process called *anodising* as described in section 4.12.

Lead

When the surface of a lead sheet is scratched, but or scraped, the newly exposed metal will have a silvery — or bluish-white — appearance. However, the fresh surface of lead quickly loses its considerable lustre which tarnishes on exposure to the atmosphere. Thus the normal appearance of lead is a characteristic dull grey colour.

This 'white' oxide film resulting from exposure to the atmosphere is very tenacious and prevents further attack. Lead is one of the most corrosion resistant of all metals and certainly the most resistant of the common base metals. Today, approximately one-third of the world's output of lead is used as a sheathing material for underground telephone and power cables. It has the property of remaining virtually incorrodible year after year. It is also used as a protective metallic coating for sheet iron and steel ('Terne-plate').

Stainless steel

Stainless steel is one of the outstanding corrosion-resistant metals, which is remarkable when it is considered that the main constituent of the alloy is iron, an element which possesses no resistance when exposed to corrosive influences. Unlike the metals so far considered, stainless steel has high structural strength as well as corrosion resistance. One familiar stainless steel is known as '18/8' because the alloy contains 18 per cent chromium and 8 per cent nickel.

As with aluminium and lead, it is the formation of a complex oxide film which protects the surface from attack. The grades containing molybdenum are recommended for architectural applications in heavily polluted areas as they will retain their polish and remain unchanged in appearance indefinitely. Stainless steels are not confined to applications requiring resistance to atmospheric corrosion; they are used extensively for chemical plant and food handling equipment where they combine corrosion resistance with non-toxic properties. They also exhibit corrosion-resistant properties at elevated temperatures. No finishing treatment is given to, or required by, stainless steel other than its initial polishing or brushing, and although expensive in first cost it is economical in the long term as minimum maintenance is required.

Nickel

This metal has already been mentioned as a constituent of stainless steel but, because of its high resistance to chemical attack, is used extensively for 'nickelplating'. When alloyed with copper, in the proportions of two-thirds nickel to one-third copper, '*Monel Metal*' is produced. This alloy is a very useful fabrication material, being ductile and readily welded. It is extremely resistant to corrosion and particularly to attack by alkalis, sea water and acids. It is widely used for appliances in the chemical manufacturing industries and for marine engineering.

Chromium

Although this metal is an important constituent of stainless steel, one of its more important uses is for electro-plating (section 4.7) articles against corrosion. It is highly resistant to corrosive influences and retains its high polish and colour for long periods.

4.4 Corrosion prevention

Methods of preventing or retarding corrosion are largely applied to iron and steel, since, as already explained, most of the common non-ferrous metals and alloys form their own protective coating when exposed to the atmosphere.

For short-term protection, oiling or greasing up a component may be adequate. However, in the long term this is not satisfactory for the following reasons:

(*a*) the protective film will dry up and no longer seal the surface from the corrosive environment;

(*b*) oils and greases absorb moisture from the atmosphere and corrosion can take place *under* the oil or grease film;

(*c*) cheap oils and greases often contain active sulphur and acid impurities that will themselves attack and corrode the very metal surface that they are supposed to be protecting.

Table 4.2 lists the more permanent methods of corrosion prevention. Care must be taken when cladding one metal with another, for when two dissimilar metals come into contact in a moist or wet corrosive (polluted) environment they behave like a simple electric cell. The currents generated cause corrosion.

Table 4.2   Corrosion-resistant coatings   87

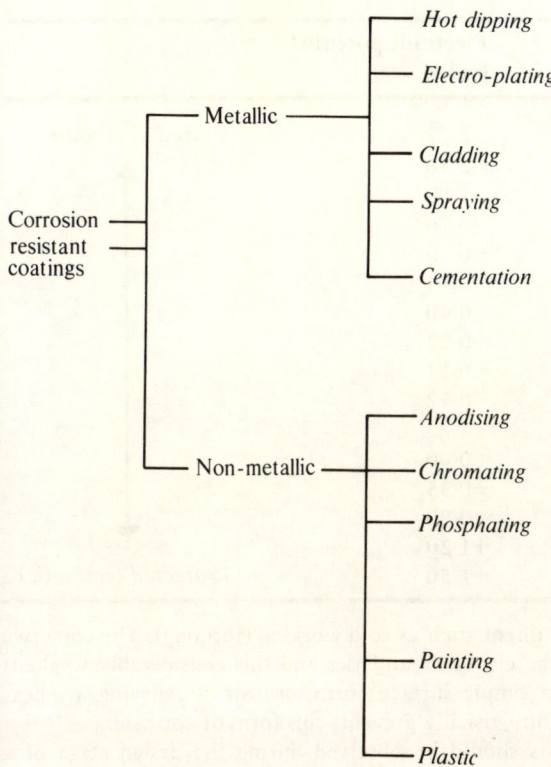

Note: That where a protective coating is applied to the metal component, this coating should be applied AFTER manufacture to ensure that it is not broken at any point.

Metals can be arranged in a special order called the *electrochemical series*. This series is given in Table 4.3 and it should be noted that, in this context, hydrogen gas behaves like a metal. If any two metals from the table come into contact in moist surroundings the more negative one will corrode the most rapidly. For example:

(*a*)   in galvanised iron (zinc-coated steel) the zinc corrodes away whilst protecting the steel. The zinc is said to be *sacrificial*;

(*b*)   in tin-plate, the mild steel is corroded if the tin coating is broken at any point. Hence cut edges should always be sealed by soft soldering or painting with a suitable lacquer, and marking out should be done with a soft pencil and not a scriber. Figure 4.4 shows what happens during electrolytic corrosion.

Chemical and electro-chemical corrosion is intensified when a metal is under stress. The stresses involved are usually internal stresses caused by some

Table 4.3   Electro-chemical series

Metal	Electrode potential (volts)	
Sodium	−2·71	*Corroded (anodic)*
Magnesium	−2·40	
Aluminium	−1·70	
Zinc	−0·76	
Chromium	−0·56	
Iron	−0·44	
Cadmium	−0·40	
Nickel	−0·23	
Tin	−0·14	
Lead	−0·12	
Hydrogen (reference potential)	0·00	
Copper	+0·35	
Silver	+0·80	
Platinum	+1·20	
Gold	+1·50	*Protected (cathodic)*

previous mechanical treatment such as cold working (forming). The corrosive attack is usually along the crystal boundaries and this considerably weakens the metal far more than simple surface corrosion. Stress relieving, by heat treatment after manufacture, usually prevents this form of corrosion.

The following points should be observed during the design stage of a component or assembly to reduce corrosion to a minimum.

1. Prevention of crevices and moisture traps.
2. Selection of a suitable material which is inherently corrosion resistant, or the close specification of an anti-corrosive treatment process for the material chosen.
3. Sealing joints if they are not continuously welded.
4. Adequate ventilation and drainage.
5. Contact with corrosive substances to be kept to a minimum.
6. Ease of washing down and cleaning.

4.5   Surface preparation

It has been firmly established that the essential and most important factor of any efficient anti-corrosion treatment is surface preparation. However carefully selected the protective process may be, it cannot fulfil its purpose if it is applied to an inadequately prepared surface. Surfaces carrying dirt, grease, corrosion or millscale are unsuitable for direct application of an anti-corrosion treatment. Table 4.4 describes how the surface may be prepared for treatment.

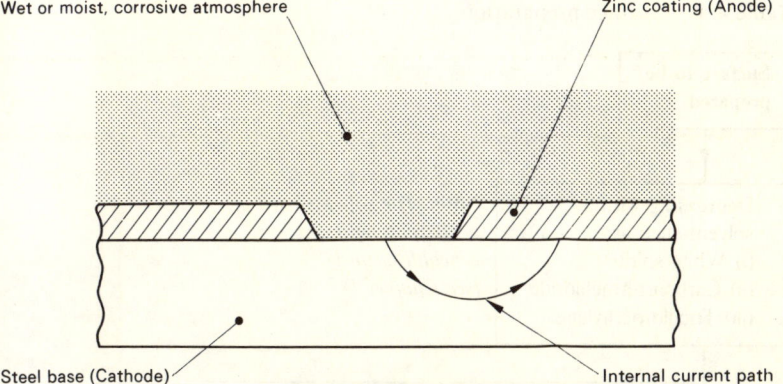

Wet or moist, corrosive atmosphere

Zinc coating (Anode)

Steel base (Cathode)

Internal current path

(a) Protection by a sacrificial coating

Coating is eaten away whilst protecting the base

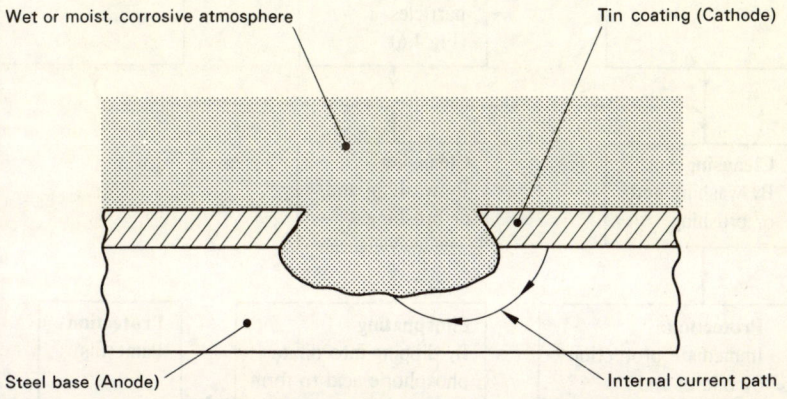

Wet or moist, corrosive atmosphere

Tin coating (Cathode)

Steel base (Anode)

Internal current path

(b) Protection by a purely mechanical coating

*Coating only protects the base if intact.
If coating is damaged, base is eaten away
quicker than if coating were not present.*

Fig. 4.4   Electrolytic corrosion

4.6   Galvanising

This is the coating of mild steel with zinc. There are two alternative processes: *hot dip galvanising*, in which the cleaned and fluxed work is dipped into a bath of molten zinc; and *electrolytic galvanising*, where the zinc is deposited electrolytically on the sheet metal base.

Table 4.4   Surface preparation

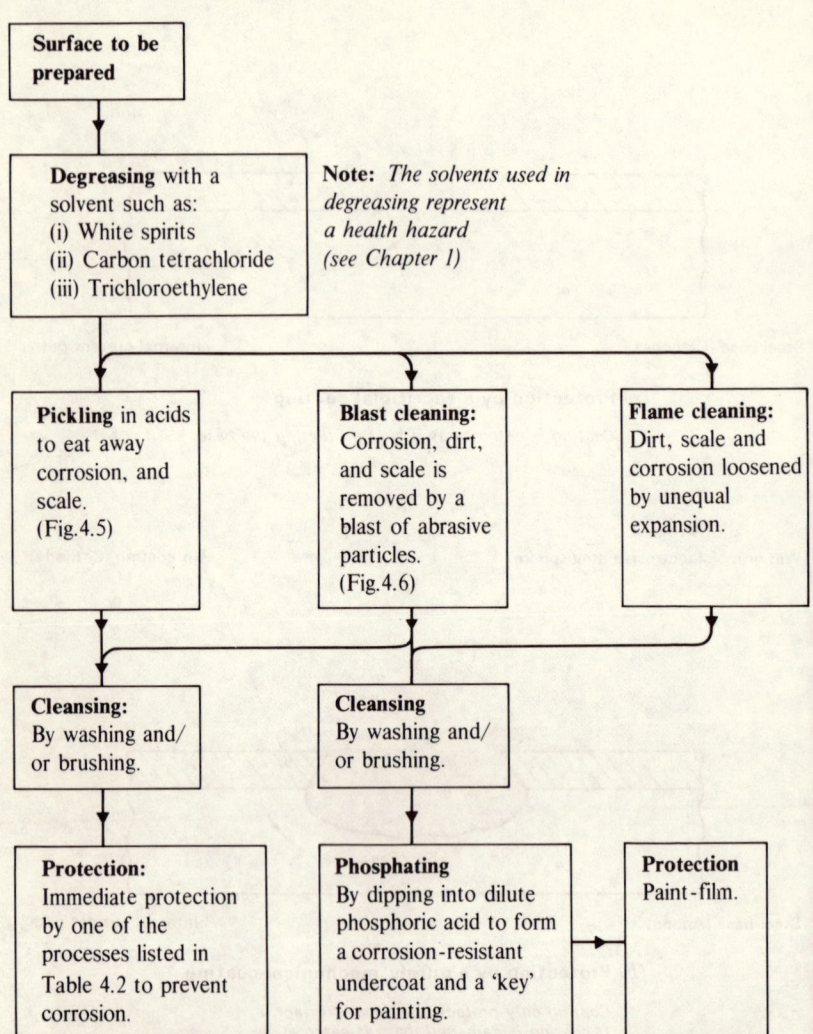

Hot dip galvanising is a very versatile process. It can be applied equally well to structural steelwork, nuts and bolts, strip, tube, wire or cast iron. It is widely used because of its reliability, its ability to withstand rough treatment, its low cost and the unique type of corrosion resistance (sacrificial coating) provided by all methods of zinc coating. Wherever possible *finished* fabricated components should be hot dip galvanised to seal raw edges and joints to avoid corrosion at these points.

The work to be galvanised is initially pickled in hot sulphuric or cold hydrochloric acid to clean the surface. It is then fluxed with a mixture of zinc

chloride ($ZnCl_2$) and ammonium chloride (NH_4Cl). Finally the work is dipped
into molten zinc to which a trace of aluminium is added to reduce the surface
oxidation of the molten zinc; that is, the formation of 'zinc ash'. The
addition of the aluminium brightens the coating and makes it more smooth
and uniform in thickness. The zinc bath is usually maintained at a temperature of 450° to 465° C. The work is then quenched in a water bath to cleanse
the surface, prevent further zinc/iron reactions and to make the work cool
enough to handle readily.

In electrolytic galvanising, the zinc is deposited on the work electrically.
This gives greater control over the thickness and uniformity of the coating.
The coating deposited by electrolytic galvanising is generally much thinner
than for hot dip galvanising and it is often used as a pre-treatment for steel
prior to painting or plastic coating.

4.7   Electro-plating

It has just been stated that in electrolytic galvanising the zinc is deposited on
the work electrically. Many other metals apart from zinc can be plated on to
components electrically and this process is called *electro-plating*. It is used to
cover components with a coating that may be decorative or protective or
both. Electro-plating is a solution treatment finishing process in which the
component(s) to be plated are immersed in a solution called the electrolyte.
The components to be plated are made the *cathode*. That is, they are
connected to the negative pole of a low voltage, heavy current, d.c. supply as
shown in Fig. 4.5. To complete the circuit *anodes* are immersed in the

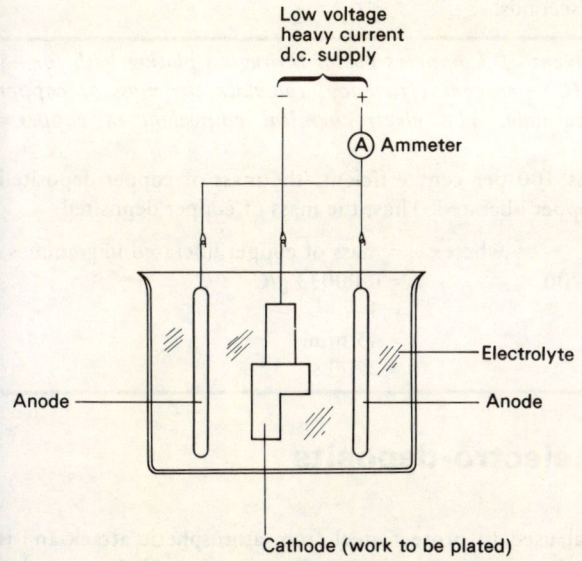

Fig. 4.5   Electro-plating

solution and connected to the positive pole of the supply.

It is the electrolyte that supplies the metal ions that are deposited on to the cathode surface (components being plated). The anodes may be *soluble*, that is they may be of the same metal as that being deposited, for example nickel, copper or zinc. Alternatively, they may be *insoluble,* as in chromium plating, in which case they only complete the circuit. Soluble anodes maintain the strength and stability of the electrolyte, whereas the strength of the electrolyte is steadily depleted when insoluble anodes are used. The rate of deposition is governed by Faraday's Laws of Electrolysis. They may be summarised as:

(a) **The mass of metal deposited is directly proportional to the magnitude of the current, and to the time for which it is passing.**

(b) **For any given quantity of electricity the mass of metal liberated from the electrolyte is proportional to the chemical equivalent of the metal.**

Note: The mass of metal deposited is only equal to the mass liberated in a plant which is 100 per cent efficient. This never happens in practice, as some metal is rejected and forms a sludge at the bottom of the bath.

The mass of metal liberated may be calculated by using the **electro-chemical equivalent.** (This is derived from (b), above, and is the mass of metal liberated by *one coulomb* of electricity passing through the cell.)

$$m = Ixt$$

where: m = mass of metal liberated in grammes.
x = electrochemical equivalent.
I = current in amperes.
t = time in seconds.

Example 4.1. *A current of 3 amperes passes through a plating bath for 45 minutes. Assuming 100 per cent efficiency, calculate the mass of copper deposited on the cathode. The electrochemical equivalent of copper =* 0.00033 *g/C.*

Since the bath is 100 per cent efficient, the mass of copper deposited equals the mass of copper liberated. Thus, the mass of copper deposited:

$m = Ixt$ where: m = mass of copper liberated in grammes
$= 3 \times 0.00033 \times 2700$ $x = 0.00033$ g/C
$= \underline{2.67\ g}$ $I = 3$ A
     $t = 45$ min.
     $= 2700$ s

4.8 Typical electro-deposits

Cadmium

This is a white metal used to protect steel from atmospheric attack and is particularly used in the electrical and aircraft industries. Its low contact resistance reduces the risk of electro-chemical corrosion when plated steel components and aluminium components are fastened together. Cadmium is toxic and should not be used where there is a possibility of contact with foodstuffs or drinking water. It is resistant to alkali attack and for this reason it is to be preferred to zinc for marine applications.

Copper

The rich colour of this metal enables it to be used for both industrial and decorative purposes. It can be used directly either as a polished or an oxidised finish, in which case it is usually lacquered to prevent a 'patina' (section 4.3) forming. Alternatively, it can be used as a 'base' prior to nickel plating. Zinc base die castings are usually copper plated before nickel plating to prevent attack on the zinc by the nickel plating solutions.

Chromium (decorative and protective)

The brilliant blue-white colour of chromium coupled with its resistance to tarnishing makes it an ideal finishing deposit over a nickel base. Only a thin film is required and this usually ranges from 0·00025 to 0·00075 mm in thickness. The chromium is not itself polished, but is applied over a bright nickel base when a bright finish is required, or over a matt or semi-bright nickel surface when a satin chrome finish is required.

Chromium (hard)

For engineering applications a thick layer of chromium is built up directly on to the component. Deposits up to 0·4 mm thick can be built up where a component is to be sized by grinding after plating. The plated surface is resistant to abrasion and has anti-friction properties. For this reason it is frequently used to protect the working faces of gauges and to build up worn gauges.

Nickel

This is widely used as an initial deposit before decorative chromium plating. It is easier to process than chromium and gives a good corrosion-resistant finish. It can be polished and has 'surface levelling' properties that causes it to build up in the hollows and improve the smoothness and appearance of poorly polished and unpolished surfaces. Unfortunately nickel has a yellowish tinge to its colour and tends to tarnish. For this reason it is frequently given a coating of chromium — a much more expensive metal — which has a more attractive appearance.

Silver

Silver has a very low electrical resistance and is frequently used for plating blade contacts. It is also used for plating high quality cutlery. Since it tends to tarnish in the sulphurous atmosphere found in urban areas its popularity is giving way in favour of stainless steel.

Tin is used as a corrosion-resistant deposit on steel and on non-ferrous metals. It is non-toxic and is widely used on food handling equipment. Tin-plate is produced in continuous strip plants where thin mild steel sheet is coated. Tin-plate is largely used in the food canning industry. Printed circuit boards and terminal tags are frequently tin-plated to facilitate soldering.

4.9   Cladding

In this process a composite billet is made up of the base metal and the coating as shown in Fig. 4.6(a). As the metal is reduced by rolling or drawing, as shown in Fig. 4.6(b), the thickness of the base and coating reduces in proportion. A typical application of this process is the cladding of steel with aluminium (ALCLAD).

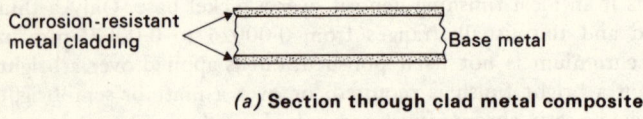

Corrosion-resistant metal cladding | Base metal

(a) **Section through clad metal composite**

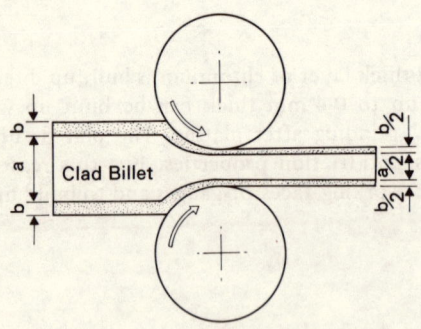

Clad Billet

(b) **Proportional reduction of clad billet**

Fig. 4.6   Cladding

4.10   Metal spraying

Metal spraying is used for a variety of purposes. The most important is the coating of ferrous metals with corrosion-resistant coatings. Other purposes include building up worn shafts, providing surfaces that are wear resistant or have improved electrical and heat conduction and for decoration.

The process consists of spraying molten or heated particles of metal on to a prepared surface. The surface should be free from grease and slightly roughened to provide a 'key' for the coating. This is usually performed by a abrasive grit blasting process. After preparation the prepared surface must be sprayed within four hours to prevent re-contamination of the surface.

The metal coating is sprayed from a 'gun' into which it is fed in the form of wire or powder. It is melted in the 'gun' by means of an oxy-fuel gas flame, an electric-arc or a plasma-arc, this latter technique being reserved for the spraying of high melting point metals. Compressed air is also fed into the gun to expel the molten metal through the nozzle.

Some typical metal spraying applications are:

Zinc

Used to rust-proof plain carbon steels. Slightly porous, so it is sealed by painting. Widely used in marine environments. This is the most commonly used deposit metal.

Aluminium

Used to rust-proof plain carbon steels in moist environments. Prevents scaling (oxidisation) at elevated temperatures. Increasingly used to protect car exhaust systems.

Copper

Used to coat and re-coat printing rolls used for high quality colour reproduction.

Used to provide high electrical conductivity coatings where electro-plating would not be thick enough to carry the required current.

Used for decorative coatings and to provide a high level of corrosion resistance.

Brass

Used for decorative and corrosion-resistant coatings; cheaper than copper.

Bronze

Used to build up bearing surfaces. Porosity improves lubrication.

Plain carbon steels

Rebuilding worn bearing surfaces of shafts.
Note: wear-resistant alloy steels can be deposited on plain carbon steels.

Stainless steel (18/8)

Extreme corrosion-resistant coatings capable of operating at very high temperatures. Architectural decoration.

4.11 Cementation processes

Three cementation processes are widely used for protecting ferrous metals. They are:

1.   Sherardising (zinc coating).
2.   Calorising (aluminium coating).
3.   Chromising (chromium coating).

1. Sherardising

In this process the articles to be coated are first cleansed by grit-blasting or acid-pickling. They are then washed and thoroughly dried. This is absolutely essential as any trace of water present will react with the zinc powder to form hydrogen gas. The accumulation of this gas would lead to an explosion. The cleansed and dried work is then placed in a rotating steel barrel containing zinc powder and heated to a temperature just below the melting point of zinc, a typical temperature being 370° C. The time taken depends upon the thickness of the coating required, but may take up to twelve hours. The zinc bonds to the ferrous work piece by diffusion and forms a hard, even layer of iron/zinc inter-metallic compounds, the slight roughness of the surface produced forming an excellent 'key' for subsequent painting processes.

2. Calorising

This is a similar process to sheradising except that the work is baked in aluminium powder at temperatures in the range 850° to 1000° C. Calorising is used to protect steel components from high temperature oxidation rather than against ambient temperature atmospheric corrosion. It also provides better protection at higher levels of humidity than sherardising.

3. Chromising

The work is baked in aluminium oxide and chromium powder to provide a chromium rich surface. This is an expensive process involving the use of high temperatures in the range of 1300° to 1400° C. Further, the process must take place in an atmosphere of hydrogen to prevent oxidation of the chromium. Because of its cost, this process is only used where extreme protection is required as, for example, in chemical plant.

4.12   Anodising

It has already been explained in section 4.3 that aluminium relies upon an oxide film on the surface of the metal to prevent corrosion. The processes of anodising artificially builds up a thick, adherent layer of aluminium oxide that is resistant to atmospheric corrosion both for interior and exterior purposes even when subjected to the pollution of urban atmospheres.

Components to be anodised are first cleaned and degreased by the use of chemical solvents, after which they are etched or polished depending upon the surface texture required. The work is then made the anode of an electrolytic cell and a direct electric current is passed through the work. The electrolyte is a dilute acid and typical acids and types of finish they provide are as follows:

Sulphuric acid

The oxide films produced by anodising in sulphuric acid are widely used for protection against general atmospheric and marine corrosion.

Oxalic acid

Oxalic acid films give a corrosion resistance comparable with sulphuric acid films and are much harder than those obtained by other general acid processes approaching the wear resistance of hard anodising. Oxalic acid is largely used on the Continent of Europe for anodising architectural metalwork. The colour obtained is the natural film colour (integral colour) and is not a dye.

Chromic Acid

This was the first anodising process to be used and is still widely used for anodising aircraft components — particularly riveted assemblies — where entrapped acid could cause corrosion. Chromic acid produces minimum corrosion under such conditions. The film produced provides an excellent 'key' for subsequent painting processes.

Selected sulphonated aromatic acids

These have been developed to provide a wide range of integral colours that are superior to dyed colours and highly resistant to weathering. The principal applications are domestic and architectural where highly decorative appearance is combined with abrasion and corrosion resistance. Typical examples being: shop fronts, windows, curtain walling and indoor metalwork such as door frames and door furniture.

Finally, the surface is sealed, usually with boiling water or steam, to improve the corrosion resistance of the coating and to minimise the absorptive properties of the oxide film. Depending upon the process used, anodised surfaces can be produced with a wide range of physical properties to suit any given set of service conditions.

4.13 Chromating

Magnesium alloys also rely upon the formation of a coherent oxide film to retard the effects of corrosion. Unfortunately, they do not respond to the anodising process and an alternative process called *chromating* (not to be confused with chromising) has to be used. In this process the components made from magnesium alloys are dipped in a solution of potassium dichromate to form a hard oxide film on the surface. Unfortunately this film is far from decorative, usually finishing grey or black. The oxidised surface should be sealed by the use of zinc chromate paint and then treated with a coloured paint to give the appropriate decorative finish.

4.14 Phosphating

Traditionally, phosphate anti-corrosion coatings were known by such names as: *Parkerising, Bonderising, Granodising* and *Walterising*. More recently the whole range of phosphate processes have been standardised under BSS 3189. It is now almost standard practice to prepare steel, zinc coated (galvanised) steel, zinc or aluminium and their alloys for painting by converting their surfaces into complex metal phosphates, chromates or oxides by immersion in chemical solutions. This not only improves their resistance to corrosion but provides a 'key' for subsequent painting processes.

Since the phosphate film has excellent anti-friction properties due to its high lubricity under extreme pressure applications, conversion coatings are also used to prepare bearing surfaces for gears, tappets, pistons, etc. Phosphate coatings are also used to prepare metal prior to cold forming operations such as wire drawing, tube drawing, sheet metal deep-drawing and cold forging. Phosphating is also used to provide high electrical resistant films on the iron laminates used for transformer cones, and the stators and rotors of electrical machines.

As an alternative to painting, phosphate coatings may be impregnated with oils, greases and waxes together with suitable stains. Such sealing methods give high degrees of corrosion resistance and were widely used from the turn of the century until quite recently when painting became more widespread.

Before components can be phosphated they have to be pickled and de-greased to provide a chemically clean surface. They are then rinsed and dried. Next they are immersed in the phosphating bath and again rinsed and dried. To give maximum corrosion resistance the components are subjected to an after treatment by immersion in a chromating solution. They are again rinsed and dried and finally sealed by oiling, waxing or painting.

4.15 Plastic coating

Plastic coatings can be both functional as well as being anti-corrosive and decorative. The wide range of plastic materials available for coating purposes provides the designer with a means of achieving:

abrasion resistance,
cushion coating (up to 6 mm thick),
electrical and thermal insulation,
flexibility over a wide range of temperatures,
non-stick, mould release properties,
permanent protection against weathering and atmospheric pollution,
reduction in maintenance costs,
resistance to corrosion by a wide range of chemicals,
the covering of welds and the sealing of porous castings.

The method of applying the plastic depends upon the type of plastic and its form of supply, but whatever the type and process, the preparation of the surface of the work is paramount in ensuring a satisfactory coating. The work must be free from grease and oxidation and, as for other coating processes, this may be carried out by grit blasting or chemical cleaning. Plastic coatings should not be applied over plated or galvanised finishes as they will affect the quality of the plastic coating. The work to be coated must be designed to suit the finish as entrapped air in badly made joints and porous castings will expand during stoving and lift the plastic off the metal sub-strata.

Fluidised bed dipping

This technique is widely used with all fluidisable powders except for epoxy resins. Figure 4.7 shows a section through a fluidised bed dipping bath. The powder has a particle size of 60–200 mesh and is supported on a bed of air passing through the porous bed from the plenum chamber. The air lifts the powder, causing it to continuously bubble up and fall back. The effect is similar in appearance to boiling water, and the diffused powder offers little resistance to the immersed, pre-heated work.

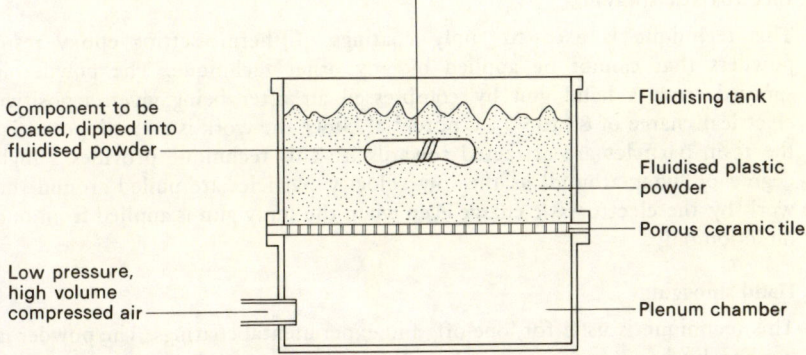

Fig. 4.7   Fluidised bed dipping

Liquid plastisol dipping

This process is limited to the use of PVC plastisol.
Note: a plastisol is a resin powder suspended in a plasticiser and no dangerous solvent is present. The process is similar to that described for the fluidised bed, except that the pre-heated work is dipped into a thixotropic (non-drip) liquid plastic instead of a powder.

Flame spraying

It is obvious that large components and structures cannot be dip coated. For site work, flame spraying is used. This process is similar to metal spraying (see section 4.10). The plastic powder is sprayed by compressed air through a suitable flame which both melts the powder and heats the air. The heated air pre-heats the work which has been previously prepared by shot or sand blasting. The molten particles of plastic impings on the workpiece where they fuse together and cool to form a coherent surface. Considerable skill is required to produce a uniform, coherent surface.

Cold powder is sprayed by means of a compressed air gun (similar to a paint spray gun) on to a pre-heated surface that has been prepared as for fluidised bed coating. It is used where the work is too large for immersion in the fluidised bed tank.

Dispersion spraying

The powder is held in suspension in a water or solvent base and then sprayed on to the cold workpiece. Stoving is usually a lengthy process as the base has to be thoroughly evaporated. This is a hazardous process and care has to be taken against the inhalation of solvent fumes and the possibility of explosions. This process is widely used to apply fluorocarbon coatings which cannot be applied by any other process.

Electrostatic spraying

This technique is used to apply coatings of thermo-setting epoxy resin powders that cannot be applied by any other technique. The powder is sprayed from a hand gun by compressed air after being given a positive electrical charge of 80 to 90 kV potential. Since the work is at earth potential the resin particles are attracted towards it. This technique provides a high degree of 'wrap-around' as the low velocity particles are pulled around the work by the electrostatic charge even when the spray gun is applied from one direction only.

Hand smogging

This technique is used for 'one-off' and experimental coatings. The powder is worked by hand (wearing suitable gloves) over the pre-heated components. No special equipment is required.

4.16   The paint film

Painting is widely used for the protection and decoration of metallic components and structures. It is the easiest and cheapest coating that can be applied with any degree of permanence and by careful choice painting can provide a wide range of protective properties.

Paints may generally be described as finely divided solids in a liquid suspension which dries or sets to provide a coherent film over the metal surface. Usually a paint is made up of three main constituents:

1. The vehicle: This contains the film-forming component or binder in a volatile solvent. The binder is a natural or synthetic resinous material and reflects the essential properties of the paint; its durability; corrosion resistance; flexibility and adhesion.

2. The pigment: This provides the paint with its opacity and colour.

Further, some pigments have special properties and act as corrosion inhibitors, fungicides, insecticides, etc.

3. The solvent or thinner: This controls the consistency of the paint and controls its application. Since the solvent evaporates once the paint has been spread, it does not form part of the final paint film. In addition, a paint may contain small quantities of a catalyst or accelerator to speed up the drying reactions, together with anti-skinning, anti-settling and thixotropic (anti-drip) agents.

A complete paint system consists of:

(a) *A primer:* This is used as an adhesive for the subsequent protective coats, and may also contain a corrosion-inhibiting pigment. Primers used on some metals, such as aluminium, contain an etching agent to produce a suitable 'key' so that the primer will adhere to the metal sub-strata.

(b) *Putties or fillers:* Those are applied by knife or spatula to fill casting porosity and larger surface blemishes such as dents in sheet metal.

(c) *Undercoats:* These are used to build up the thickness of the paint film, cover the primer and filler and give opacity to the colour of the finish coats, as well as providing a smooth surface for their application where a high gloss is required (e.g. motor car bodywork).

(d) *Finish or top coats:* These are both decorative and provide most of the corrosion resistance. They also have the mechanical strength to withstand abrasive damage.

4.17   Types of paint

Paints may be broadly classified by the manner in which they dry.

Group 1

Atmospheric oxygen reacts with the binder, causing it to polymerise into a solid film. This reaction can be speeded up by *forced drying* at 70° C. Paints that dry by oxidation include the traditional linseed oil-based paints, the oleo-resinous paints, and the modern, general purpose air drying paints based on oil-modified alkyd resins.

Group 2

These paints are based on amino alkyd resins which do not cure (set) at room temperature and which have to be 'stoved' at 110° to 150° C. When set, such paints are tougher and more resistant to abrasion that air-drying paints. Because of their increased toughness these paints are now widely used for motor car body finishes during initial manufacture.

Polymerisation is caused by the addition of an activator or hardener. Such paints are known as 'two-pack' paints and the two components are mixed together immediately before use. Polymerisation commences immediately but slowly in the can but, after spreading, a solvent evaporates causing an increase in the concentration of the catalyst. The polymerisation process speeds up at this point and the paint is soon 'touch-dry'. However, it does not attain its full mechanical properties and resistance to damage until after a few days. Paints in this category are based on polyester, polyurethane and epoxide resins.

Group 4

Paints that dry by the evaporation of a volatile solvent. Suitable resins are nitro-cellulose and thermosetting acrylic. As the solvent evaporates the resins form a solid film. Paints which dry by the evaporation of a volatile solvent are known as *lacquers*.

4.18   Application of paints (preparation of surface)

As for any of the protective coatings described so far, the success of a print film largely depends upon satisfactory preparation of the metal sub-strata prior to the application of the paint. As described in section 4.5, this may consist of simple purging of the surface by mechanical or chemical means to remove scale, rust, grease and dirt. It may, alternatively, include more sophisticated pre-treatment such as galvanising, sherardising or phosphating, for example. The purpose of the surface preparation is not only to provide a satisfactory 'key' so that the paint film can adhere strongly to the surface of the work, but to ensure that this surface is chemically neutral so that no reactions will take place beneath the paint film, the products of which might 'lift' the paint film and break down its protection.

4.19   Application of paints (spraying)

This is one of the most versatile methods of coating surfaces with paint. Originally introduced for finishing mass produced motor cars, it is now used for large panels and small components alike. There are three basic techniques.

1. Conventional spraying (compressed air).
2. Airless spraying.
3. Electrostatic spraying (conventional and airless).

1. Conventional spraying

Compressed air is used to atomise the paint in a spray gun and project it on to the component being coated. It is a quick and relatively simple process requiring comparatively low-cost equipment. Further, it is versatile and can accommodate frequent colour changes. It gives consistently high standards of finish but paint and solvent wastage is high due to overspray and bounce.

2. Airless spraying

In this process the paint is not atomised by compressed air but is pumped through a fine jet. Airless spraying gives less overspray and bounce, resulting in less hazardous spray-dust than with conventional spraying. As a result the process is more suitable for use outside the spray booth, for example, when used for large structures and for maintenance on-site. Airless spraying can give better penetration into the corners of awkwardly shaped components and the rate of covering is greater, but the film thickness is difficult to control.

3. Electrostatic spraying

In this process the paint emerging from a conventional spray gun, or an airless spray gun, is given an electric charge of up to 150 kV. Since like charges repel each other, the spray droplets disperse and form a cloud. This gives uniformity to the film thickness and helps to prevent runs and 'teardrops' forming. The ionised spray droplets are strongly attracted towards the work which is earthed. This not only removes wastage by overspray and bounce-back but produces a 'wrap-around' effect. That is, spray droplets are deposited on the back as well as the front surfaces of components and edge coverage is greatly improved.

4.20   Application of paints (dipping)

In this process the work to be coated is suspended in paint and then lifted out to drain off. Usually the drying process is completed by stoving. In order to maintain the consistency of the paint bath it is neither usual or desirable to use air drying paints. Dipping plants usually operate on a continuous conveyor system where the components pass through the paint bath then over a drainage area and finally through a stoving tower before cooling and being off loaded from the conveyor. Dip painting is highly productive and the labour costs are low, but close control is required for consistent results. The process is as equally applicable to small components as it is to motor car bodies.

4.21   The hazards of spraying and stoving

The hazards of industrial paint application falls into two main categories:

1. Explosion and fire hazards resulting from the use of flammable solvents and the formulation of flammable dust particles as the spray mist dries in the atmosphere.
2. Toxic and irritant effects due to the inhalation of paint mist (wholly or partially solidified) and solvent fumes.

These hazards are particularly related to spraying and stoving processes, but the storage of paints also presents special problems. The local factory inspector and the fire authorities should be consulted before painting on an industrial scale is undertaken.

It is essential when spray painting to provide an efficient means of extraction to remove the excess spray mist and solvent fumes. Spray booths serve the double function of removing spray mist and fumes from the working area and then treating the exhausted air so that it is cleansed before being released back into the atmosphere. They can be divided into three main types.

1. Dry back spray booths: These consist of three-sided roofed canopies made from sheet steel and provided with an electrically driven exhaust fan. The air drawn from the booth passes through a disposable filter before passing to the atmosphere.

2. Down draught water washed spray booths: These have floor gratings through which the air in the booth is exhausted. This system has the advantage that large components can be accommodated more easily than in any other type. The air drawn down from the booth passes through water sprays and is washed clean before being returned to the atmosphere. The dross removed from the exhausted air forms a floating spongy mass that is easily removed.

3. Wet screen water washed spray booths: These have a vitreous enamelled back screen over which water cascades in a continuous water fall. Spraying takes place towards this screen and excess spray mist droplets are carried down into a tank where separation takes place.

The lighting and other electrical installations associated with spray painting booths have to be to Buxton Approved Standards for flame and explosion proof fittings.

Stoving

The main hazards associated with stoving ovens result from the use of unsuitable paints having volatile and flammable solvents and from the accumulation of explosive dusts and gases in the fume extraction ducts. Stoving ovens and their extractors should have pressure relief vents so that any explosion is carried upwards and clear of the working area.

Problems

Section A

1. The 'rusting' of steel is caused by the action of: (a) air (oxygen) alone; (b) moisture alone; (c) air and moisture together; (d) industrial pollutants.

2. The most effective method of preventing the atmospheric corrosion of steels is by: (a) galvanising; (b) anodising; (c) greasing; (d) painting.

3. Copper is protected from atmospheric corrosion by the formation of a surface film called: (a) verdigris; (b) copper oxide; (c) copper sulphate; (d) patina.

4. Aluminium alloys have their corrosion resistance improved by: (a) galvanising; (b) chromising; (c) calorising; (d) anodising.

5. Phosphate anti-corrosion coatings are used to prepare metals for: (a) painting; (b) plastic coating; (c) metal spraying; (d) electroplating.

Section B

6. (a) Zinc is used to galvanise steel components. Explain what is meant by the zinc coating being 'sacrificial'.

   (b) Calculate the mass of nickel deposited on the cathode of an electroplating cell if a current of 5 amperes flows for 3 hours and the cell is 85 per cent efficient. (The electro-chemical equivalent of nickel is 0·304.)

7. Describe the precautions that must be taken when: (a) paint spraying; (b) stove enamelling.

8. Paints are classified by the way in which they dry or harden. Describe the four different systems available and state a typical application of each system.

9. State **six** important reasons for coating metallic components with a plastic coating. Describe with the aid of sketches the following plastic coating processes: (a) fluidised bed dipping; (b) flock spraying; (c) hand smogging.

10. (a) Explain with the aid of a diagram how the gases given off by factory chimneys form corrosive pollutants in the atmosphere and list the more destructive of them.

    (b) Explain why it is always necessary to carefully prepare a metal surface before applying anti-corrosion treatment.

    (c) Describe how structural steelwork can be prepared prior to painting.

Chapter 5

Fastening and joining

5.1 Screwed fastenings

Various types of screwed fastenings are used where components must be assembled and dismantled regularly. Screwed fastenings are proportioned so that it requires approximately the same force to make them fail in the following ways:

(a) by the head pulling off;

(b) By the thread stripping. (This assumes that the nut thickness is equal to the minor diameter of the thread. For example: A nut size M12 has a thickness of 9·5 and 10·0 mm, which compares with the minor diameter of an M12 bolt of 9·8 mm. If the bolt screws directly into a component the same proportions apply.);

(c) By failure of the shank, in tension or shear, across the core diameter (minor diameter) of the thread.

Figure 5.1 shows some typical applications of screwed fastenings.

5.2 Locking devices

In order to prevent screwed fastenings slackening off due to vibrations, various locking devices are employed. A selection are shown in Fig. 5.2. It

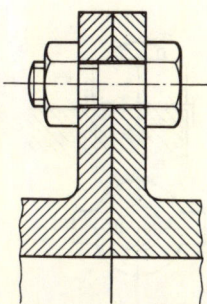

(a) Section through a bolted joint

Plain shank extends beyond joint face

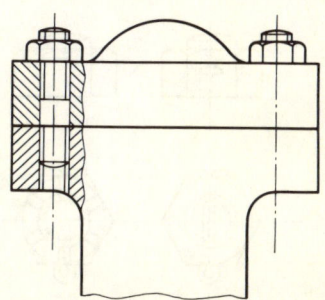

(b) Stud and nut fixing for an inspection cover

This type of fixing is used where a joint is regularly dismantled. The bulk of the wear comes on the stud which can eventually be replaced cheaply. This prevents the wear falling on the expensive casting or forging

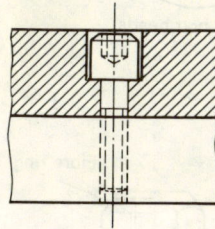

(c) Cap head socket screw

Although much more expensive than the ordinary hexagon head bolt, the socket screw is made from high tensile alloy steel. They are heat treated to make them very strong, tough and wear resistant. They are widely used in the manufacture of machine tools. The above example shows how the head is sunk into a counterbore to provide a flush surface

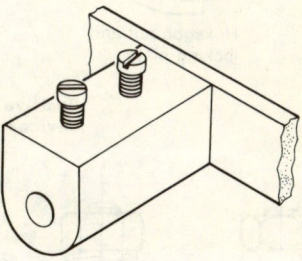

(d) Cheese head brass screws

These are used in small electrical appliances for clamping cables into terminals

Fig. 5.1 Use of screwed fastenings

will be seen that they can be divided into two categories: those where the locking action is positive, and those where the locking action is frictional. Positive locking devices are more time consuming and difficult to fit, but they are essential for critical joints where failure could cause serious accidents, as in the clutch and brake controls of machine tools.

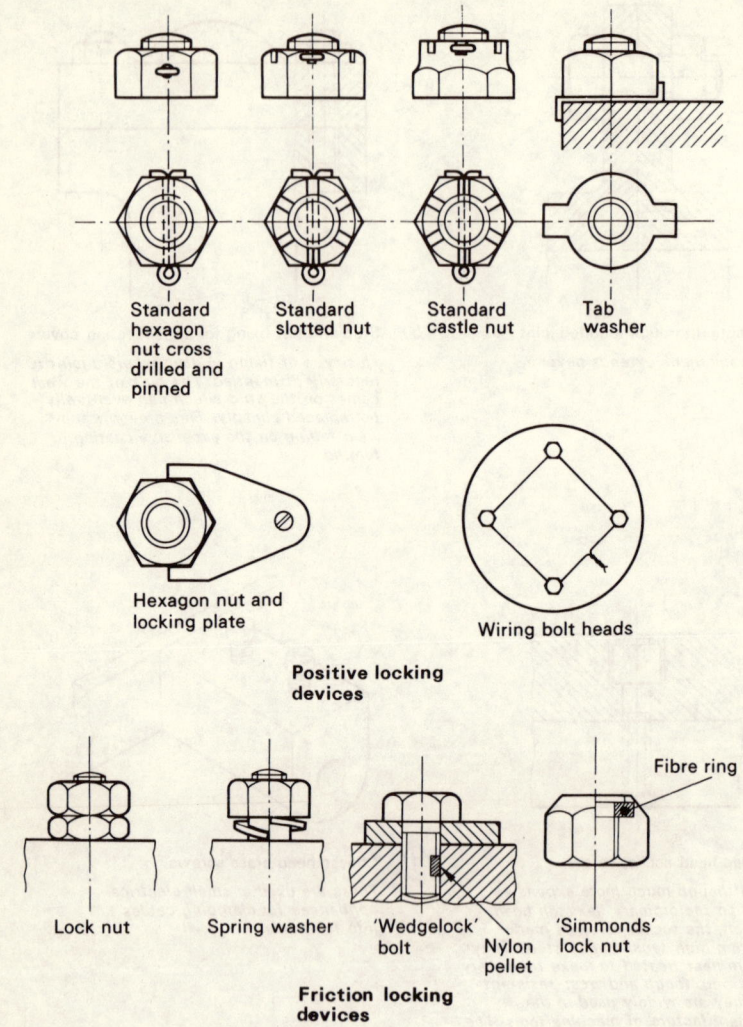

Standard
hexagon
nut cross
drilled and
pinned

Standard
slotted nut

Standard
castle nut

Tab
washer

Hexagon nut and
locking plate

Wiring bolt heads

**Positive locking
devices**

Lock nut

Spring washer

'Wedgelock'
bolt

Nylon
pellet

Fibre ring

'Simmonds'
lock nut

**Friction locking
devices**

Fig. 5.2   Locking devices

5.3 Spanners

Spanners are used to tighten screwed fastenings. They are carefully proportioned so that their length enables a man of average strength to fully tighten
the fastening correctly. A selection of spanners is shown in Fig. 5.3.

In order to ensure that critical screwed fastenings are tightened correctly
a *torque spanner* should be used. Figure 5.4 shows such a spanner. Torque
spanners fall into two categories. One type uses a pre-set slipping clutch. This

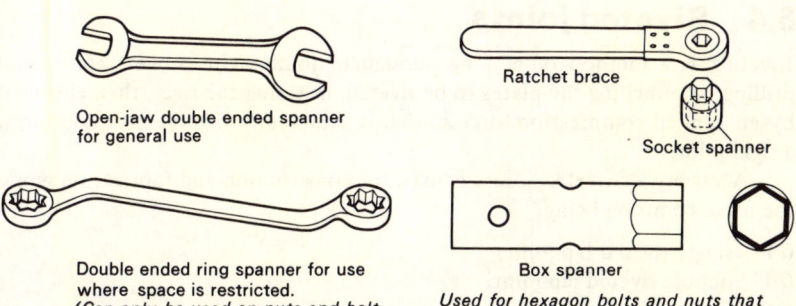

Open-jaw double ended spanner
for general use

Ratchet brace

Socket spanner

Double ended ring spanner for use
where space is restricted.
*(Can only be used on nuts and bolt
heads that are in very good condition.)*

Box spanner

*Used for hexagon bolts and nuts that
are recessed into a casting. The spanner
is turned by a tommy bar inserted
through holes in the spanner body.*

Fig. 5.3   Spanners

can be set to slip when the nut is correctly tightened and prevents over-tightening. The other type has a spring-loaded handle. A pointer indicates on a scale the torque being exerted on the nut. This latter type relies heavily upon the skill of the fitter for its correct functioning. However, it is much cheaper and simpler in construction.

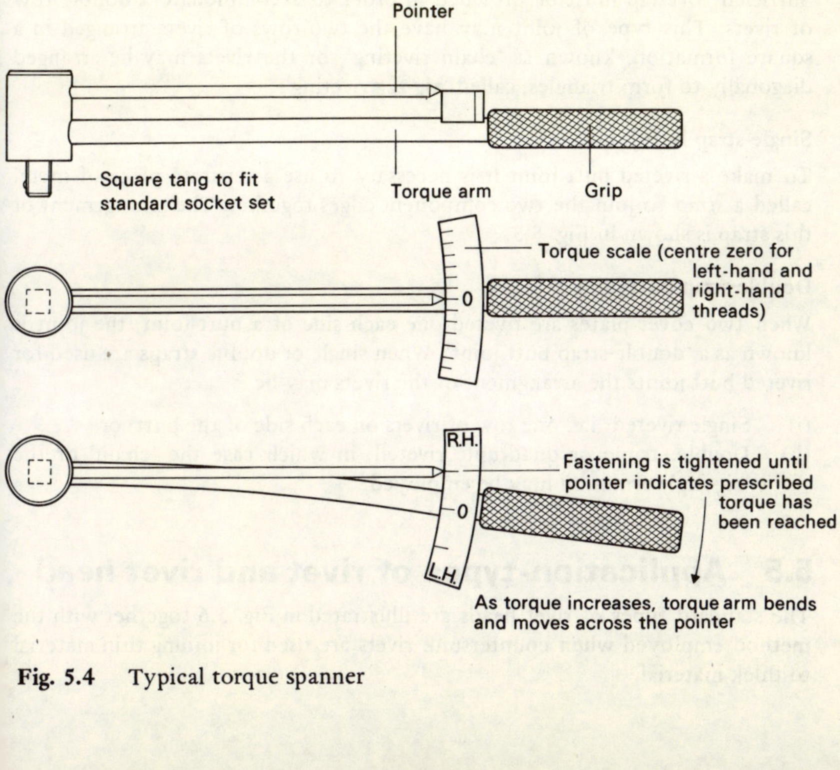

Pointer

Square tang to fit
standard socket set

Torque arm

Grip

Torque scale (centre zero for
left-hand and
right-hand
threads)

R.H.

0

L.H.

Fastening is tightened until
pointer indicates prescribed
torque has
been reached

As torque increases, torque arm bends
and moves across the pointer

Fig. 5.4   Typical torque spanner

5.4 Riveted joints

Riveting is a method of making permanent joints. The process consists of drilling or punching the plates to be riveted, inserting the rivet, then closing it by an applied compression force so that it completely fills the hole and forms a rigid joint.

A variety of riveted joints is used for construction and fabrication work, the more common being:

(a) single riveted lap joint;
(b) double riveted lap joint;
(c) single-strap butt joint;
(d) double-strap butt joint.

These are illustrated in Fig. 5.5.

Single riveted lap joint

This is the simplest of all riveted joints and is extensively used for joining both thick and thin plates. The plates to be joined are overlapped a short distance and a single row of rivets, conveniently spaced along the middle of the lap, completes the joint.

Double riveted lap joint

A lap joint having two rows of rivets is known as a 'double riveted lap joint'. Sufficient overlap must be provided in order to accommodate a double row of rivets. This type of joint may have the two rows of rivets arranged in a square formation, known as 'chain riveting', or the rivets may be arranged diagonally to form triangles, called 'zig-zag riveting'.

Single-strap butt joint

To make a riveted butt joint it is necessary to use a separate piece of metal called a *strap* to join the two component edges together. The arrangement of this strap is shown in Fig. 5.5.

Double-strap butt joint

When two cover plates are riveted one each side of a butt joint, the joint is known as a 'double-strap butt joint'. When single or double straps are used for riveted butt joints the arrangment of the rivets may be:

(i)   Single riveted: i.e. one row of rivets on each side of the butt; or
(ii) Double, triple or quadruple riveted; in which case the 'chain' or the 'zig-zag' formation may be employed.

5.5 Application-types of rivet and rivet head

The standard types of rivet heads are illustrated in Fig. 5.6 together with the method employed when countersunk rivets are used for joining thin material to thick material.

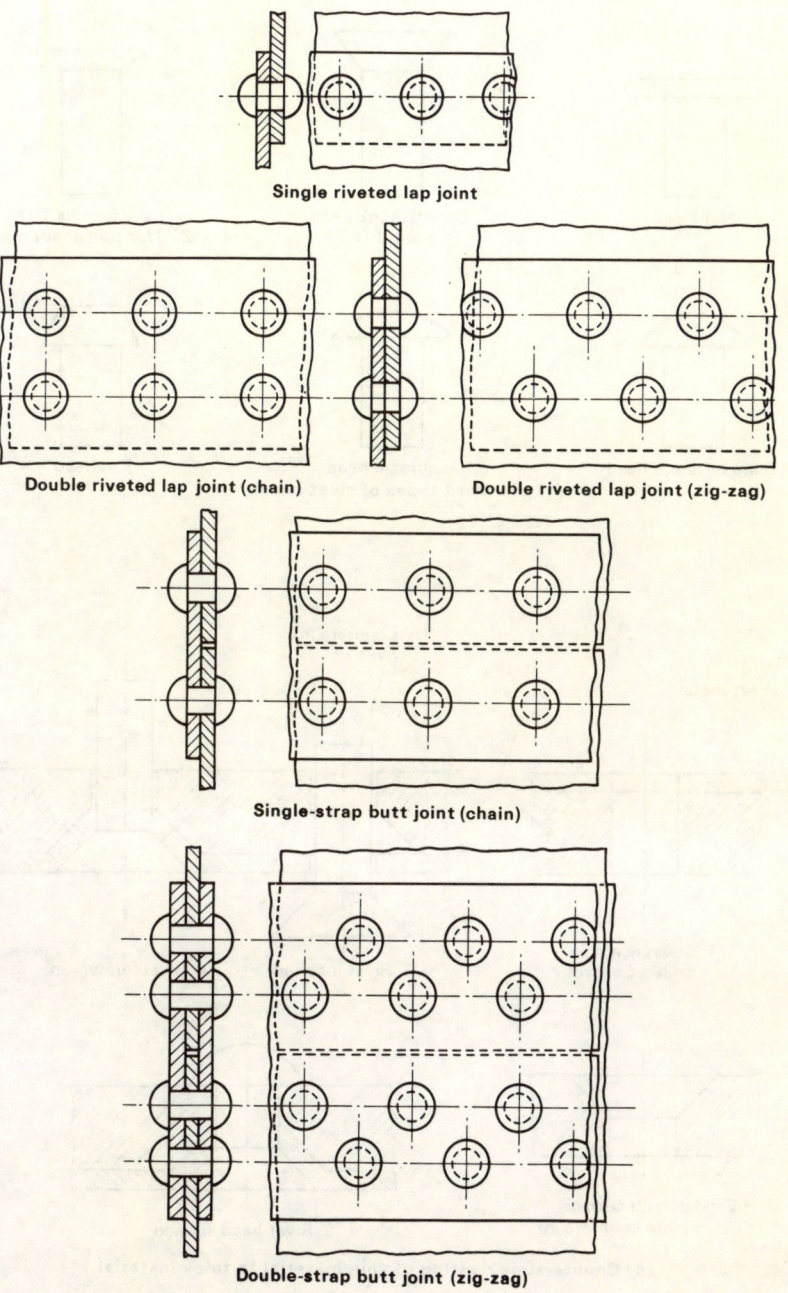

Single riveted lap joint

Double riveted lap joint (chain)

Double riveted lap joint (zig-zag)

Single-strap butt joint (chain)

Double-strap butt joint (zig-zag)

Fig. 5.5 Types of riveted joint

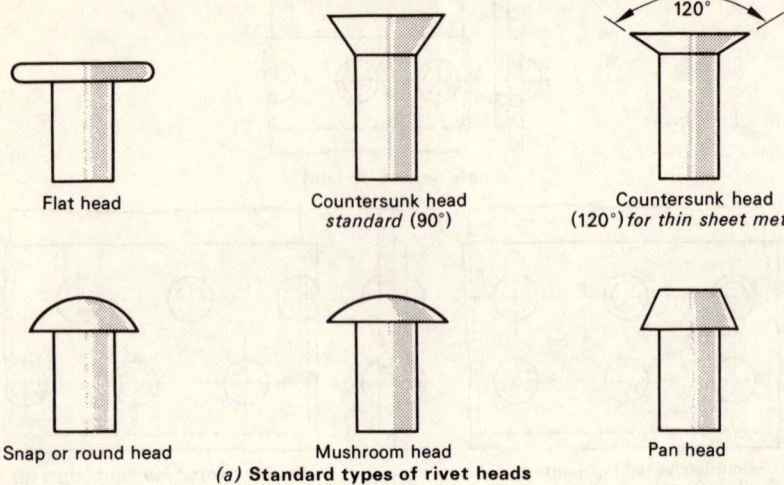

Flat head

Countersunk head
standard (90°)

Countersunk head
(120°) *for thin sheet metal*

Snap or round head

Mushroom head

Pan head

(a) **Standard types of rivet heads**

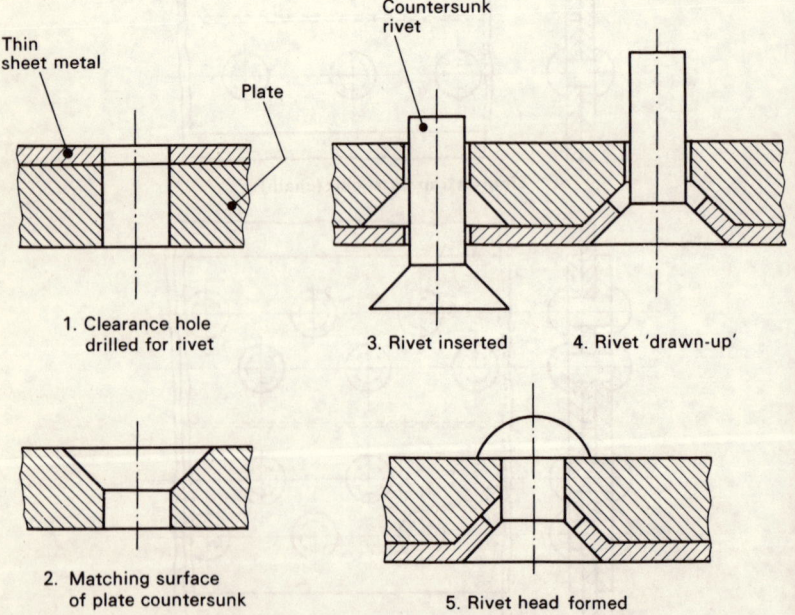

Thin
sheet metal

Plate

Countersunk
rivet

1. Clearance hole
drilled for rivet

3. Rivet inserted

4. Rivet 'drawn-up'

2. Matching surface
of plate countersunk

5. Rivet head formed

(b) **Countersunk riveting of thin material to thick material**

Fig. 5.6 Rivet heads and applications

'Tinners' and flat-head rivets are used in most general sheet metal fabrications, where the metal is very thin and little strength is required. The countersunk head is used when a flush surface is required, and the roundhead or snaphead is most widely used where the joint must be as strong as possible. Mushroom head or 'knobbled rivets', as they are termed in the steel construction industry, are used where it is necessary to curtail the height of the rivet head above the metal surface, as, for example, on the outer fuselage 'skins' of aircraft in order to decrease 'drag', or in the case of steel chutes and bunkers, to reduce obstruction on the inside surfaces. Pan head rivets are very strong, and are, therefore, widely used for girders and heavy constructional engineering.

'Tinners' are similar to flat-head rivets. They are made of soft iron and are usually coated with tin to prevent corrosion and to make them easier to soft solder — hence their name 'tinners'.

Always use the correct rivet for a particular metal to be riveted. When riveting aluminium, for example, use aluminium rivets; and when riveting copper use copper rivets.

5.6   Defects in riveted and bolted joints

When making connections which require the use of rivets or bolts, the following are a few of the basic details which should be considered in order to prevent many common defects:

(a)   when marking out, use the correct allowance for edge clearance and pitch;

(b)   all drilled or punched holes should be made to the correct clearance size to suit the bolt or rivet diameter, or as specified on the drawing;

(c)   remove any 'burrs' from around the edges of all holes before final assembly of the parts to be joined;

(d)   ensure that holes are correctly lined and matched before inserting the bolt or rivet;

(e)   use the proper type of rivet or bolt as specified on the drawing;

(f)   use rivets or bolts of the correct length;

(g)   where bolts are employed always use a suitable washer under the nut;

(h)   when using bolts do not overtighten the nut;

(i)   when inserting bolts or rivets, do not attempt to force or drive them into the hole;

(j)   always use the correct tools for the job.

Some of the common forms of defects associated with riveted connections are illustrated in Fig. 5.7.

Figure 5.8 shows some of the common faults that occur with bolted connections.

(a)   Unless a suitable washer is fitted under the nut damage will be caused to the contact surface of the plate or structural member.

As the nut is tightened it will tend to bite into the metal and spin up an

CAUSE OF RIVETING DEFECT	RESULTANT EFFECT
Insufficient hole clearance	Rivet not completely 'drawn through' Not enough shank protruding to form head  Original head of rivet 'stands proud', the formed head is weak and mis-shaped
Hole too large for rivet	Hole not filled 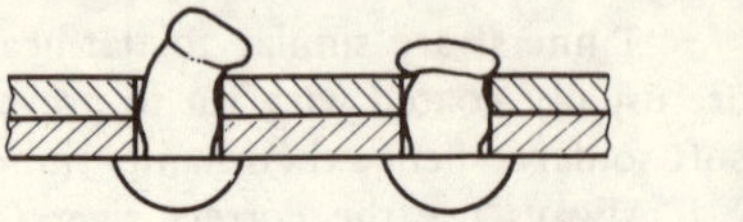 Rivet tends to bend and deform. Head weak and poorly shaped
Rivet too short	Not enough shank protruding to produce a correctly shaped head 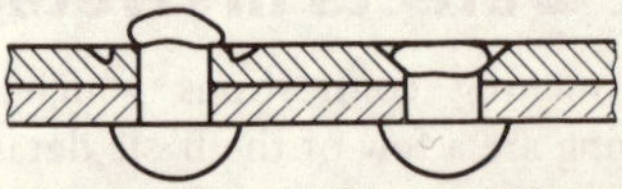 Plate surface damaged Countersinking not completely filled
Rivet too long	Too much shank protruding to form required head 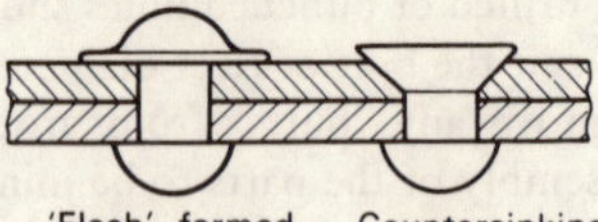 'Flash' formed around head (Jockey cap) Countersinking over-filled
Rivet set or dolly not struck square	Badly shaped head - off centre 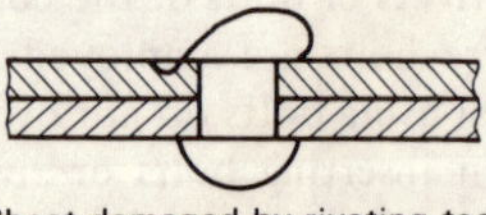 Sheet damaged by riveting tool
Drilling burrs not removed	Not enough shank protruding to form the correct size head 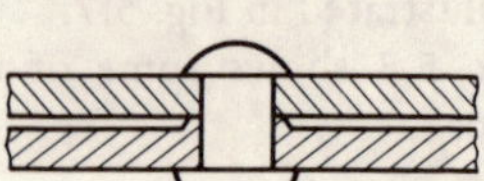 Plates or sheets not closed together. Unequal heads

Fig. 5.7   Common defects in riveting

CAUSE OF RIVETING DEFECT	RESULTANT EFFECT
Sheets not closed together — **rivet not drawn up sufficiently**	Weak joint. Rivet shank swells between the plates  Not enough shank protruding to form correctly shaped head
Rivet holes not matched	Weak mis-shapened head 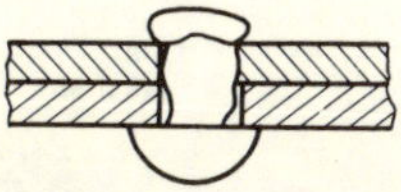 Rivet deformed and does not completely fill the hole

Fig. 5.7 (*continued*)

annular bead as it penetrates the bearing surface. The bolt will always tend to become slack, and any subsequent tightening of the nut will cause it to bite still further into the metal. Eventually, when the whole of the threaded portion of the bolt has completely penetrated the nut, no further tightening can take place and the joint becomes weak.

(*b*) With any bolted joint at least two screw threads on the bolt should protrude through the nut. Therefore, a bolt of the correct length should be used. If a bolt is too short the nut cannot obtain sufficient grip on the screw thread of the bolt, and there is always a tendency for it to work loose. Any attempt to overtighten the nut generally results in stripping the thread, especially when a bolt with a fine thread is used. No part of the thread on a bolt should be in the hole.

(*c*) The contact surfaces of a nut or bolt must be square to the hole.

5.7 Self-secured joints

These joints, as the name implies, are formed by folding and interlocking thin sheet metal edges together in such a manner that they are made secure without the aid of any additional jointing process. Their use is confined to fabrications or components constructed with light gauge sheet metal less than 1·6 mm thick. Figure 5.9 shows a selection of the most commonly used self-secured joints.

Very often self-secured joints are employed in the manufacture of containers which have to hold foodstuffs and liquids.

In this case it is necessary to seal the joints and seams. Use is made of *latex rubber* inserts, and *soft solder* with a high tin content.

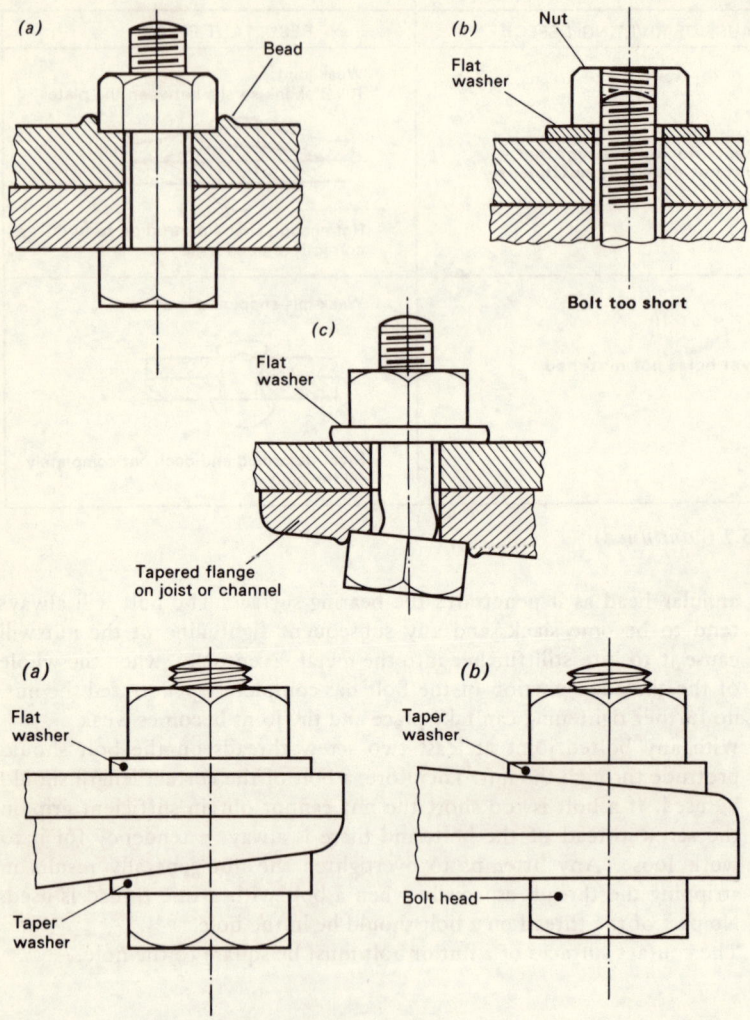

Fig. 5.8   Defects in bolted connections

5.8   Soft soldering

Soft soldering is a low temperature thermal process in which the metal of the components being joined is not melted. The process involves the use of a suitable low-melting-temperature alloy of *tin* and *lead* which is 'bonded' by the application of heat and a suitable flux to an unmelted parent metal. Therefore, soft solders must have a lower melting point than the metals they join.

It is an essential feature of a soft soldered joint that each of the joint

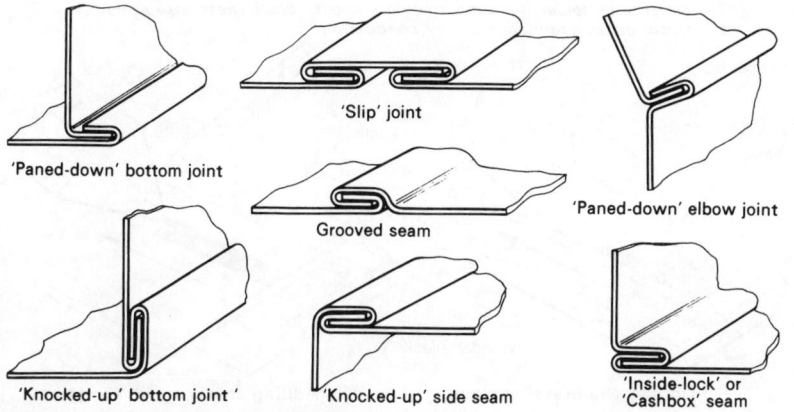

Fig. 5.9 Self secured joints

surfaces is 'tinned' by a film of solder and that these two films of solder are made to 'fuse' with the solder filling the space between them.

Basically the action of a soft solder, when applied to a prepared metal joint which has been heated to the required temperature, is to:

(a) flow between the parent metal surfaces which remain unmelted;
(b) completely fill the space between the sufaces;
(c) adhere thoroughly to the sufaces;
(d) solidify.

In a correctly soldered joint, examination under a microscope has shown that in the action of tinning metals such as brass, copper and steel a definite chemical reaction takes place. The metal surface and the tin in the solder react together to form an 'intermetallic compound' which acts as a 'key' for the bulk of the solder in the joint.

The intermetallic compound layer will continue to grow in thickness the longer the joint is kept at soldering temperature. Solder cannot be completely wiped or drained off the parent metal when in the molten state. The metal surface, therefore, remains permanently wetted or 'tinned' by a film of solder. This film of solder cannot be mechanically prised off, leaving the surface of the metal bare and in its original state.

In general there are two stages in making a soft-soldered joint:

1. Tinning the metal surfaces.
2. Filling the spaces between the tinned surfaces with solder.

These stages are shown in Fig. 5.10. The metal to be soldered is supported on a wood or asbestos block (heat insulator) to prevent unnecessary heat loss from the joint by conduction.

Table 2.18 gives a selection of typical soft solders in general use, together with their properties and uses. The complete list is to be found in BS 219 : 1959.

118

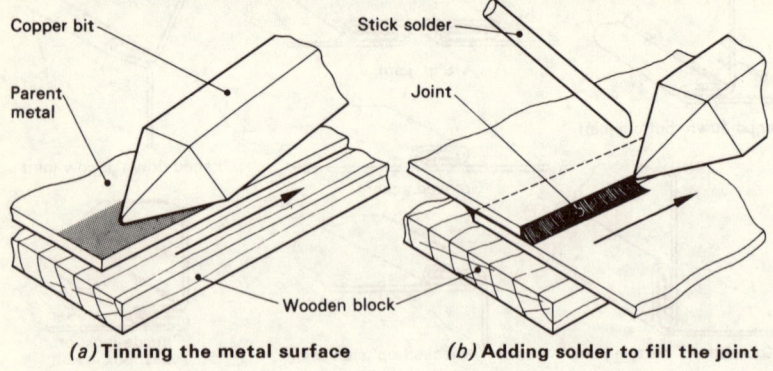

*The metal to be soldered is supported on a wooden block (**heat insulator**) to prevent unnecessary heat loss by **conduction***

Copper bit

Parent metal

Stick solder

Joint

Wooden block

(a) **Tinning the metal surface**   *(b)* **Adding solder to fill the joint**

Fig. 5.10   Basic stages in soft soldering

5.9 Preparation of the joint

The 'tinning action' of solder cannot take place unless the two surfaces to be joined are *chemically clean*. That is the mating surfaces must not only be free from dirt and grease, they must also be free of any *oxide film*.

Prior to any thermal joining process, including soft soldering, brazing or welding, it is necessary to expose bare metal at the mating or joining surfaces. Dirt, grease, oils or oxides are themselves unsolderable and act as barriers between the molten solder and the metal surfaces to be joined.

5.10 Soldering fluxes

When a metal is exposed to the air at 'room temperature' it acquires a thin oxide coating within a few minutes of being cleaned. Since oxygen combines with metals even more rapidly at high temperatures, it is very important that not only is the oxide film removed, but that it is prevented from reforming as the metal is reheated to the soldering temperature. A soldering *flux* is a substance used for this purpose.

The requirements of a good flux are:

(*a*)   it must remain liquid at the soldering temperature;

(*b*)   in its liquid state it must act as a cover over the joint and exclude the air;

(*c*)   it must dissolve any oxide film present on the surfaces being joined;

(*d*)   it should be readily displaced from the joint surfaces by the molten solder.

Figure 5.11 shows the essential functions of a soldering flux: how it removes the oxide, protects the surface from atmospheric attack and then gives way for the 'wetting' action of the molten solder. A molten solder is

said to 'wet' a surface when it leaves a continuous permanent film on the 119
surface of the parent metal instead of rolling off in the form of globules.

A molten solder is said to 'wet' when it leaves a continuous permanent film on the surface of the parent metal instead of rolling over it.

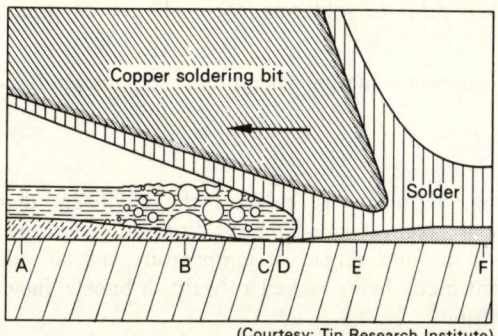

(Courtesy: Tin Research Institute)

Diagrammatic representation of the displacement of flux by molten solder.
A Flux solution lying above oxidised metal surface.
B Boiling flux solution removing the film of oxide (e.g. as chloride).
C Bare metal in contact with fused flux.
D Liquid solder displacing fused flux.
E Tin reacting with the basis metal to form compound.
F Solder solidifying.

Fig. 5.11   The essential functions of a soldering flux

Fluxes used for soft soldering operations may be classified as active and inactive (passive).

The active fluxes quickly dissolve the oxide film on a metal and, at the same time, act as a barrier to prevent further oxidation. Unfortunately, all active fluxes leave a corrosive residue which will be present along the edges of the joint after soldering. These apparently dry flux residues are hygroscopic and rapidly absorb water from the atmosphere. As soon as the residue becomes damp it exhibits corrosive properties and will attack the metal. If an active flux has to be used the workpiece must be washed clean after soldering.

Because of the difficulty of avoiding residual corrosion the use of active fluxes is banned in electrical and other work that cannot be washed effectively. In these situations passive fluxes are used. These prevent oxidation during soldering, but will not remove oxide already present on the joint faces, which have to be prepared by mechanical cleaning.

5.11   Types of soldered joints

Soft soldering, as a joining method, relies almost entirely on *adhesion* for its strength. Soft-soldered joints are at their strongest at room temperature, their mechanical strength decreasing rapidly as the temperature increases. The strength of a soft soldered joint is determined by three basic factors.

1. The strength of the solder itself — this is governed by its composition (i.e. ratio of **tin/lead**). Surplus solder does not add strength to the joint.
2. The strength of the bond between the bulk of the solder and the surfaces which it 'tins' (i.e. the joint interfaces).
3. The design of the joint. Where strong joints are required the joint edges may be interlocked prior to soldering. (See self-secured joints, section 5.7.)

The three most common soldering operations are: *tacking, sweating* and *floating*.

Tacking

This is used for joining long seams. Tack first at the middle and then at each end of the seam. Finally tack at equal intervals on alternative sides of the middle. This ensures a uniform spread of temperature and so reduces any tendency by the parent metal being tacked to warp or buckle due to uneven expansion and contraction.

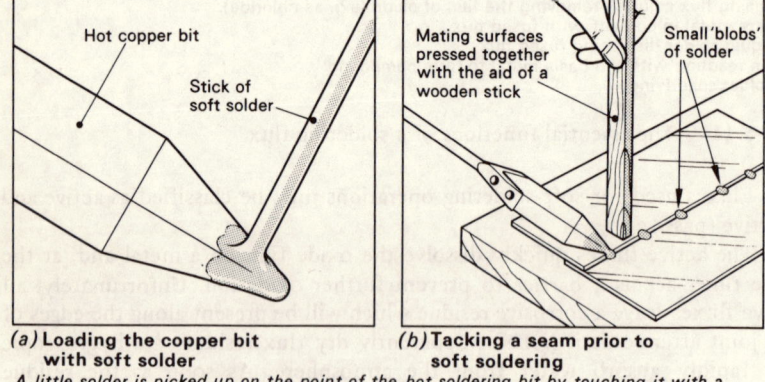

(a) **Loading the copper bit with soft solder**

(b) **Tacking a seam prior to soft soldering**

A little solder is picked up on the point of the hot soldering bit by touching it with a stick of solder. Any excess solder may be removed by simply shaking or flicking the bit.

Fig. 5.12   Soft soldering processes — tacking

The process of tacking is shown in Fig. 5.12 and consists of spot soldering at regular intervals instead of making a continuous run. The tip of a hot copper bit is used and the local heating effect causes the small amount of solder carried on the bit to penetrate the joint at the point of contact. During tacking, the mating surfaces of the seam are pressed tightly together by means of a wooden stick or the 'tang' of an old file.

Sweating

This produces a continuous joint as shown in Fig. 5.13. A thin coating of a suitable flux is applied to the mating surfaces which will have been previously cleaned. A uniform coating of solder is applied to the surfaces to be joined.

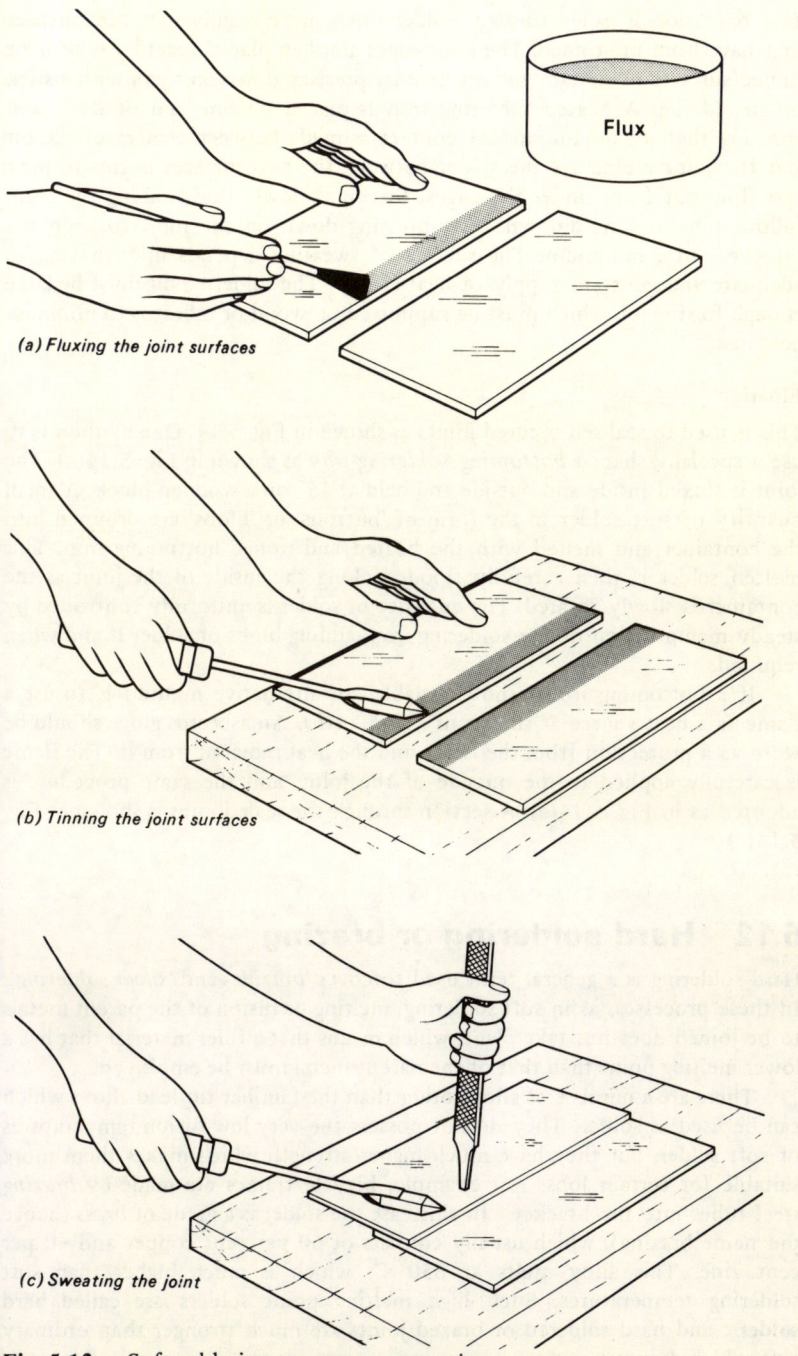

(a) Fluxing the joint surfaces

(b) Tinning the joint surfaces

(c) Sweating the joint

Fig. 5.13 Soft soldering processes — sweating

This operation is called *tinning*. Solder flows more readily between surfaces that have been pre-tinned. The joint edges are then placed together with their tinned surfaces in contact and are held by pressing down on them with a stick or an old file. A heated soldering iron is placed on one end of the seam, ensuring that maximum surface contact is made between a facet of the bit and the joint metal. As the solder between the two surfaces begins to melt and flow out from under the edges, the bit is slowly drawn along the seam followed by the hold-down stick pressing down on the joint to keep the thickness to a minimum. The success of sweating depends upon having an adequate and constant supply of heat energy. The soldering bit must be large enough for the job which must be supported on wood or asbestos to minimise heat loss.

Floating

This is used to seal self-secured joints as shown in Fig. 5.14. One method is to use a specially shaped *bottoming soldering iron* as shown in Fig. 5.14(a). The joint is fluxed inside and outside and held at 45° on a wooden block. A small quantity of soft solder in the form of 'buttons' or 'blobs' are dropped into the container and melted with the heated and tinned bottoming iron. This molten solder is then carefully flooded along the inside of the joint as the container is slowly rotated. The quantity of solder is uniformly controlled by steady manipulation of the soldering iron, adding blobs of solder if and when required.

If a bottoming iron is not available, an alternative method is to use a flame as a heat source as shown in Fig. 5.14(b). An asbestos glove should be worn as a protection from the flame and the heat radiated from it. The flame is carefully applied to the outside of the joint, and the same procedure is adopted as in Fig. 5.14(a). A section through the sealed joint is shown at Fig. 5.14(c).

5.12 Hard soldering or brazing

Hard soldering is a general term used to cover '*brazing*' and '*silver soldering*'. In these processes, as in soft soldering, melting or fusion of the parent metals to be joined does not take place, which means that a filler material that has a lower melting point than that of the parent metal must be employed.

There are a number of alloys other than the familiar tin/lead alloys which can be used as solder. They do not possess the very low fusion temperatures of soft solder, but they have much higher strength which makes them more suitable for certain jobs. For example, bicycle frames are made by *brazing* steel tubes into the brackets. In this case the solder is a grade of *brass* (hence the name brazing), which usually consists of 60 per cent copper and 40 per cent zinc. This alloy melts at 850° C, which is much higher than soft soldering temperatures. Such high melting point solders are called **hard solders,** and hard soldered or brazed joints are much stronger than ordinary soft soldered ones.

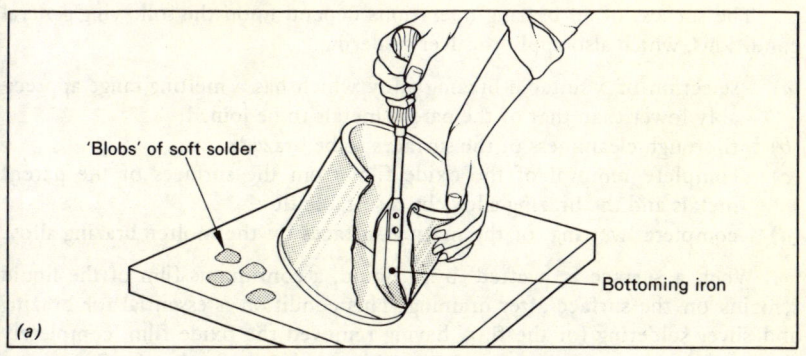

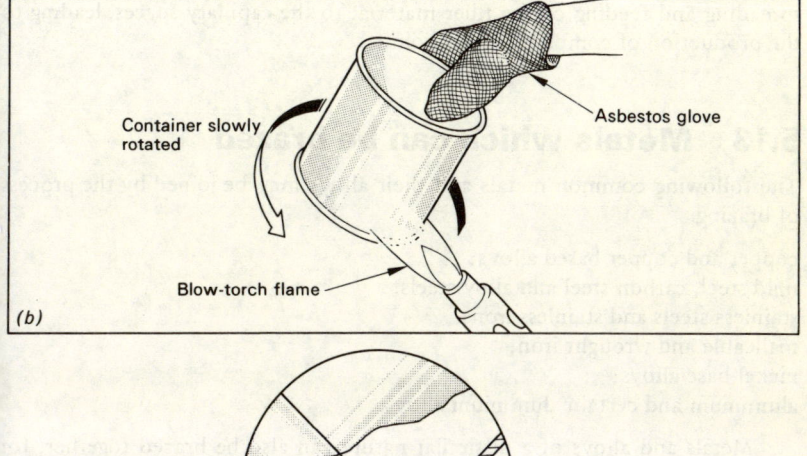

(c) Section through a knocked-up bottom

Fig. 5.14 Soft soldering processes — floating

Brazing is defined as:

'A process of joining metals in which molten filler metal is drawn by capillary attraction into the space between closely adjacent surfaces of the parts to be joined.'

In general the melting point of the filler metal lies above 500° C but below the melting point of the parent metals being joined. In this respect, and broadly in its application, brazing lies between soft soldering and fusion welding.

The success of all brazing operations depend upon the following general conditions, which also apply to silver soldering:

(a) selection of a suitable brazing alloy which has a melting range appreciably lower than that of the parent metals to be joined;

(b) thorough cleanliness of the surfaces to be brazed;

(c) complete removal of the oxide film from the surfaces of the parent metals and the brazing alloys by a suitable flux;

(d) complete 'wetting' of the mating surfaces by the molten brazing alloy.

When a surface is 'wetted' by a liquid, a continuous film of the liquid remains on the surface after draining. This condition is essential for brazing and silver soldering for the flux, having removed the oxide film, completely 'wets' the surfaces of the joint faces. This wetting action by the flux assists spreading and feeding of the filler material to the capillary spaces, leading to the production of completely filled joints.

5.13 Metals which can be brazed

The following common metals and their alloys may be joined by the process of brazing:

copper and copper based alloys;
mild steel, carbon steel and alloy steels;
stainless steels and stainless irons;
malleable and wrought iron;
nickel-base alloys;
aluminium and certain aluminium alloys.

Metals and alloys of a dissimilar nature can also be brazed together, for example: copper to brass; copper to steel; brass to steel; cast iron to mild steel; mild steel to stainless steel.

5.14 Filler materials

For most metals the brazing alloys used are normally based on copper–zinc alloys and are dissimilar in composition to the parent metals to which they are applied. Copper is commonly used as a brazing material for the flux-free brazing of mild steel in reducing atmosphere furnaces. Brazing alloys may be classified into three main types:

1. Silver-bearing brazing alloys or 'silver solders'.
2. Brazing alloys containing phosphorous.
3. Brazing brasses or 'spelters'.

Silver solders

These are more expensive than the normal brazing alloys because they contain a high percentage of silver, but they offer the advantages of produc ing very strong and ductile joints at much lower temperatures. Silver solders are very free-flowing at brazing temperature. Brazing temperatures required have very little heat-effect upon the properties of the parent metals.

By using silver solders it is possible to increase the speed of brazing and to eliminate or to limit 'finishing' operations. One of its main applications is for delicate work in which small neat joints are essential.

Borax type fluxes are not suitable for silver soldering, and a suitable proprietary flux should be used.

An efficient flux should melt at a temperature at least 50° C lower than the melting point of the brazing material and retain its activity at a temperature of at least 50° C above the melting temperature of the brazing alloy being used.

Brazing alloys containing phosphorous

Filler materials which contain phosphorous are usually referred to as 'self-fluxing brazing alloys'. These alloys contain silver, phosphorous and copper or copper and phosphorous, the former possessing a lower melting range. The outstanding features of these alloys is their ability to braze copper in air without the use of a flux. When melted in air the products of oxidation form a fluid compound which acts as an efficient flux. A suitable flux is required when brazing copper based alloys.

These brazing alloys find their greatest application in the resistance brazing operations, in refrigerator manufacture, electrical assemblies (electric motor armatures), for brazing seams and fittings in domestic copper hot water cylinders and in plumbing.

Brazing brasses

The oldest and best known method of brazing involves the use of brazing brasses or 'brazing spelters', using borax as a flux. These alloys melt at much higher temperatures than the silver solders and the phosphorous-containing brazing alloys, but produce stronger joints. Increasing the zinc content decreases the melting range, and this makes it possible to make a joint in 60/40 brass using a 50/50 brass as the brazing alloy. Conversely, it is important that a brass to be joined by brazing should have a high copper content compared with the brazing alloy used.

5.15 Aluminium brazing

There is a distinction between the brazing of aluminium and the brazing of other metals. For aluminium, the filler material is one of the aluminium alloys having a melting point below that of the parent metal.

Borax-base fluxes are unsuitable for brazing aluminium and its alloys, but many proprietary fluxes are available, and these are basically mixtures of the alkali metal chlorides and fluorides. A standard aluminium brazing flux contains essentially chlorides of sodium, potassium and lithium. The melting points of the parent metals which can be aluminium brazed range between

590° C and 660° C. Extreme care must be taken when aluminium brazing, because of the very small margin temperature which permits the joint to be made without melting and the possible collapse of the parent metal during the operation.

5.16 Flame or torch brazing

Flame brazing may be used to fabricate almost any assembly, and gives particular advantage where the joint area is small in relation to the bulk of the assembly. The torch is the most common method of heating. The brazing spelter may be applied in one of two ways.

1. By carefully loading the joint with granulated spelter which has been 'calcined' by boiling it in a borax solution. The calcined granules are mixed with borax paste. As heat is applied from the torch and the joint brought uniformly up to brazing temperature the spelter melts. Borax is usually sprinkled on the joint which is tapped with an iron spatula in order to assist the flow of the molten spelter throughout the joint.
2. In wire or strip form. The end of the spelter is heated and dipped into the flux which adheres to it. When the joint has been brought to the brazing temperature the flux end of the filler metal is applied and flows into the joint.

5.17 Furnace brazing

Furnace brazing is used extensively when:

(a) the parts to be brazed can be pre-assembled or jigged to hold them in position;
(b) the brazing filler material can be pre-placed as shown in Fig. 5.15(a);
(c) when a controlled atmosphere is required. The components to be brazed are assembled with the filler alloy in the required position and fluxed if necessary.

Pre-placed brazing alloy inserts are available in a variety of forms, such as wire rings, bent wire shapes, washers and foils.

The method of heating varies according to the application, and in general muffle furnaces are used in order that the flame does not impinge directly on the parts being brazed. Some furnaces are heated by gas or oil, but the majority are electrically heated. The furnaces can be classified into two types

(a) the batch type with either an air or controlled atmosphere;
(b) the conveyor type with a controlled atmosphere.

Figure 5.15 shows some aspects of furnace brazing. The schematic layouts of these furnaces are shown in Fig. 5.15(b).

Individual assemblies or components for batch brazing (in trays) are passed through the furnace on a conveyor system. Inert or reducing atmospheres can be used to protect the work from oxidation.

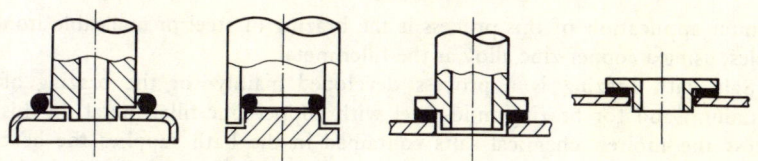

(a) **Use of prepared filler metal**

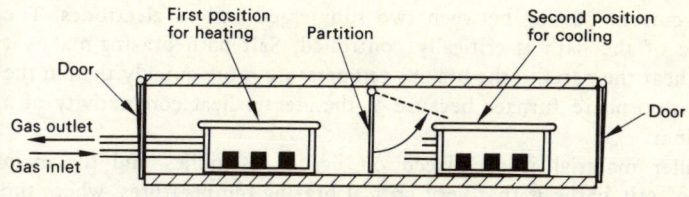

First position for heating — Partition — Second position for cooling

Door

Gas outlet

Gas inlet

Door

Schematic layout of batch brazing in a sealed container

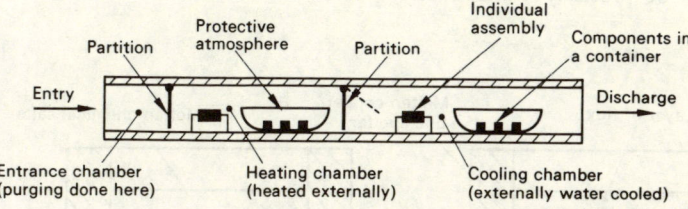

Partition — Protective atmosphere — Partition — Individual assembly — Components in a container

Entry

Discharge

Entrance chamber (purging done here)

Heating chamber (heated externally)

Cooling chamber (externally water cooled)

(b) **Schematic layout of continuous brazing furnace**

Fig. 5.15   Furnace brazing

5.18   Dip brazing

Two methods of heating are employed:

(*a*)   molten filler metal dip heating;
(*b*)   molten chemical flux bath dip heating.

Instead of a furnace, the parts to be brazed may be submerged in a bath of molten filler metal. A crucible, usually made of graphite, is heated externally to the required temperature to maintain the brazing alloy in fluid form. A cover of flux is maintained over the molten brazing alloy, and in practice the assembled parts to be joined should be raised and lowered two or three times in order that flux and filler metal may flow into the joint clearances by capillary attraction. The size of molten bath and the heating method must be such that the immersion of the parts to be brazed will not lower the temperature of the bath below that necessary for brazing. Large articles are generally pre-heated before being placed in the bath. The most

128 common application of this process is the brazing of steel or malleable iron articles, using a copper-zinc alloy as the filler metal.

Salt bath brazing is a process developed mainly for the brazing of aluminium, and for brazing mild steel with copper-zinc filler metal. In this process the molten chemical salts contained in the bath supplies the heat required to raise the parent metal to brazing temperature. The chemical salts are contained in a metal or ceramic container (the bath) which may be heated externally by gas or oil, or heated internally by an electric low voltage alternating current passing between two submerged carbon electrodes. The temperature of the bath is critically controlled. Salt bath brazing makes it possible to heat the parts to the brazing temperature more quickly than in the controlled-atmosphere furnace because of the greater heat conductivity of a liquid medium.

The filler material is pre-placed on jigged assemblies and the main advantage of salt baths is that very critical brazing temperatures, where the difference in melting points between the parent metal and filler metal are slight, can be precisely controlled. Figure 5.16 shows two methods of dip brazing.

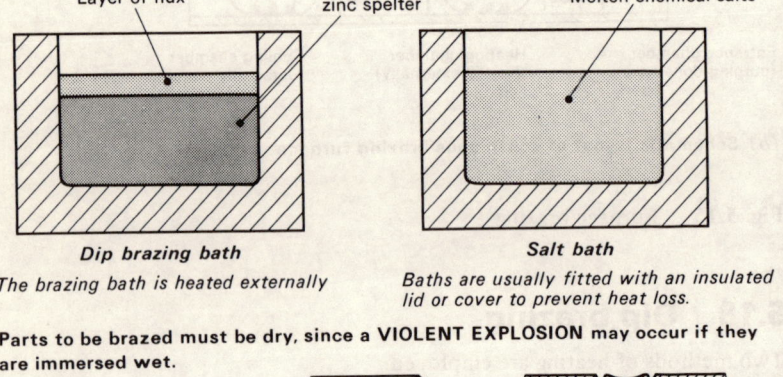

Layer of flux   Molten copper/zinc spelter   Molten chemical salts

Dip brazing bath
The brazing bath is heated externally

Salt bath
Baths are usually fitted with an insulated lid or cover to prevent heat loss.

Parts to be brazed must be dry, since a **VIOLENT EXPLOSION** may occur if they are immersed wet.

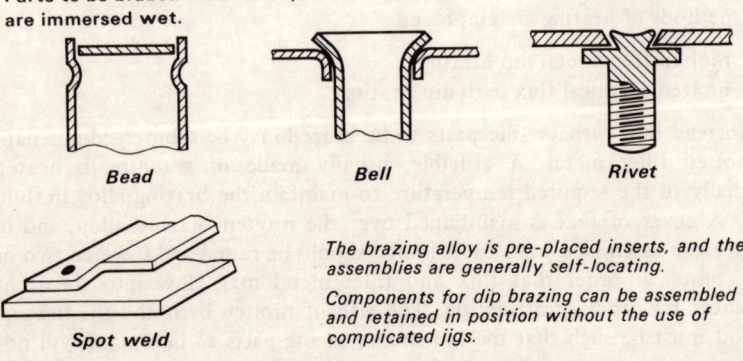

Bead   **Bell**   **Rivet**

Spot weld

The brazing alloy is pre-placed inserts, and the assemblies are generally self-locating.

Components for dip brazing can be assembled and retained in position without the use of complicated jigs.

Fig. 5.16   Two methods of dip brazing

5.19 Electric induction brazing

Induction heating is used extensively on parts which are self-jigging and where very rapid heating is required. The component to be brazed is placed in the magnetic field of a coil carrying high frequency alternating electric current. Parts may be assembled in jigs provided effective heating is not reduced by their use and that the jigs themselves are not made of an electrically conducting material. In practice the operator loads one jig whilst another assembly is being brazed.

In this process, which is normally operated continuously, the heat required for brazing is generated within the material itself from a high-frequency electric current passing through a high conductivity water-cooled copper tube coil encircling the joint. Most of the intense heat generated by this method is relatively near the surface, the interior being heated by thermal conduction from the hot surface. As a rule, the higher the frequency the shallower the heating. Internal and external coils may be used, as shown in Fig. 5.17, and several different coil designs may be used. A paste or powder flux is commonly used, and the filler material is pre-placed. Silver solders find an extensive use in this process.

5.20 Electric resistance brazing

In this process the heat required for brazing is developed by:

(a) resistance at the joint interface, as in resistance welding;
(b) resistance in carbon electrodes which conduct heat to the joint faces.

The basic principle of resistance heating is that a heavy electrical current at low voltage is passed through the assembly in such a way that a hot spot is generated at the joint suitable for brazing operations. Heating can be precisely localised and this ensures no general loss of mechanical properties throughout the parent metal. One main advantage of resistance brazing is normally that no jigs are required, for the electrodes themselves act as jigs and hold the components in correct relationship. There are two methods of heating, either direct or indirect, and these are shown schematically in Fig. 5.18.

5.21 Fusion welding

In the soldering and brazing processes described earlier in this chapter, the joints are formed by a thin film of metal that has a lower melting point and inferior strength to the metals being joined. In *fusion welding* any additional metal added to the joint has a similar composition and strength as the metals being joined. Figure 5.19 shows the principle of joining two pieces of metal by fusion welding where not only the filler material but also the edges of the components being joined are melted. The molten metals fuse together and, when solid, form a homogeneous joint whose strength is equal to the metal of the components being joined.

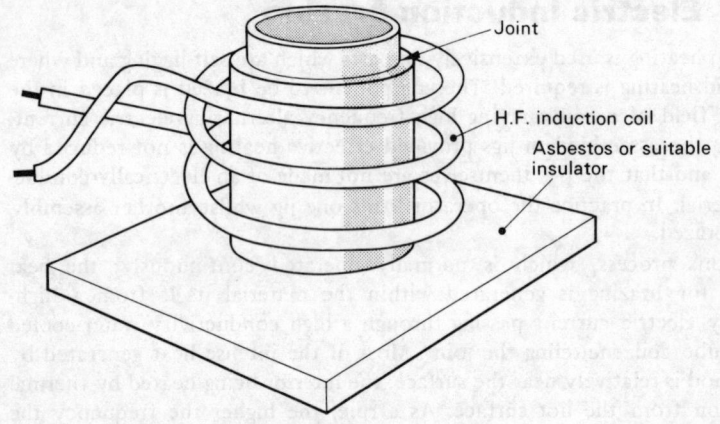

*Solid copper coils are also sometimes employed. The coils
are normally insulated with glass fibre. In practice the work
to be brazed is brought to the coils.*

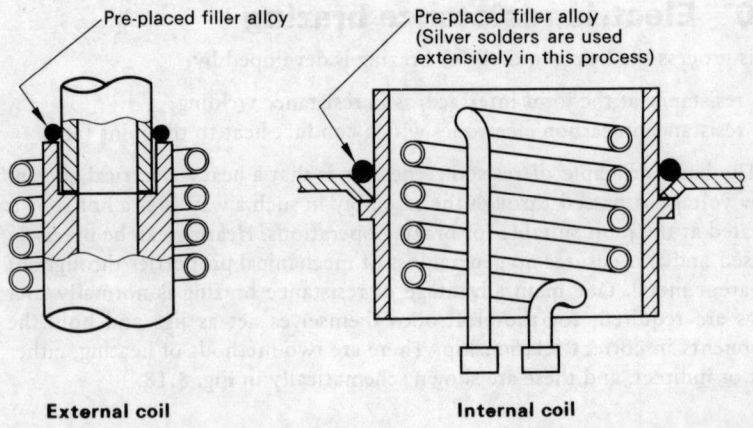

*It is usual for induction coils to be designed to surround the joint,
but internal coils can be used for certain applications.*

Fig. 5.17   Electric induction brazing

Oxy-acetylene welding

In this process the heat source is a mixture of oxygen and acetylene gases
burning to produce a flame whose temperature can reach 3250° C, and this is
above the melting point of most metalas. Figure 5.20 shows a typical set of
gas welding equipment. Since the gases are stored under very high pressures

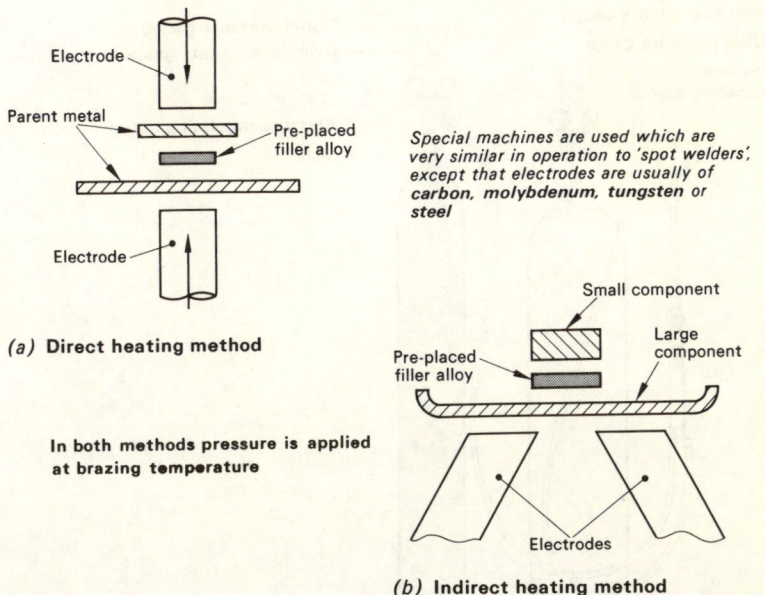

(a) **Direct heating method**

Special machines are used which are very similar in operation to 'spot welders', except that electrodes are usually of **carbon, molybdenum, tungsten** or **steel**

In both methods pressure is applied at brazing temperature

(b) **Indirect heating method**

Fig. 5.18   Electric resistance brazing

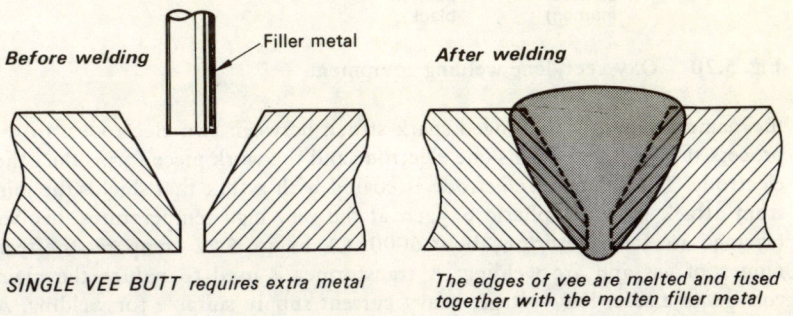

SINGLE VEE BUTT requires extra metal

The edges of vee are melted and fused together with the molten filler metal

Fig. 5.19   Fusion welding

and form highly flammable and even explosive mixtures the equipment must be handled with great care. This equipment must only be used by persons who have been fully instructed in the operating and safety procedures recommended by the Home Office and the equipment suppliers. Figure 5.21 shows the two basic techniques for fusion welding using an oxy-acetylene torch.

Metallic arc welding

This is a fusion welding process where the energy required to melt the edges of the components and the filler rod is provided by an electric arc. The *arc* is

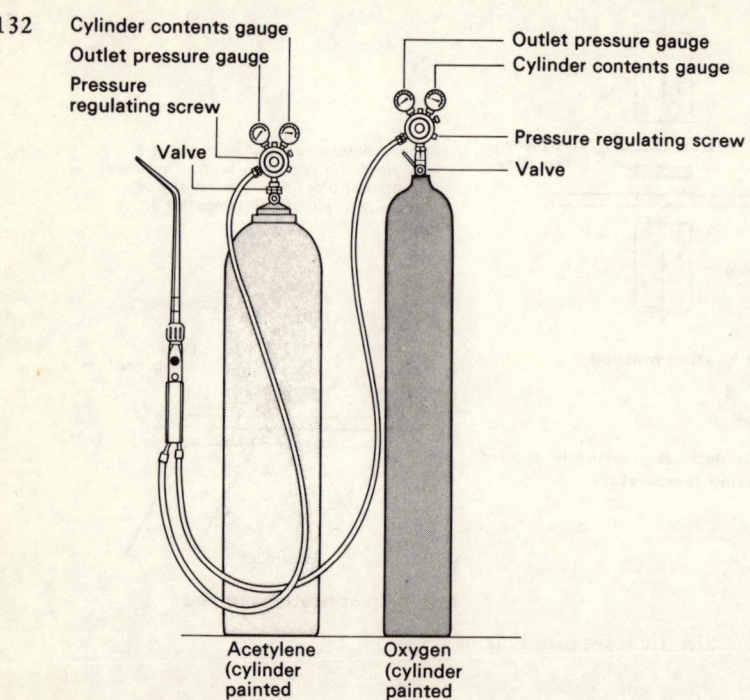

Cylinder contents gauge
Outlet pressure gauge
Pressure regulating screw
Valve

Outlet pressure gauge
Cylinder contents gauge
Pressure regulating screw
Valve

Acetylene
(cylinder
painted
maroon)

Oxygen
(cylinder
painted
black)

Fig. 5.20   Oxy-acetylene welding equipment

the name given to a prolonged spark struck between two electrodes. In this process the filler rod forms one electrode and the work piece forms the other electrode. The filler rod/electrode is coated with a flux that shields the joint from attack by atmospheric oxygen at the very high temperatures involved (average arc temperature is about 6000° C). Figure 5.22 compares the techniques of gas and arc welding. A transformer is used to reduce the mains voltage to a safe, low voltage, heavy current supply suitable for welding. As with gas welding, arc welding equipment must not be used by unskilled persons, except under the closest supervision. The dangers associated with the process arise not only from the very high temperatures and the magnitude of the heat energy involved, but also from the possibility of electric shock if the equipment is not correctly installed and used. Figure 5.23 shows the general arrangement of a metallic arc welding installation.

5.22   Use of adhesives

Traditionally, adhesives fell into two categories:

1.   *Glues:*   These were made from the bones, hooves and horns of animals and the bones of fishes. Derivatives of milk and blood were also used.

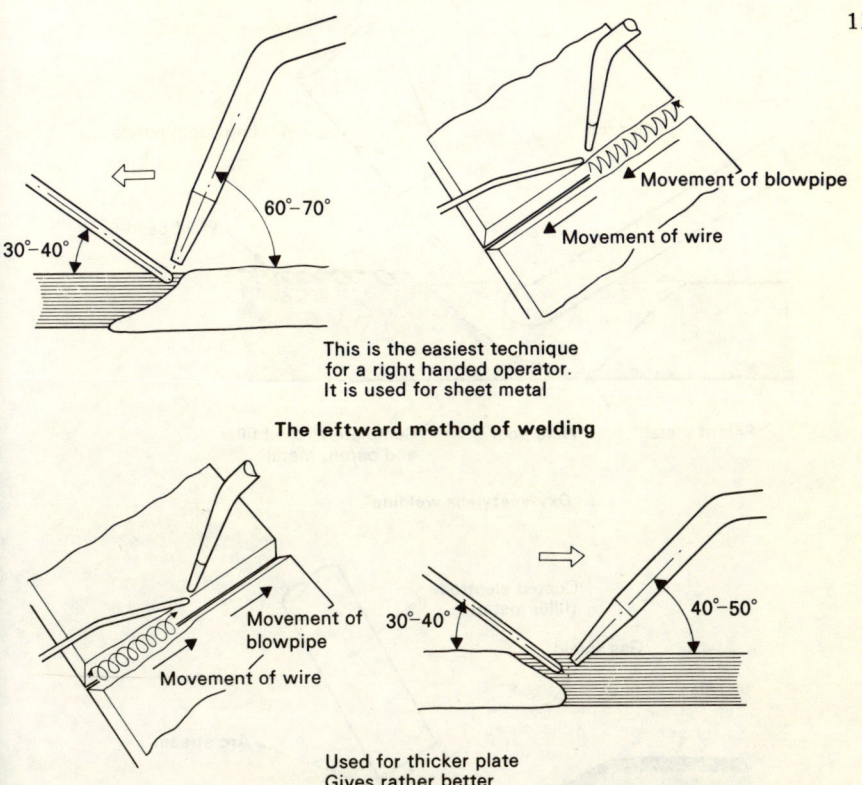

This is the easiest technique
for a right handed operator.
It is used for sheet metal

The leftward method of welding

Used for thicker plate
Gives rather better
penetration

The rightward method of welding

Fig. 5.21   Gas welding techniques

Glues were largely used for jointing wood and were particularly important in furniture and toy manufacturing industries.

2. *Gums:*   These were made from vegetable matter. Resin and rubber being extracted from trees, and starches being extracted from the byproducts of flour milling.

Although natural glues and gums are still widely used for low strength applications, they are being increasingly supplanted by high strength synthetic adhesives. Shortly before the Second World War, the growth of the high polymer (plastics) industry led to the development of many new synthetic materials suitable for high strength adhesives. These new adhesives, which are a considerable advance on the traditional natural glues and gums, are used extensively for bonding together materials in a wide variety of industries.

Table 5.1 lists some of the more important advantages and limitations of

134

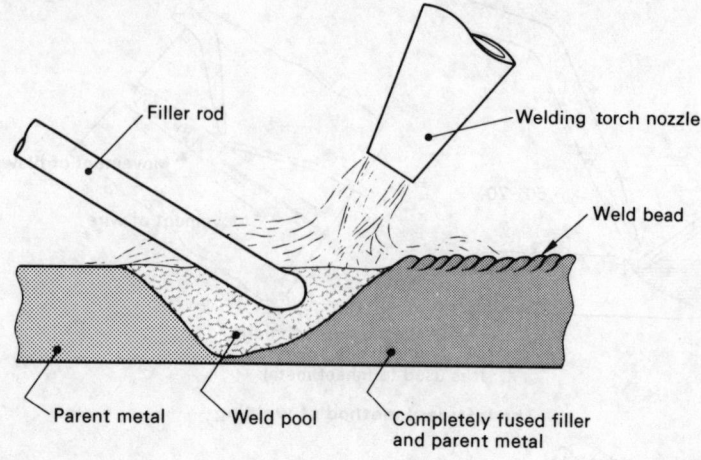

(a) **Oxy-acetylene welding**

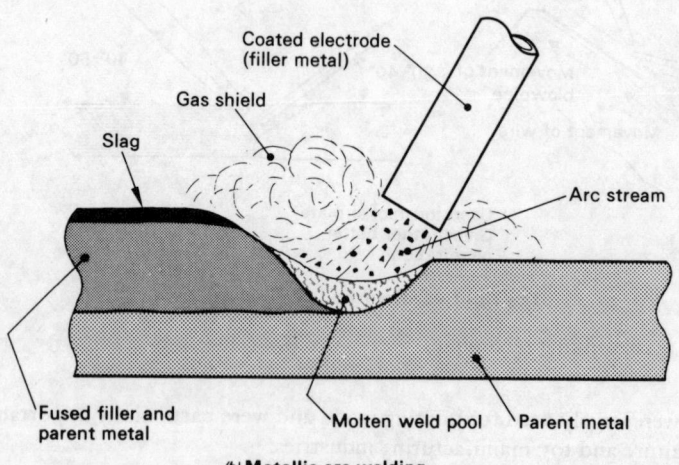

(b) **Metallic arc welding**

Fig. 5.22 Comparison of oxy-acetylene and metallic arc welding

adhesive bonding as compared with the mechanical and thermal jointing
processes discussed earlier in this chapter.

5.23 The adhesive bond

Figure 5.24(a) shows a typical bonded joint and explains the terminology
used for the various features of the joint. The strength of the bond depends
upon two factors:

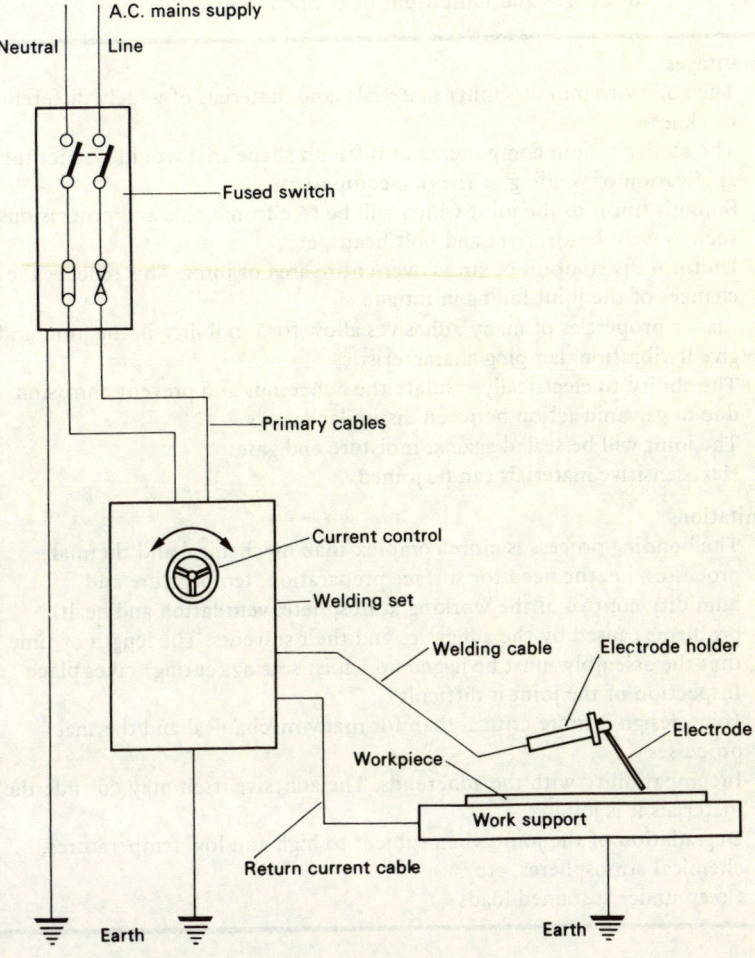

Fig. 5.23 Manual metal-arc welding circuit diagram

(a) adhesion;
(b) cohesion.

Adhesion is the ability of the bonding material (adhesive) to stick (adhere) to the materials being jointed (adherends). There are two ways in which the bond can occur and these are shown in Fig. 5.24(b).

Cohesion is the ability of the adhesive to resist the applied forces within itself.

Table 5.1 Advantages and limitations of bonded joints

Advantages
1. The ability to join dissimilar materials, and materials of widely different thicknesses
2. The ability to join components of difficult shape that would restrict the application of welding or riveting equipment
3. Smooth finish to the joint which will be free from voids and protrusions such as weld beads, rivet and bolt heads, etc
4. Uniform distribution of stress over entire area of joint. This reduces the chances of the joint failing in fatigue
5. Elastic properties of many adhesives allow for flexibility in the joint and give it vibration damping characteristics
6. The ability to electrically insulate the adherends and prevent corrosion due to galvanic action between dissimilar metals
7. The joint will be sealed against moisture and gases
8. Heat-sensitive materials can be joined

Limitations
1. The bonding process is more complex than mechanical and thermal processes, i.e. the need for surface preparation, temperature and humidity control of the working atmosphere, ventilation and health problems caused by the adhesives and their solvents. The length of time that the assembly must be jigged up whilst setting (curing) takes place
2. Inspection of the joint is difficult
3. Joint design is more critical than for many mechanical and thermal processes
4. Incompatibility with the adherends. The adhesive itself may corrode the materials it is joining
5. Degradation of the joint when subject to high and low temperatures, chemical atmospheres, etc
6. Creep under sustained loads

5.24 Failure in adhesive bonds

Figure 5.25 shows three ways in which a bonded joint can fail. These failures can be prevented by careful design of the joint, correct selection of the adhesive, careful preparation of the joint surfaces and control of the working environment (cleanliness, temperature and humidity).

5.25 The strength of bonded joints

No matter how effective the adhesive is and how carefully it is applied, the joint will be a failure if it is not correctly designed and executed. It is bad practice to apply adhesive to a joint that was originally proportioned for

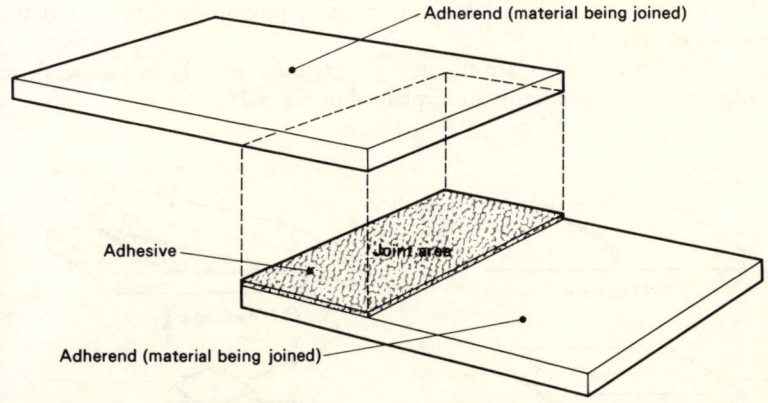

Adherend (material being joined)

Adhesive

Joint area

Adherend (material being joined)

(a) **Elements of the bonded joint**

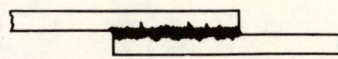

A simple cemented joint in which the adhesive penetrates the pores of the adherends to form the bond. This occurs with rough or porous surfaces.

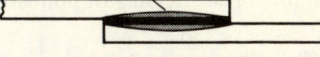

The molecules of the adhesive and the adherend diffuse and mingle together in this zone.

The adhesive and the adherends react together chemically so that an intermolecular bond is formed.

(b) **Types of bond**

Fig. 5.24   The bonded joint

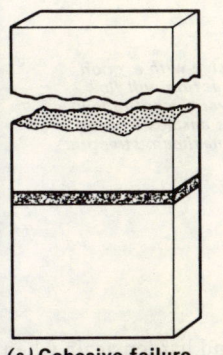

(a) **Cohesive failure of the adherend**

(over-strong adhesive)

(b) Cohesive failure of the adhesive

(weak adhesive)

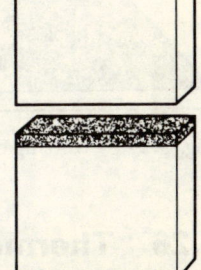

(c) **Adhesive failure**

(inadequate preparation of the joint faces resulted in a poor bond)

Fig. 5.25   Adhesive and cohesive failure

bolting, riveting or welding. The joint must be proportioned to exploit the properties of adhesives.

Most adhesives are relatively strong in tension and shear, but weak in cleavage and peel; these terms are explained in Fig. 5.26.

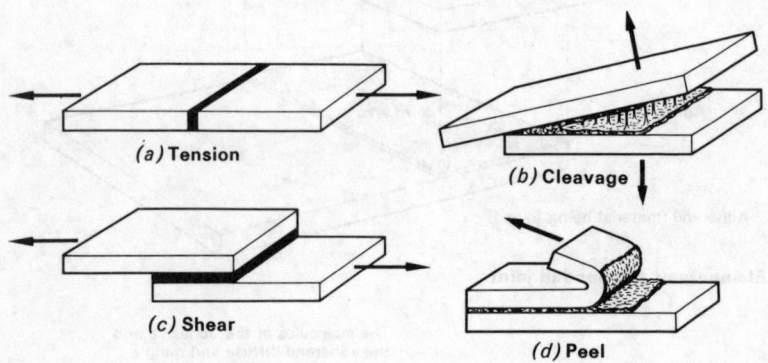

Fig. 5.26   The stressing of bonded joints

The adhesive must 'wet' the joint surfaces thoroughly, otherwise voids will occur and the bonded area will be considerably less than the theoretical maximum: this will weaken the joint considerably. Fig 5.27 shows the effects of wetting on the formation of the joint.

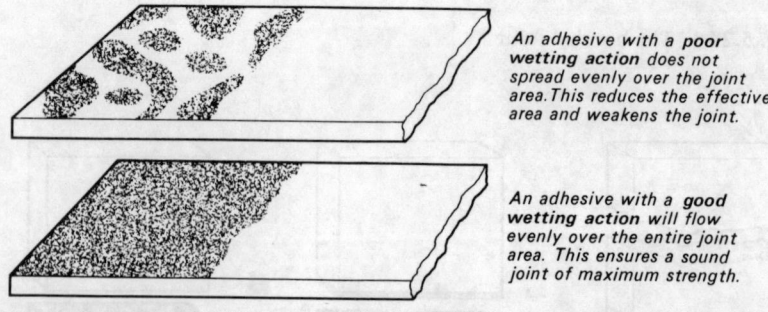

An adhesive with a **poor wetting action** does not spread evenly over the joint area. This reduces the effective area and weakens the joint.

An adhesive with a **good wetting action** will flow evenly over the entire joint area. This ensures a sound joint of maximum strength.

Fig. 5.27   Wetting capacity of an adhesive

5.26   Thermo-plastic adhesives

These are materials that soften when they are heated and harden again when cooled. They may be applied to the joint in three ways:

1. *Heat activated:*   The adhesive is softened by heating until it is fluid enough to spread freely over the whole joint and adhere to the materials being joined. Upon cooling to room temperature a bond is achieved.

2. *Solvent activated:* The adhesive is softened by a suitable solvent and a bond is achieved by the solvent evaporating. Because evaporation is essential to the setting of the adhesive, a sound bond is almost impossible to achieve at the centre of a large joint area as shown in Fig. 5.28. This is particularly the case when joining non-absorbent materials.

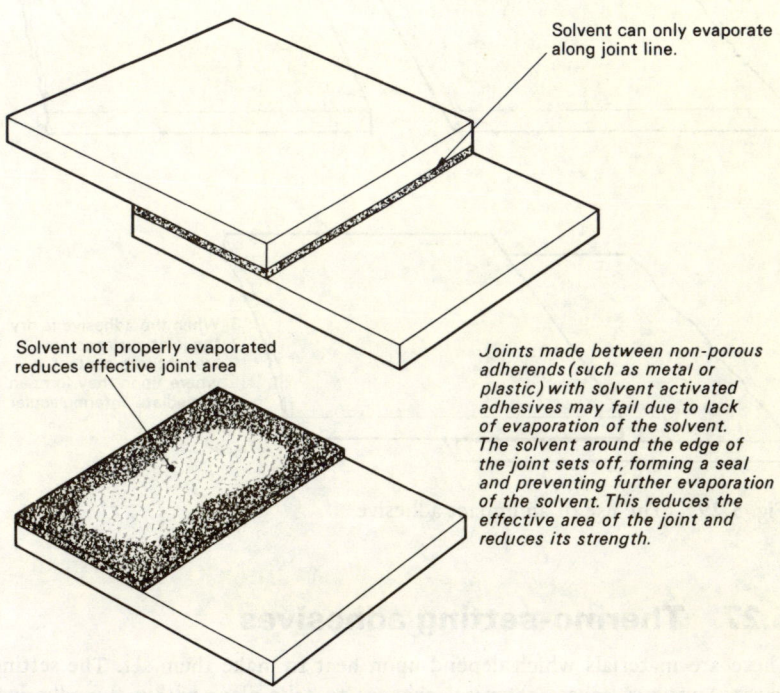

Solvent can only evaporate along joint line.

Solvent not properly evaporated reduces effective joint area

Joints made between non-porous adherends (such as metal or plastic) with solvent activated adhesives may fail due to lack of evaporation of the solvent. The solvent around the edge of the joint sets off, forming a seal and preventing further evaporation of the solvent. This reduces the effective area of the joint and reduces its strength.

Fig. 5.28 Solvent activated adhesive fault

3. *Impact adhesives:* These are solvent activated adhesives which are spread separately on the two joint faces and then left to dry by evaporation. When dry, the treated faces are brought together whereupon they instantly bond together by intermolecular attraction. This enables non-absorbent materials of a large area to be successfully joined. Figure 5.29 shows the steps in making an impact joint.

Thermo-plastic adhesives are based upon synthetic materials such as polyamides, vinyl and acrylic polymers and cellulose derivatives. They are also based on such natural materials as resin, shellac, mineral waxes and rubber. They are not so strong as thermo-setting plastics but, being more flexible, are more suitable for joining non-rigid materials.

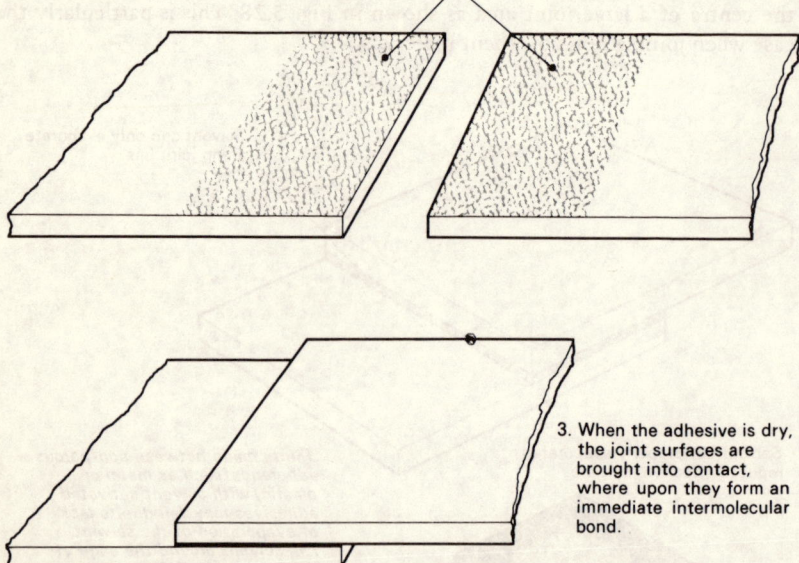

1. The impact adhesive is spread thinly and evenly on both joint surfaces.
2. The adhesive is then left to dry by evaporation. This avoids the problem in Fig. 5.28.

3. When the adhesive is dry, the joint surfaces are brought into contact, where upon they form an immediate intermolecular bond.

Fig. 5.29 The use of an impact adhesive

5.27 Thermo-setting adhesives

These are materials which depend upon heat to make them set. The setting (curing) process causes chemical changes to take place within the adhesive. Once cured they cannot be softened again by the re-application of heat. This makes them less heat sensitive than thermo-plastics.

The heat necessary to cure the adhesive can be applied externally as when phenolic resins are used or internally by adding a chemical hardener as when epoxy resins are used. The hardener is a chemical that reacts with the adhesive to generate heat (exo-thermic reaction). Since the setting process is a chemical reaction and not dependent upon evaporation, the area of the joint is unimportant. Thermo-setting adhesives are extremely strong and are used for making structural joints in high strength materials such as metals. The body shells of motor cars and stressed members of aircraft are increasingly dependent upon these adhesives for their joints in place of spot welding and riveting. The stresses are more uniformly distributed and the joints are sealed against corrosion. Further, the low temperatures involved do not affect the crystallographic structure of the metal. Thermo-setting adhesives tend to be brittle when cured and, therefore, are not suitable for flexible (non-rigid) materials.

5.28 Safety

One great advantage of natural gums and glues is that they are non-toxic. Therefore, they are widely used in the labelling and packaging of foodstuffs. Most synthetic adhesives and their solvents, hardeners, catalysts, etc. are highly toxic and must be used under carefully controlled conditions. In addition, the solvents used in thermo-plastic and impact adhesives are highly flammable. They must be stored and used in well ventilated conditions and the working area must be declared a **no-smoking** zone.

The health hazards presented by these materials range from dermatitis and sensitisation of the skin to permanent damage to the internal organs of the body if inhaled or accidently swallowed.

Precautions

(a) Use only in well ventilated areas.
(b) Wear protective clothing appropriate to the process, no matter how inconvenient.
(c) Use a barrier cream.
(d) After use, wash thoroughly in soap and water; do not use solvents.

5.29 Pipe jointing

Machine tools are increasingly using pneumatic and hydraulic services in their operation. Thus the engineer is frequently called upon to design and install high and low pressure fluid services.

Tapered threads are self-seal when pulled up tight and are used with fittings threaded parallel internally.

To ensure pressure tightness, the joint is usually sealed by brushing a 'jointing compound' on the threads before screwing the joint together. There are many proprietary makes of jointing compound on the market. Linseed-oil based, putty-type compounds are suitable at low temperatures and pressures. They have the advantage that the joint may be easily broken out if a fitting has to be altered or replaced. At high pressure and temperatures, a hard setting synthetic resin-based compound should be used.

Although the screwed joints are simple to make and neat in appearance, they have some fundamental limitations:

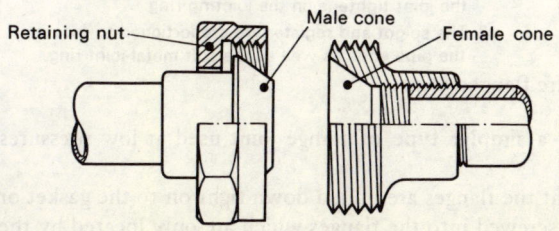

Fig. 5.30   Union joint

(*a*) valves and other fittings cannot be readily broken out of the pipe run for maintenance or replacement;

(*b*) if the pipe is cranked, a large amount of room is required to swing it round as it is screwed home.

To overcome these limitations a number of union joints (Fig. 5.30) have to be left in the pipework and these are a source of weakness.

Flanged joints overcome the disadvantages of the plain screwed joint, but are:

(*a*) more costly to produce;

(*b*) more bulky.

Figure 5.31 shows a BSS 778 : 1966 high pressure flanged joint with an all-metal seal.

It will be seen that the screw threads only locate the flanges and in no way assist in sealing the joint. The flanges do not touch, and so do not require machining on their faces.

It is usual to spot face the nut and bolt seatings.

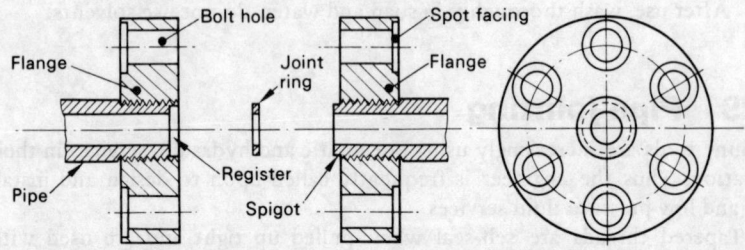

B S S Hydraulic pipe flange joint - 17·25 MN/m² (172·5 bar)

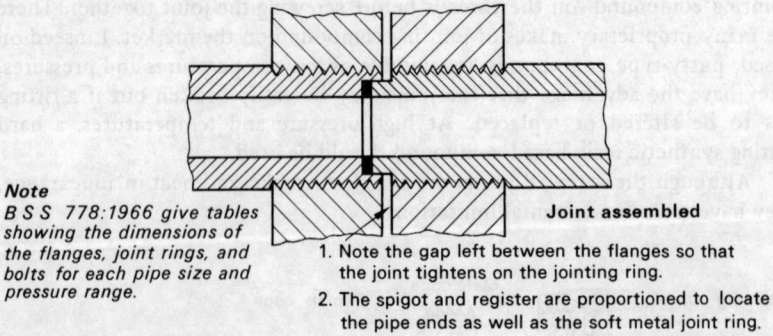

Note
*B S S 778:1966 give tables
showing the dimensions of
the flanges, joint rings, and
bolts for each pipe size and
pressure range.*

Joint assembled

1. Note the gap left between the flanges so that the joint tightens on the jointing ring.
2. The spigot and register are proportioned to locate the pipe ends as well as the soft metal joint ring.

Fig. 5.31 High pressure flanged joint

Figure 5.32 shows a simpler type of flange joint used at low pressures and temperatures.

In this type of joint the flanges are pulled down tight on to the gasket or packing. The pipes are screwed into the flanges which are only located by the retaining bolts, there being no spigot and register.

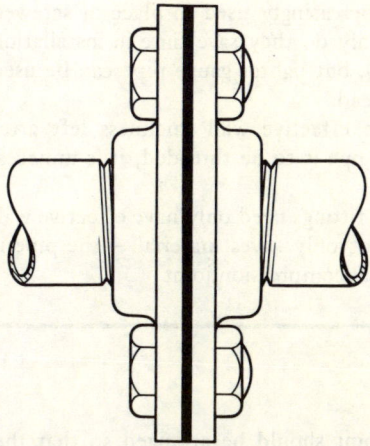

Fig. 5.32 Low pressure flanged joint

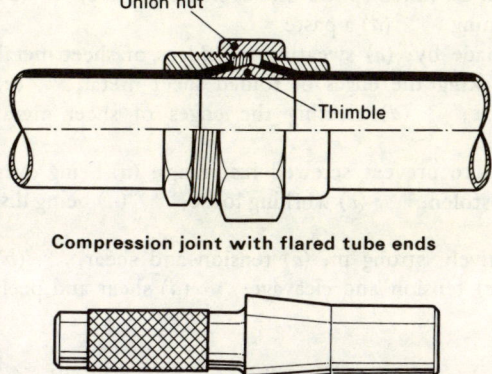

Union nut

Thimble

Compression joint with flared tube ends

Steel drift for flaring tube ends

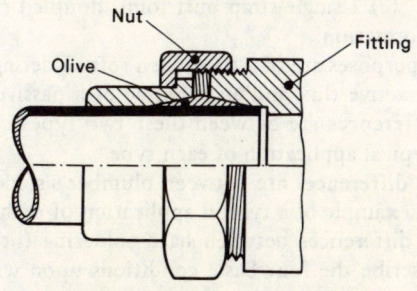

Nut

Fitting

Olive

Compression joint with plain tube ends

Fig. 5.33 Compression joints

Compression joints are becoming increasingly used in place of screwed joints for the smaller size pipes. Not only do they save time in installation (the pipe does not have to be threaded), but lighter gauge pipe can be used since it is not weakened by the screw thread.

The tube is only as strong as the effective wall thickness left after screwing. Therefore, except where the pipe is to be threaded, it is unnecessarily thick. This is a waste of material.

A pipe used with compression joint fittings need only have effective wall thickness throughout its length. This not only saves material — the pipe is easier to bend. Figure 5.33 shows a typical compression joint.

Problems

Section A

1. For maximum strength a riveted joint should be arranged so that the rivet is in: (*a*) shear;　(*b*) tension;　(*c*) compression;　(*d*) no way stressed.
2. The soldering flux known as 'killed spirits' is: (*a*) non-corrosive;　(*b*) inactive;　(*c*) self cleaning;　(*d*) a paste.
3. Self-secured joints are made by: (*a*) sweating the edges of sheet metal together;　(*b*) interlocking the edges of folded sheet metal;　(*c*) using an impact adhesive;　(*d*) riveting the edges of sheet metal together.
4. Locking devices are used to prevent screwed fastenings: (*a*) being over tightened;　(*b*) being stolen;　(*c*) working loose;　(*d*) being dismantled.
5. Most adhesives are relatively strong in: (*a*) tension and shear;　(*b*) cleavage and peel;　(*c*) tension and cleavage;　(*d*) shear and peel.

Section B

6. Describe, with the aid of sketches, how the following joints are made: (*a*) a grooved seam between two pieces of tin plate;　(*b*) a bolted joint between two 12 mm thick plates in which one of the plates has an M10 × 1·5 tapped hole;　(*c*) a single-strap butt joint, doubled riveted with the rivets in 'zig-zag' formation.
7. (*a*) Explain what the main purposes are of a flux when soft soldering.
   (*b*) Give an example of an active flux and an example of a passive flux. Explain what the main differences are between these two types of flux and give an example of a typical application of each type.
   (*c*) Explain what the main differences are between plumber's solder and tinman's solder and give an example of a typical application of each type.
8. (*a*) Describe the essential differences between hard soldering (brazing) and soft soldering, and describe the four basic conditions upon which a successful brazed joint depends.
   (*b*) Explain, with the aid of sketches, how a collar may be brazed on to a shaft using electric induction heating.

9. (*a*) Sketch a section through a corner joint between two pieces of sheet
metal that has been designed for adhesive bonding.
(*b*) Describe how the joint surfaces should be prepared, and what
precautions should be taken to ensure a sound joint.
(*c*) Select a suitable adhesive for a high strength joint between the two
pieces of metal and give reasons for your choice.
10. Describe under what circumstances the following joints would be used:
(*a*) a screwed joint;   (*b*) a riveted joint;   (*c*) a self-secured joint;
(*d*) a soft soldered joint;   (*e*) a hard soldered joint;   (*f*) an adhesive bonded joint.
Your answer should take into consideration the following factors: the conditions under which the joint has to function, the properties of the materials
being joined and the effectiveness of the joint.

Chapter 6

Working in plastics

6.1 The joining of plastics

Depending upon their composition, plastic materials may be joined by one of the following techniques.

1. Heat welding.
2. Solvent welding.
3. Adhesive bonding.

They may also be joined by mechanical methods such as riveting, the use of nuts and bolts or by the use of self-tapping screws. Screwed connections are used when only a semi-permanent joint is required. Screwed and riveted joints were discussed in Chapter 5.

6.2 Heat welding

Heat welding can only be used to join thermoplastic materials since only these plastic materials soften upon heating. Obviously such processes cannot be applied to thermosets as, once cured, they do not soften when heated. A distinction is made in the joining of plastics between welding and sealing. The term 'sealing' is reserved for the thermal joining of thin films and foils (plastic bags containing food, e.g. potato crisps), whereas the term 'welding' is applied to the joining of relatively heavy gauge (thick) sheet plastic components.

Since thermoplastic materials soften like metals when heated, they can be joined by a technique similar to oxy-acetylene gas welding. They cannot be arc welded as the temperature would be too high and, in any case, plastic materials are insulators and an arc cannot be struck. The temperature of an ordinary gas welding torch is much too high for direct application to the plastic material and would cause it to degrade. The low thermal conductivity and softening temperatures of plastic materials necessitates the use of a low welding temperature so that the heat can penetrate into the body of the plastic before the surface degrades.

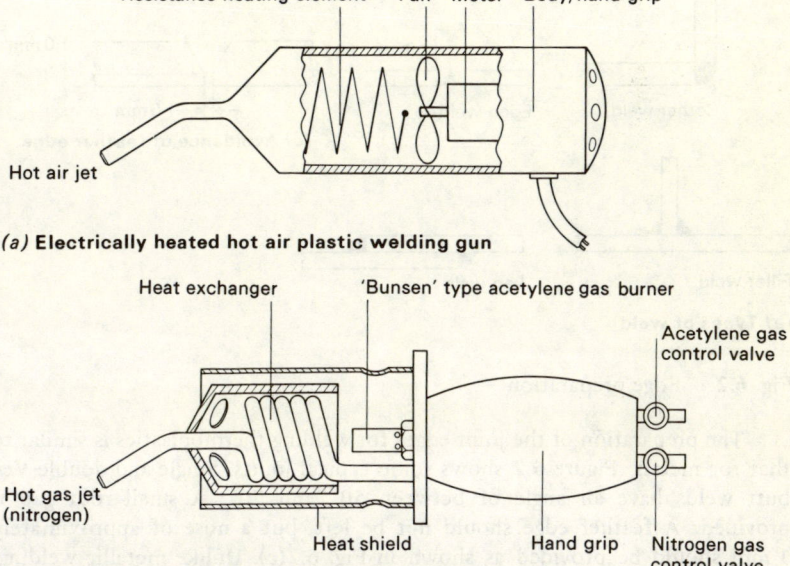

(a) Electrically heated hot air plastic welding gun

(b) Gas heated - hot nitrogen plastic welding gun

Fig. 6.1   Plastic welding guns

Heat is normally applied to the joint by a welding 'gun'. This consists of an electrically operated blower that directs a jet of hot air or hot nitrogen gas into the weld zone. Figure 6.1 shows the principle of such a gun. Nitrogen gas is used to prevent oxidation of the plastic material which would weaken the joint. The welding gun supplies heat to the jet of welding gas by means of either an electric resistance element (rather like a hair dryer) or by means of a gas flame directed upon a heat exchanger (see Fig. 6.1(b)).

Unlike metals which have a fairly sharply defined melting point, thermoplastics usually have a wide range between the softening temperature and the temperature at which they degrade. The easiest plastics to weld are polymerised vinyl chloride (PVC) and polyethylene (PE) as they have a wide

148   softening range. The basic technique is to apply a jet of heated air or nitrogen into the joint so that the edges of the parent plastic sheet are softened. Filler material, in the form of a rod of the same material as that being welded, is added into the joint in much the same way as when welding metals. Some degradation inevitably occurs, so that the strength of the joint is slightly below that of the surrounding material.

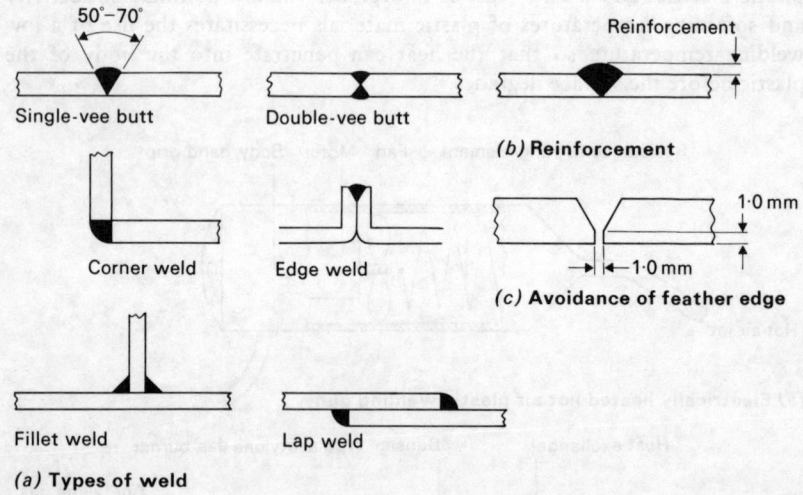

Fig. 6.2   Edge preparation

The preparation of the joint edges for welding thermoplastics is similar to that for metals. Figure 6.2 shows some typical joints. Single and double-Vee butt welds have an angle of between 50° and 70°. A small root gap is provided. A feather edge should not be left, but a nose of approximately 1 mm should be provided as shown in Fig. 6.2(c). Unlike metallic welding, the bead or 'reinforcement' standing above the surface of the sheet should not be removed as it can increase the joint strength by as much as 20 per cent. (See Fig. 6.2 (b).)

Unlike metal welding, the plastic filler rod is softened but not melted. It is presented to the work at right angles as shown in Fig. 6.3, the welding jet being kept between 10 and 20 mm from the joint. To prevent overheating and to ensure uniform softening the jet is moved with a weaving motion between rod and sheet. At the end of the weld, the downward pressure should be maintained on the filler rod until the joint has cooled. Surplus filler rod is removed with a knife. When welding polyolefins the filler rod should be presented to the work at 45°.

In addition to hot gas welding, heated tool welding is also used. In these processes the surfaces to be joined are softened against a heated surface and brought together under pressure to make them bond. A variety of heat sources may be used, some specially designed for the job and others adapted

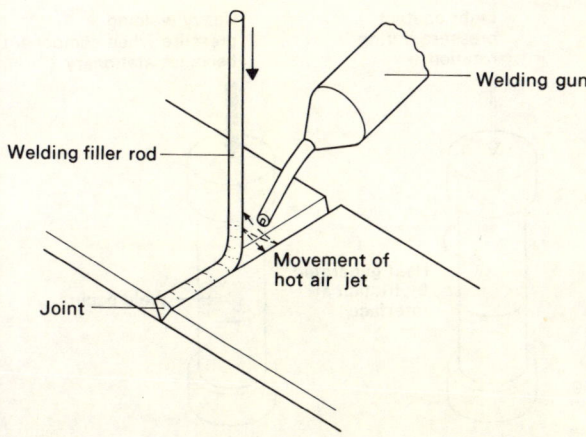

Fig. 6.3 Weld technique

from existing equipment. Examples of heat sources are: hot knives, soldering irons, hot plates, hot wires and strip heaters. To prevent the heated plastic adhering to them, the surfaces of the heat source should be coated with a non-stick material such as 'Teflon'. When welding thick sheet (6 mm or thicker) great care must be taken to avoid trapping air in the joint; thus forming discontinuities resulting in considerable weakening. For thick material it is preferable to shear the edges together to ensure perfect contact free of cavities.

6.3 Friction welding

When two surfaces are rubbed together without a lubricant the friction between them results in a conversion from mechanical energy to heat energy at the interface. This is exploited in friction welding (also called *spin welding*) by rotating one component against a stationary component until the joint faces reach their welding temperature. At this point the rotation ceases and the axial pressure is increased to cause welding to take place (see Fig. 6.4).

Obviously the heating effect will not be uniform. Overheating tends to occur at the periphery where the maximum relative velocity between the two surfaces exists, and lack of heating at the centre where the relative motion is minimal. Figure 6.5 shows some typical friction joint forms. It will be seen that the centre of the joint is either removed (Fig. 6.5(a)) so that the strength of the joint can be calculated more easily, or the centre of the joint faces form a mechanical joint or key. For the reasons already explained minimal or no welding occurs at the centre of the joint. Both the pressure and the rotational velocity must be controlled to suit the material to be friction welded.

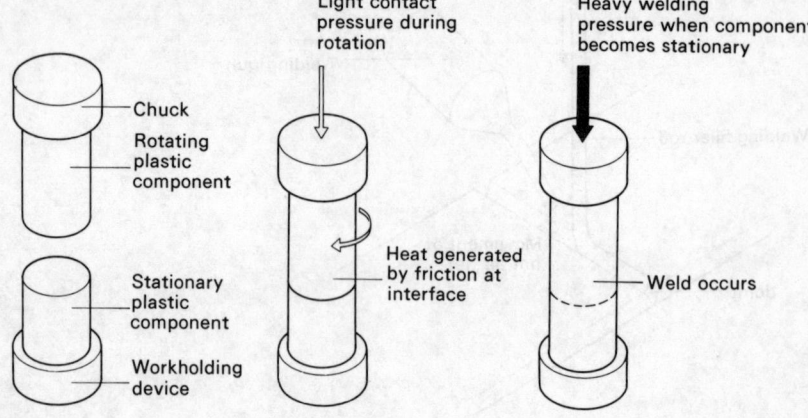

Fig. 6.4 Friction welding — principle

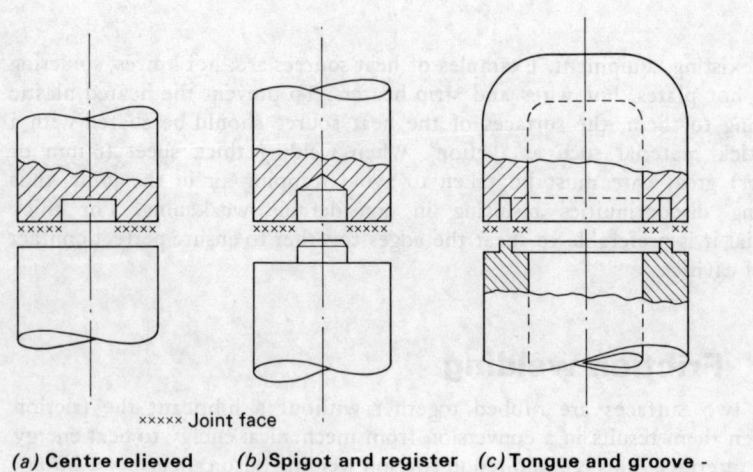

××××× Joint face

(a) Centre relieved *(b)* Spigot and register *(c)* Tongue and groove -
cylindrical components

Fig. 6.5 Types of friction weld

6.4 Hot wire (resistance) welding

The problems of butt welding solid rod by friction heating have already been
discussed. Further, external heating is equally ineffectual since all plastic
materials are poor conductors of heat. A solution to the problem is shown in
Fig. 6.6. A resistance mat of nichrome wire is laid in the joint and connected
to an external power source. As soon as the joint faces have softened the
supply is disconnected and pressure is applied. The resistance mat remains
permanently embedded in the welded joint.

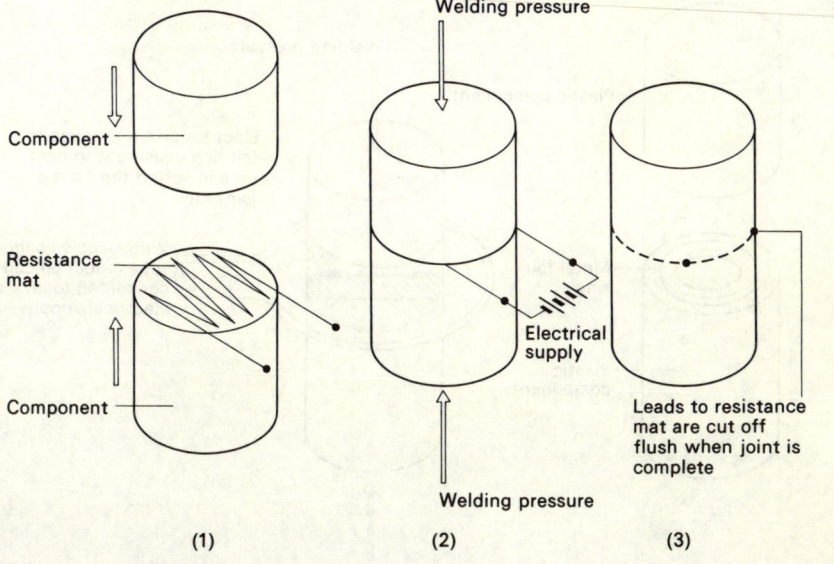

Fig. 6.6　Resistance welding

6.5　Induction welding

Since plastic materials are insulators it may seem strange to consider the possibility of using electric induction heating to form a weld. The solution is quite simple. A ring of metal foil is laid in the joint and an induction coil is placed around the joint as shown in Fig. 6.7. A high frequency alternating current is passed through the coil and this induces a current in the metal ring embedded in the joint. The ring heats up and softens the joint faces. Pressure is applied and the weld occurs, leaving the ring embedded in the joint. Obviously components with this type of joint must never be used in alternating current fields as heat will again be generated and the joint will fail.

6.6　Di-electric welding

This process exploits the insulating properties of plastic material. In this instance the plastic to be joined forms the di-electric of a capacitor placed between two electrodes as shown in Fig. 6.8. A high frequency alternating current of about 30×10^6 Hz is applied to the electrodes and the plastic commences to heat up from the inside. When the required temperature is reached the current is switched off and the electrodes apply the pressure to make the weld. The advantage of this method is the fact that the heating effect occurs in the joint itself and overcomes the problem of externally heating materials that have poor thermal conductivity.

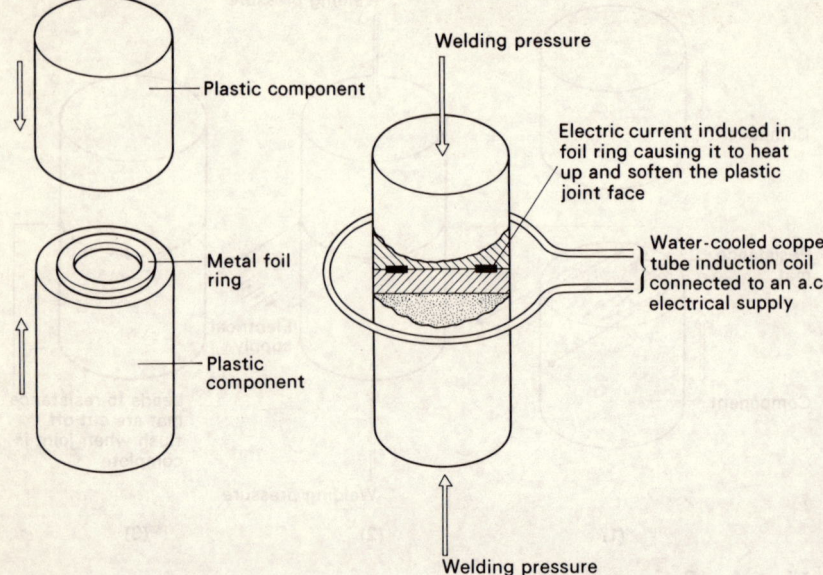

Fig. 6.7   Induction welding

Not all plastic materials have the electrical properties to lend themselves to this process. The following plastic materials cannot be welded by di-electric techniques:

polyethylene (PE);
polypropylene (PP);
polycarbonate (PC);
polytetrafluoroethylene (Teflon) (PTFE);
polystyrene (PS).

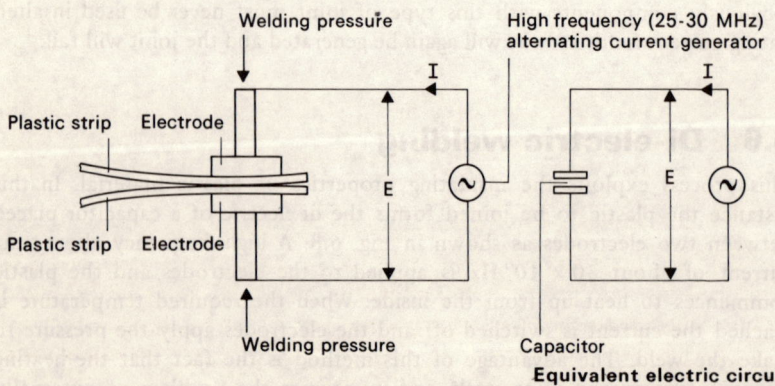

Fig. 6.8   Di-electric welding

6.7 Ultrasonic welding

Ultrasonics is the technology of very high frequency sound waves. Usually a frequency of 20×10^3 Hz or higher is used, which is above the range of normal hearing for an adult. Ultrasonic techniques induce heat in the joint by friction as the surfaces vibrate rapidly together.

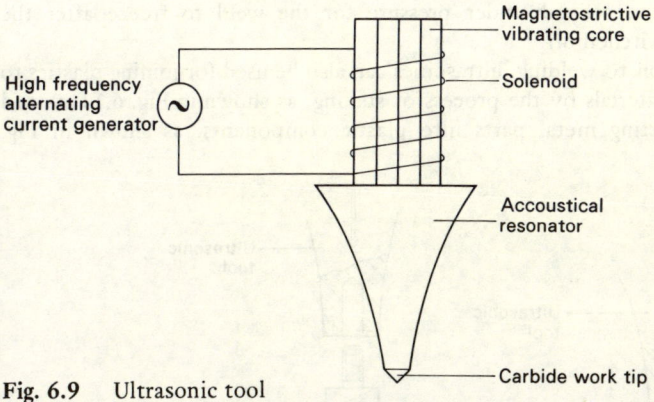

Fig. 6.9　Ultrasonic tool

Figure 6.9 shows the principal components of an ultrasonic welding tool. The horn is designed to match the characteristics of the magnetostrictive core to the work head (tip). That is, the horn acts as an accoustical tuning device. Magnetostriction is the property of the core material to expand and contract in sympathy with the alternating electromagnetic field generated by the solenoid. The work head or tip is contoured to match the joint being made and is usually of a wear- and heat-resistant material such as tungsten carbide or titanium. Its mass must be kept very small so that it does not damp the high frequency vibration.

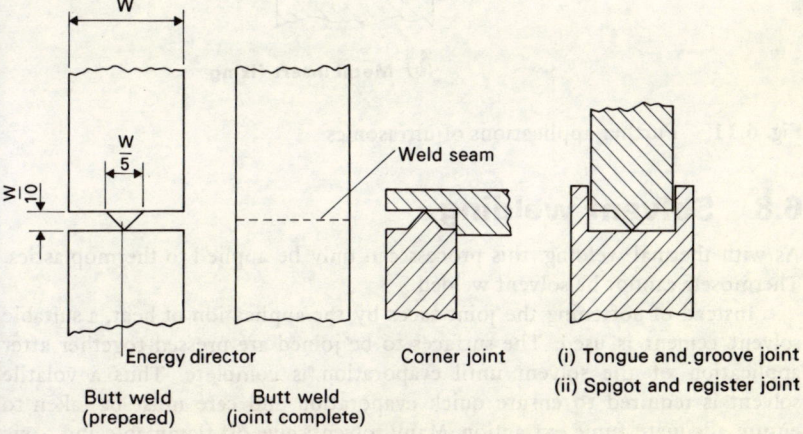

Fig. 6.10　Weld preparation

Most thermoplastics can be welded by ultrasonic techniques; the exceptions are the vinyls and the cellulosics. Figure 6.10 shows some typical joints and it will be seen that they are all formed with *energy directors* to localise the heat generated. This restricts the energy used and the volume of plastic softened. The small volume of softened plastic spreads out into the joint under pressure and produces the bond. The joint should be held for between a half and one second under pressure for the weld to freeze after the generator is switched off.

In addition to welding, ultrasonics can also be used for joining plastics to non-plastic materials by the process of staking, as shown in Fig. 6.11(*a*), and also for inserting metal parts into plastic components, as shown in Fig. 6.11(*b*).

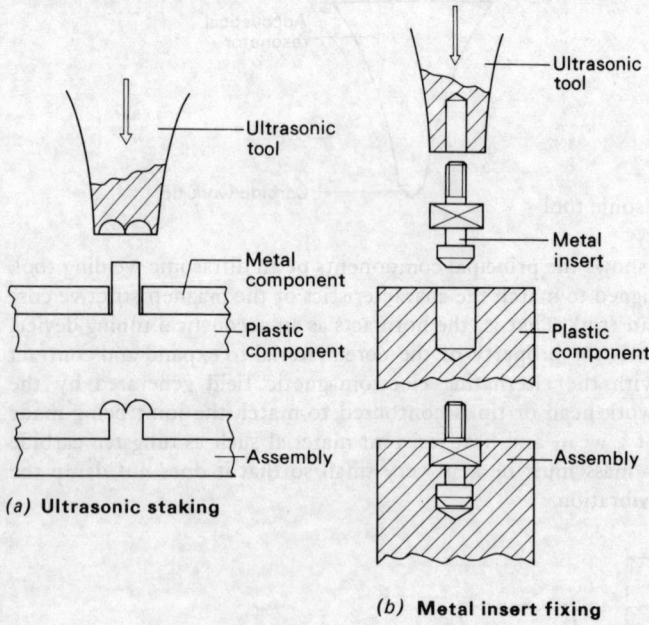

(a) **Ultrasonic staking**

(b) **Metal insert fixing**

Fig. 6.11 Further applications of ultrasonics

6.8 Solvent welding

As with thermal welding, this process can only be applied to thermoplastics. Thermosets cannot be solvent welded.

Instead of softening the joint faces by the application of heat, a suitable solvent cement is used. The surfaces to be joined are pressed together after application of the solvent until evaporation is complete. Thus a volatile solvent is required to ensure quick evaporation and care must be taken to ensure adequate fume extraction. Many solvents give off flammable and toxic fumes and great care must be taken in their storage and use.

Bodied cements: These are used where gaps occur in the joint faces, since a volatile solvent cannot fill gaps when it evaporates. A typical bodied cement is used for constructing model aeroplanes and is made by dissolving cellulose nitrate in amyl acetate. All bodied cements are made by dissolving some of the parent material in the solvent before applying the solvent to the joint faces.

Monomeric cements: These are made from the same monomers as the thermoplastics being joined. For example, methylmethacrylate may be used to join any proprietary PMMA material. However, a catalyst is added to the solvent immediately before application and the bond is produced by polymerisation rather than by evaporation.

Joints made with solvent cements are much weaker than those produced by thermal welding because of the shrinkage stresses set up by evaporation of the solvent. Further, evaporation of the solvent may also create voids in the joint. Another problem is the difficulty in ensuring complete evaporation when large surface areas are being joined.

6.9 Adhesive bonding

The principles involved in the use of adhesives have already been discussed in Chapter 5. When used to join plastic components care is required to match the adhesive carefully to the material being bonded. the surface texture of many plastic materials prevents an adequate *key* being formed between the adhesive and the parent material. Further, the adhesive must match the rigidity of the components being joined.

(*a*) *Elastomeric cements* are used to join flexible materials.

(*b*) *Thermo-setting cements* are used to join rigid materials. These cements are generally stronger than elastomeric cements.

Elastomeric adhesives are usually based on natural, synthetic or re-claimed rubber, which is dissolved in a solvent. Neoprene rubber is used for impact adhesives that form a bond on contact.

Thermo-setting adhesives are insoluble in most solvents and are hardened by the addition of a catalyst. Two methods of application are employed. Resin and catalyst are mixed prior to use and curing starts immediately. Thus the resin must be applied quickly before polymerisation advances too far and a poor joint occurs. This technique is employed for moulding glass fibre reinforced components. The second method consists of applying resin to one component and the catalyst to the mating component. Curing does not occur until contact is made between the two surfaces. This enables work to proceed at a more leisurely pace and is useful when manipulating large components and panels into position.

Epoxy resins are the strongest (and most expensive) of the thermo-setting adhesives. They have incomparable strength and weathering character-istics and can be used to bond ill-fitting joints as very little shrinkage occurs and they combine the properties of adhesive and void filler. Epoxy resins will be considered further in section 6.12.

6.10 Heat bending techniques

Simple, straight bends in thermoplastic materials follow the principles laid down for sheet metal working in Chapter 10. The only difference being that the plastic material requires to be heated before bending and bending must take place whilst the material is still hot. For this reason the bending jig must be faced with materials having a low thermal conductivity such as wood or tufnol.

Strip heaters may be used to ensure that heating is localised along the line of the bend. This makes the plastic sheet easier to handle. Rapid cooling is required immediately after the bend is complete to avoid loss of shape and degradation of the plastic. Thick sheet should be heated on both sides prior to bending because of the low thermal conduction of plastic materials.

Like metals, some spring-back occurs when the plastic is removed from the bending jig and some degree of *overbend* will be required. This has to be determined by trial and error as it depends upon the plastic, its thickness, temperature, radius of the bend and rate of cooling.

Other heat bending techniques such as vacuum forming, blow forming and pressing are outlined in Fig. 6.12. In all these examples the thermoplastic sheet is preheated before forming.

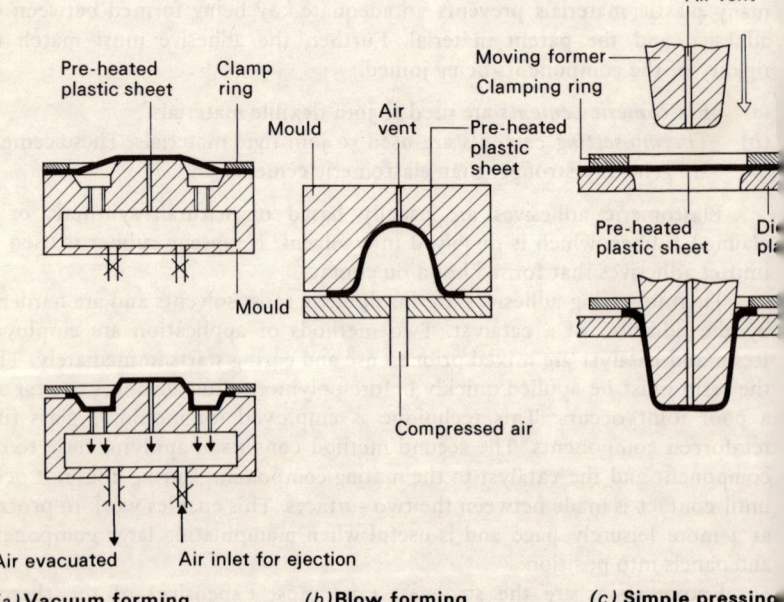

(a) Vacuum forming   *(b)* Blow forming   *(c)* Simple pressing

Fig. 6.12   Forming plastic sheet material

6.11   The machining of plastics

Although plastic components are usually moulded to their finished shape, it is sometimes necessary to resort to machining processes. For example:

(a)   drilling holes from the solid where coring is inconvenient or the quantity does not warrant the use of complex moulds;

(b)   sizing cored holes where the shrinkage after coring is not acceptable;

(c)   tapping holes;

(d)   turning bushes, screws and other small components from the bar where the quantity involved does not warrant the manufacture of a mould;

(e)   the machining, by all the normal workshop methods, of engineering components from *tufnol* (reinforced plastic bars, rods, tubes, sheets);

(f)   the *in situ* erection of ducting and equipment cabinets from sheet plastic.

The problems associated with the machining of plastics are as various as the materials themselves. Within a broad framework of recommendations, each material requires treating individually and the optimum machining conditions have to be determined by trial and error.

Just as plastic materials can be divided into the two main groups of thermoplastics and thermo-setting plastics, so the machining characteristics divide into two main categories.

Thermoplastics

(a)   Heat generated during cutting tends to build up due to poor thermal conduction. This softens the material being cut so that it may collapse and the swarf clog the tools. Cooling by air blast is recommended, together with vacuum extraction of the swarf. The cutting speed may have to be reduced to avoid overheating and softening despite the machine being worked below its optimum capacity.

(b)   Thermoplastics have low moduli of elasticity and high elastic recovery, and will therefore deform under the tool forces produced by heavy cuts.

From the above characteristics, it is obvious that tooling with free-cutting characteristics are required. Particular attention should be paid to the flow of the chips produced, as thermoplastics are usually ductile and produce a continuous, ribbon chip.

(a)   A high rake angle is required and this prohibits the use of carbide and ceramic tooling. Super-high speed steel is suitable for most applications but stellite is to be preferred for long production runs.

(b)   Cutting tools should be lapped to a high surface finish after grinding to reduce friction and the consequent possibility of the swarf adhering to the rake face.

Thermo-setting plastics

(a)   Poor thermal conductivity results in a tendency to overheating but softening does not occur.

158   (b)   Thermo-setting plastics tend to be rigid and brittle compared with the thermoplastics previously described and machine more like cast iron, producing a granular, discontinuous chip. This required vacuum chip disposal facilities.

(c)   The filler materials used with thermo-setting plastics results in high abrasion and wear of the cutting tools. Fillers such as glass fibre, glass, paper, mica, clay and paper are particularly abrasive.

From the above characteristics it will be seen that zero or negative rake carbide tooling is essential for reasonable tool life. Since low rake tooling increases the temperature at the cutting zone, the cutting speed has to be kept relatively low (30 to 50 m/min).

The drilling of plastic materials is considered in detail in Chapter 11.

6.12   Casting plastic materials

Metals are cast by raising them above their fusion temperatures, pouring them into moulds and waiting for them to freeze into solid components. The casting of plastics is entirely different. Basically, a low viscosity resin is mixed with a catalyst and poured into a mould that has been treated with a suitable release agent. After curing has taken place, the mould is opened and the casting is removed. Three groups of plastic materials (epoxides, phenolics, polyesters) are used for casting.

1. *Epoxides:*   Where maximum strength is required for tooling or where maximum electrical insulation is required for insulators and encapsulating electrical components.

2. *Phenolics:*   Where appearance is important and the inherent brittleness is relatively unimportant (e.g. billiard and snooker balls).

3. *Polyesters:*   For hobby casting as well as for a range of industrial and construction industry products, such as synthetic marble, stone and brick veneers used for interior trim, shop and bar fitting. The high viscosity, high exotherm polyesters developed for glass fibre moulding is not suitable for casting. Special low viscosity, low exotherm polyesters have been formulated for cast applications.

Since this chapter is concerned mainly with the engineering applications of plastic materials, cast epoxides will now be considered in rather greater depth. Epoxy resins have oxygen and aromatic rings in the polymer chain and this results in the superior properties of these resins. They are available in a wide range of formulations giving an equally wide range of properties and curing characteristics, leading to an equally wide range of applications. Since no water or volatile substances are produced during the curing cycle, little shrinkage occurs and epoxide castings have high dimensional stability. They also make excellent gap filling adhesives and have been considered as such in *tooling epoxides* and the *encapsulation* of electrical and electronic components.

(*a*) *Tooling epoxides* are usually reinforced with aluminium powder, and are cast to make large and complex forming dies used for short run sheet metal pressings. Aluminium-filled epoxy resins are easier to machine and lighter to handle than tool steels. Neither is costly heat treatment of large dies necessary. Tooling epoxides are also readily repaired and can be easily altered by building up additional layers of resin.

Other applications are casting patterns and models for three dimensional copy milling where their light weight and impact resistance facilitates handling and their corrosion resistance facilitates storage. Lead time between initial design and production is greatly reduced by using tooling epoxides.

(*b*) *Encapsulating electrical and electronic components* is now widely used especially for equipment designated for tropical and other electrically hostile environments. Epoxy resins are rapidly replacing traditional insulating materials because of their improved electrical and mechanical properties. For example, epoxy resins are replacing Chatterton's compound for 'potting' or encapsulating transformers and chokes. They are also replacing ceramic (porcelain) insulators for heavy duty power transmission equipment. The epoxy insulators having greater impact strength.

For the impregnation of transformers, or motor and generator stators and rotors unfilled expoxides are used, specially formulated to have very low viscosities. They are solvent free and have good wetting properties so that voids are unlikely to occur and the electrical function is improved. Two-part systems that cure, on mixing, at ambient temperatures are used for casting and encapsulation. For impregnation the epoxides are formulated to withstand a pot temperature of 20° C for up to fourteen days. The components are dipped at 20° C which causes full penetration and drives off any entrapped moisture. Curing takes upwards of two hours at 120° C and is achieved by stoving the dipped components in heated ovens.

When making a casting or encapsulating electrical components, the resin must be poured into the mould so as not to entrap air and create voids. The thermoset should be poured slowly and steadily down one side of the mould (which must be adequately vented). The mould should be vibrated to assist air bubbles to rise to the surface where they can be pricked and burst.

Problems

Section A

1. Thermo-plastic materials can be welded using: (*a*) an oxy-acetylene torch; (*b*) arc welding equipment; (*c*) a hot air gun; (*d*) a brazing torch.
2. When drilling plastic materials the feed of the drill should be: (*a*) slow but steady; (*b*) fast but steady; (*c*) fast with a woodpecker action; (*d*) very slow with a woodpecker action.

3. A suitable coolant when machining plastics is: (*a*) compressed air; (*b*) suds; (*c*) water; (*d*) paraffin.

4. Carbide-tipped tools are used when machining thermo-setting plastics because: (*a*) of the high temperature involved; (*b*) the fillers used are abrasive; (*c*) they chip less readily; (*d*) they are easy to regrind.

5. To ensure that sheet plastic takes a permanent set when bent to shape it should be: (*a*) scored with a knife along the bend line; (*b*) painted with a solvent prior to bending; (*c*) cooled prior to bending; (*d*) heated prior to bending.

Section B

6. With the aid of sketches describe how **four** of the following joining techniques are applied to plastic materials: (*a*) hot gas welding; (*b*) friction welding; (*c*) hot wire (resistance) welding; (*d*) induction welding; (*e*) di-electric welding; (*f*) ultra-sonic welding.

7. (*a*) Epoxides are an important industrial group of plastic materials. Describe in detail two applications of cast epoxides.
   (*b*) Epoxides also form the basis of high strength adhesives. Describe how a joint should be prepared and completed using epoxide adhesives.

8. (*a*) Explain how the machining of thermoplastic materials varies from the machining of metals.
   (*b*) Explain what precautions must be taken when machining thermo-setting plastics to ensure a reasonable tool life.

9. With the aid of sketches, describe how thermoplastic materials may be formed by: (*a*) vacuum forming; (*b*) blow forming; (*c*) pressing.

10. Explain the meaning of the following terms as applied to the solvent welding and adhesive bonding of plastic materials; (*a*) bodied cement; (*b*) monomeric cement; (*c*) elastomeric cement; (*d*) impact adhesive; (*e*) thermo-setting cement.

Chapter 7

Measurement

7.1 Units of linear measurement

From the earliest days of trading, unscrupulous merchants have tried to give 'short measure' to make more profit. To combat their attempts, governing bodies have always tried to lay down the 'correct' weights and measures to be used in trading. These 'correct' weights and measures are referred to as **legal standards**.

In this chapter we are only interested in the **standard of length** upon which all our measurements of dimension are based.

The rapid advances made in engineering during the nineteenth century were due to the improved materials available and the more accurate measuring techniques that were developed.

This demand for more accurate measurement stimulated the British and French governments into producing new and more accurately read standards of length, as shown in Fig. 7.1.

1. The imperial standard yard British (1855).
2. The international prototype metre French (1872).

The British imperial standard yard was made from a simple bronze alloy that has continually shrunk at the rate of one-millionth of an inch per year for the past 50 years.

An attempt was made in 1922 to overcome this shrinkage by measuring

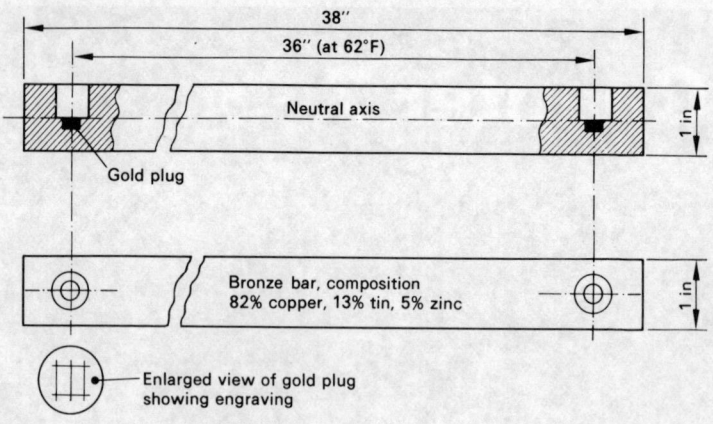

(a) The international prototype metre (1872)

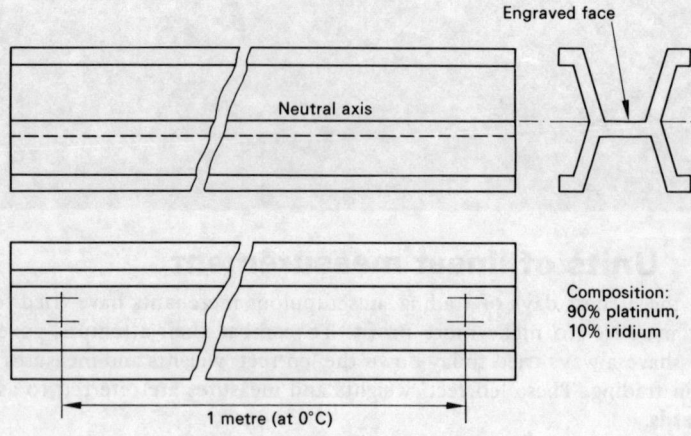

(b) The imperial standard yard (1855)

Fig. 7.1 Material standards of length

the yard/metre relationship in that year and specifying it as a legal size.

1 yard = 0·914 398 41 metre (1922).

The increasing demand for accurate, internationally interchangeable engineering devices and the problems arising from the varying ratio between the imperial standard yard, the American standard yard and the prototype metre were resolved in 1960.

The United Kingdom, America, Canada, Australia, New Zealand and South Africa agreed to adopt an international yard based upon the metre:

1 international yard = 0·9144 metre.

The Weights and Measures Act of 1963 laid down that this was to become the legal standard yard from 31 January 1964.

Further, the Eleventh General Conference of Weights and Measures held in Paris in 1960 defined the metre as equal to 1 650 763·73 wavelengths of the orange radiation of Krypton isotope 86 gas as shown in Fig. 7.2.

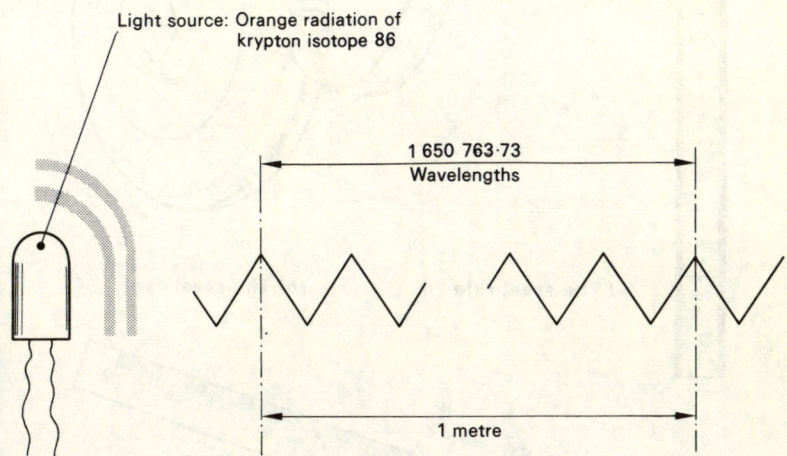

Fig. 7.2 The light standard of length

Such a 'natural' standard has several advantages over the previous standard 'bars'.

1. It does not change length.
2. If destroyed it can easily be replaced.
3. Identical 'copies' can be kept in all standard rooms and physical laboratories.
4. It can be used for making comparative measurements of a much higher accuracy than was possible with the older material standards.

For most workshop purposes the metre is too large a unit and most engineering drawings are dimensioned in millimetres (mm). One metre equals 1000 millimetres (1 mm = 0·001 m).

7.2 Workshop standards of length

Accurate measurement is the basis of good engineering practice. It is a significant fact that the craftsmen in many other trades are content to use a wooden rule, whereas the engineer always uses a precision engraved steel rule. Further, the engineer only uses the rule for his least important measurements. Figure 3.7(*a*) shows a typical steel rule, whilst Fig. 7.3(*b*) shows a typical steel tape.

For more accurate measurements the micrometer caliper and the vernier

164

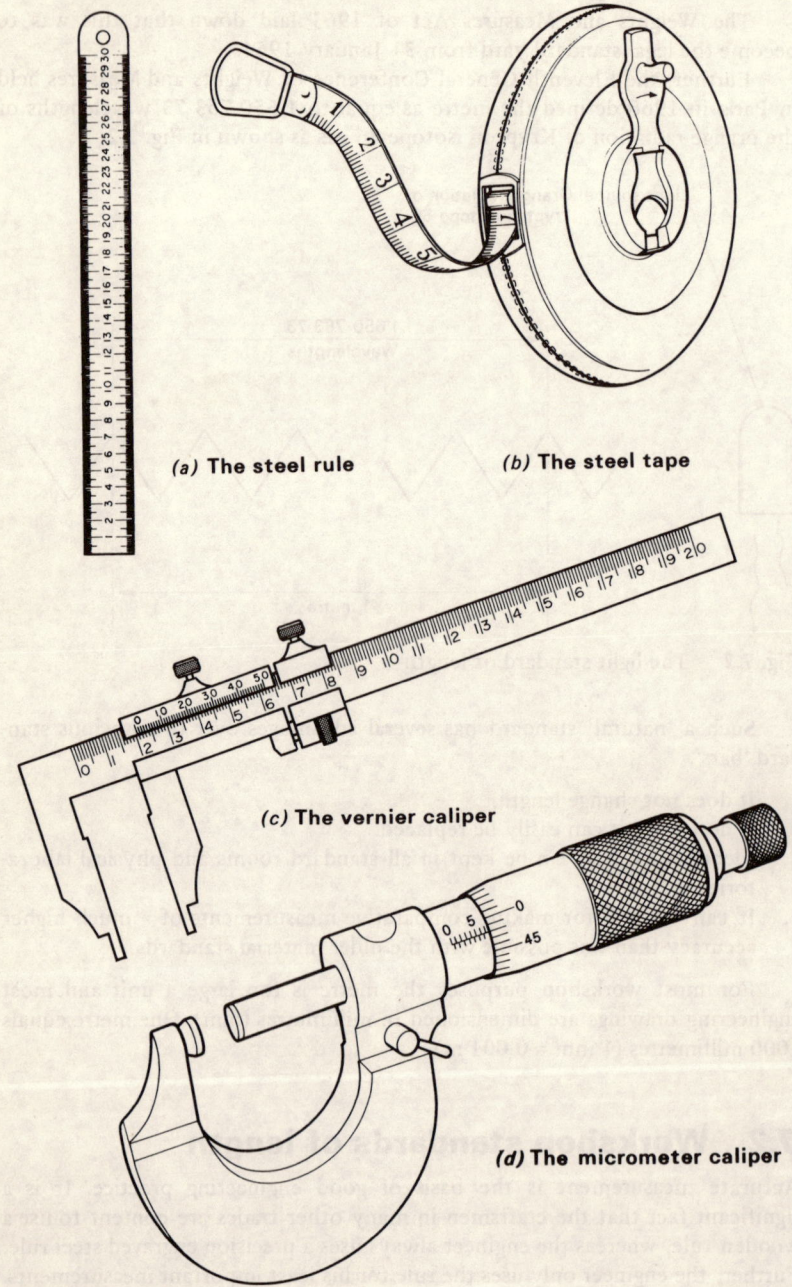

(a) The steel rule

(b) The steel tape

(c) The vernier caliper

(d) The micrometer caliper

Fig. 7.3 Measuring instruments

caliper are used. Figure 7.3(c), (d) shows examples of these instruments.

For the most accurate workshop measurements slip gauges are used. These are made to very high accuracy and must be used with extreme care.

Table 7.1 Workshop standards of length

Name	Range (in mm)	Reading accuracy
Steel rule	150 to 1000	0·5 mm
Vernier caliper	0/150 to 0/2000	0·02 mm
Micrometer caliper	0/25 to 1800	0·01 mm
Slip gauges	1·0025 to 327	0·0025 mm
	(105-piece set)	

Table 7.1 gives the accuracy and range of size in which the above workshop standards of length are available.

Note: The accuracy given is the reading accuracy. Whether or not this accuracy will be achieved will depend upon the care with which the measurement is made.

The workshop standards are themselves based upon the national and international standards of length. Thus the components produced in one factory will be interchangeable with similar components produced in another factory.

7.3 The engineer's rule

The steel rule is frequently used in the workshop for measuring components of limited accuracy quickly. The quickness and ease with which it can be used, coupled with its low cost, makes it a popular and widely used measuring device. Steel rules may be 'rigid' or 'flexible' depending upon their thickness and the 'temper' of the steel used in their manufacture. They may be obtained in lengths from about: 150 mm to 1000 mm (1 m). Until the complete changeover to metric measurement is made, and this is still many years away, a rule should be chosen with both millimetre and inch units engraved on it.

When choosing a steel rule the following points should be looked for. It should be:

1. Made from hardened and tempered spring steel.
2. Engine divided; that is, the graduations should be precision engraved into the metal.
3. Ground on the edges so that it can be used as a straight edge when scribing lines or testing a surface for flatness.
4. Satin chrome finished so as to reduce glare and make it easier to read, also to prevent corrosion.

166   A good rule should be looked after carefully to prevent damage to its edges and to the datum end. It should never be used as a scraper or a screwdriver, and it should never be used to remove swarf from machine-tool table 'Tee' slots. After use the rule should be wiped clean and lightly oiled to prevent rusting. Dulling of the surface will make it difficult to read.

All measurements made with the rule are of limited accuracy by reason of the difficulty of sighting the graduations in line with the feature being measured. Despite the claims of many craftsmen, the rule should not be trusted for measurements where an accuracy of better than 0·5 mm is required.

In addition to being used for the measurement of linear distances, the rule can be used as a straight edge for testing components for flatness.

The engineer's rule, used for making direct measurements, depends upon **visual alignment** of a mark or surface on the work to be measured against the nearest division on its scale. This may appear to be a relatively simple exercise, but in practice errors can very easily occur, as shown in Fig. 7.4.

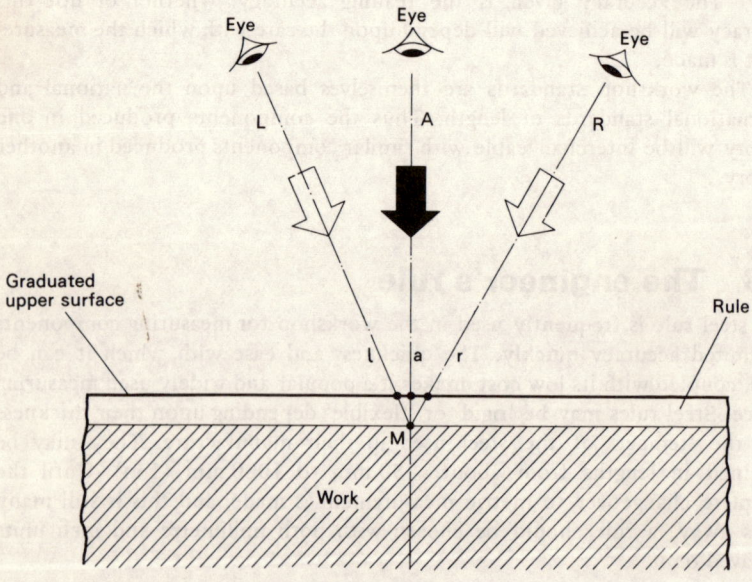

Fig. 7.4   Possible errors with direct eye measurement

These errors can be minimised by using a rule whose thickness is as small as possible — this emphasises the importance of using a thin steel rule.

It is important when making measurements with an engineer's rule to have the eye directly opposite and at 90° to the mark on the work, otherwise there will be an error — known as '**parallax**' — which is the result of any sideways positioning of the direction of sighting.

1. 'M' represents the mark on the work whose position is required to be measured by means of a rule laid alongside it. The graduations of measurement are on the upper face of the rule, as indicated.
2. If the eye is placed along the sighting line A—M, which is at 90° to the work surface, a **true reading** will be obtained at 'a', for it is then directly opposite 'M'.
3. If, however, the eye is not on this sighting line, but displaced to the right, as at 'R', the division 'r' on the graduated scale will appear to be opposite 'M' and an **incorrect reading** will be obtained. Similarly, if the eye is displaced to the left, as at 'L', an incorrect reading on the opposite side, as at 'l' will result.

7.4   Line and end measurement

Distances have sometimes to be measured between two lines and sometimes between two surfaces, or a combination of line and surface. When the length being measured is expressed as the distance between two lines, this is referred to, obviously, as *'line measurement'*. When the length being measured is expressed as the distance between two surfaces, this is referred to as *'end measurement'*. It is very difficult to convert between end systems of measurement and line systems of measurement. For example, a rule (which is a line-measuring device) is not convenient for the direct measurement of distances between two edges. Similarly the micrometer, which is an end-measuring device, would be equally inconvenient to measure the distance between two lines. The measuring device must be selected to suit each particular measuring situation.

7.5   Use of calipers

Calipers are used to extend the application of the rule and increase the reading accuracy. They are used for transferring the distance between the faces of a component to the rule in such a way as to reduce sighting errors. That is, to convert an end measurement situation to the line system of the rule.

Firm-joint calipers are usually used in the larger sizes. Spring joint calipers are usually used for fine work, as in instrument-making and tool-making. Some uses of calipers are shown in Fig. 7.5.

The accurate use of calipers depends upon a highly developed sense of feel that can only be acquired by practice. When using calipers, the following rules should be observed:

(a)   hold the caliper gently and near the joint;
(b)   hold it square to the work;

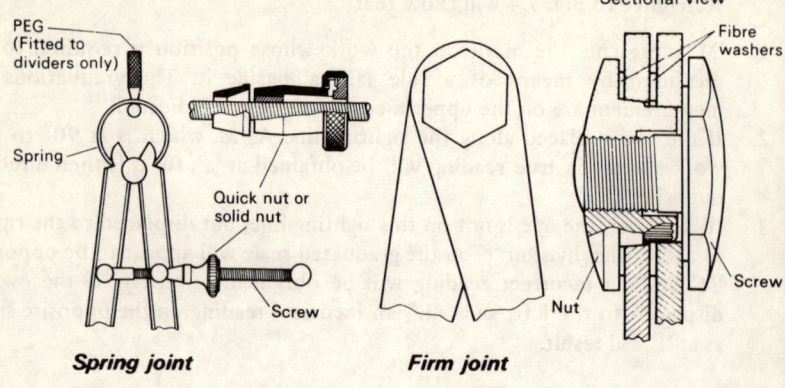

Spring joint   **Firm joint**

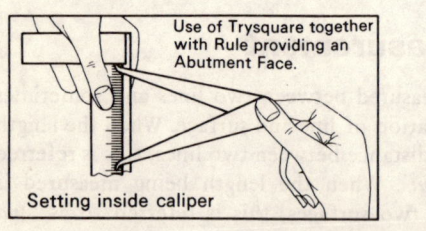

Use of Trysquare together with Rule providing an Abutment Face.

Setting inside caliper

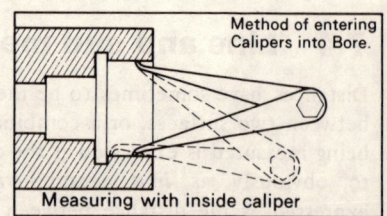

Method of entering Calipers into Bore.

Measuring with inside caliper

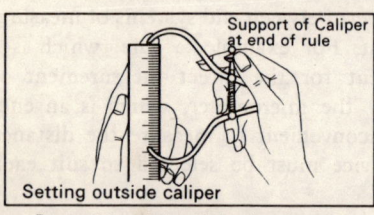

Support of Caliper at end of rule

Setting outside caliper

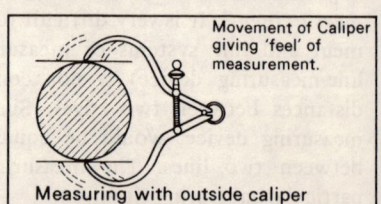

Movement of Caliper giving 'feel' of measurement.

Measuring with outside caliper

By careful setting and developed sense of touch or Caliper 'feel', a surprising degree of accuracy is attained.

Care being taken at all times, to see that in measurement or setting, Calipers are held square to rule or work.

(Courtesy of Moore & Wright Ltd)

Fig. 7.5   Construction and use of calipers

(c)   no force should be used to 'spring' it over the work. Contact should only just be felt;

(d)   the caliper should be handled and laid down gently to avoid disturbing the setting;

(e)   lathe work should be **stationary** when taking measurements. This is essential for **safety** and **accuracy**.

7.6   The micrometer caliper

The equipment so far described has only limited accuracy. As already stated, the unaided rule should only be trusted to 0·5 mm.

Most engineering work has to be measured to a much greater accuracy than this, especially where component parts must fit together as, for instance, a shaft and its bearing. To achieve this greater precision, measuring equipment of a greater accuracy and sensitivity must be used.

One of the most familiar precision measuring devices found in the workshop is the **micrometer caliper**. Figure 7.6 shows the construction of a typical micrometer caliper and names the more important parts.

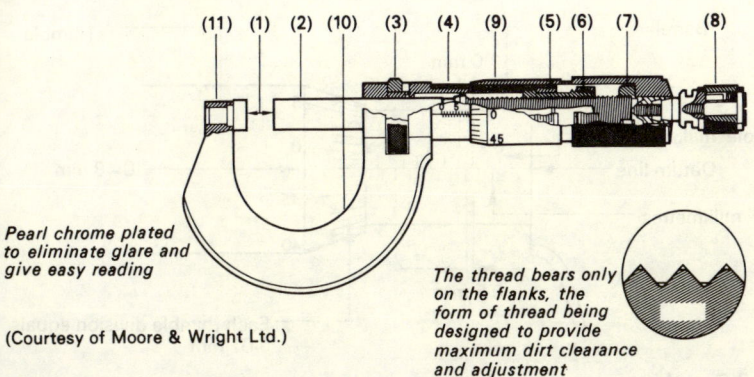

Pearl chrome plated to eliminate glare and give easy reading

(Courtesy of Moore & Wright Ltd.)

The thread bears only on the flanks, the form of thread being designed to provide maximum dirt clearance and adjustment

(1) **Spindle and anvil faces** — Glass hard and optically flat, also available with **Tungsten carbide** faces
(2) **Spindle** — Thread ground, and made from alloy steel, hardened throughout, and stabilised
(3) **Locknut** — effective at any position. Spindle retained in perfect alignment
(4) **Barrel** — Adjustable for zero setting. Accurately divided and clearly marked. Pearl chrome plated
(5) **Main nut** — Length of thread ensures long working life
(6) **Screw adjusting nut** — For effective adjustment of main nut effective
(7) **Thimble adjusting nut** — Controls position of thimble
(8) **Ratchet** — Ensures a constant measuring pressure
(9) **Thimble** — Accurately divided and every graduation clearly numbered
(10)**Steel frame** — Drop forged. Marked with useful decimal equivalents
(11)**Anvil end** — Cutaway frame facilitates usage in narrow slots

Fig. 7.6   Construction of the micrometer caliper

The operation of the micrometer depends upon the principle that the distance moved by the nut along the screw is proportional to the number of revolutions made by the nut. Therefore by controlling the number of revolutions and fractions of a revolution made by the nut, the distances it moves along the screw can be accurately predicted. This principle forms the basis of a number of measuring devices.

Note: The movement of the screw and nut are relative, i.e. the same arguments apply if the nut is fixed and the screw rotates.

To apply this principle to a practical measuring device, one needs:

(*a*)   a precision screw;
(*b*)   a means of counting the 'whole' revolution of the screw;
(*c*)   a means of measuring the extent of the partial revolutions.

Examination of any micrometer measuring device will show how these principles are incorporated. The screw thread is rotated by the thimble which indicates the 'partial' revolution, the 'whole' revolutions being 'counted' on the barrel of the instrument.

The screw has a lead of 0·5 mm, and the thimble and barrel are graduated as shown in Fig. 7.7.

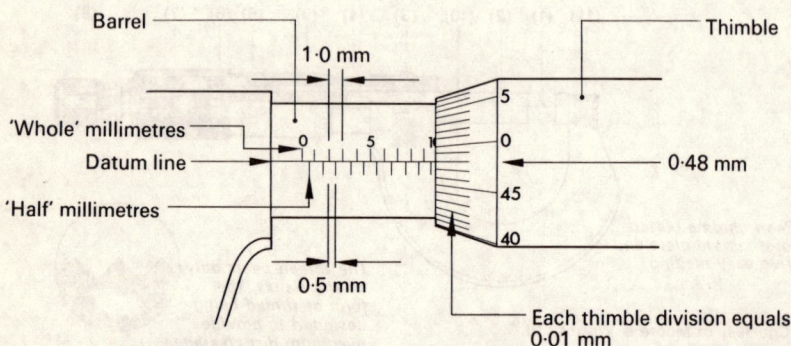

Fig. 7.7   Micrometer scales (metric)

Since the lead of the screw of a standard metric micrometer is 0·5 mm and the barrel divisions are 0·5 mm apart, one revolution of the thimble and therefore the screw moves the thimble along a distance of one barrel division. (The barrel divisions are placed on alternate sides of the datum line for clarity.) Further, since the thimble has 50 equal divisions and one revolution of the thimble equals 0·5 mm, then a movement of one thimble division equals:

$$\frac{0·5}{50} = 0·01 \text{ millimetre (mm)}$$

The micrometer reading equals:
   The largest visible 'whole' millimetre +
   The largest visible 'half' millimetre +
   The thimble division coincident with the datum line.

The reading in Fig. 7.7 is as follows:

9 'whole' millimetres	=	9·00
1 'half' millimetre	=	0·50
48 hundredths of a mm	=	0·48
		9·98 mm

Unless a micrometer is properly looked after it will soon lose its high initial accuracy. To maintain this accuracy the following precautions should be observed:

(*a*)   wipe the anvils and the work to be measured perfectly clean before making a measurement;

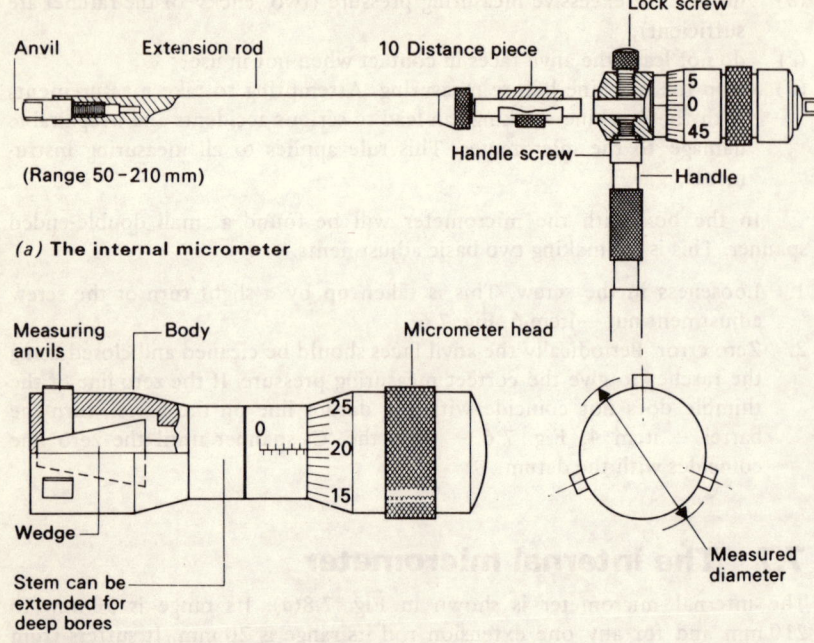

Anvil　　Extension rod　　　　　10 Distance piece

(Range 50–210 mm)

Lock screw

Handle screw

Handle

Measured diameter

(a) **The internal micrometer**

Measuring anvils　　　Body　　　　Micrometer head

Wedge

Stem can be extended for deep bores

(b) **The micrometer cylinder gauge**

Fig. 7.8　Internal measuring devices

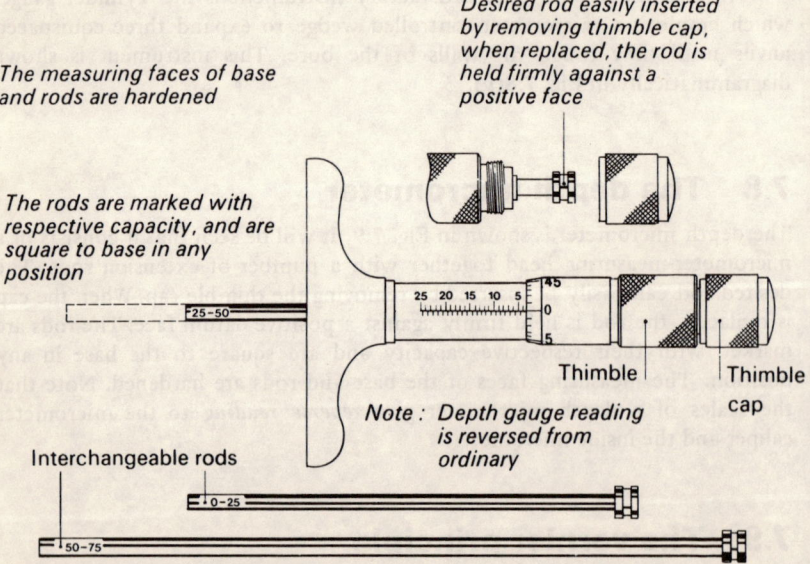

The measuring faces of base and rods are hardened

Desired rod easily inserted by removing thimble cap, when replaced, the rod is held firmly against a positive face

The rods are marked with respective capacity, and are square to base in any position

Interchangeable rods

Thimble

Thimble cap

Note : Depth gauge reading is reversed from ordinary

Fig. 7.9　Micrometer depth gauge

accurately the magnitude of the 'bit'.

In engineering it is not good enough to say that it looks as if the datum is 'roughly half way' between divisions one and two.

Therefore in Fig. 7.10(b) an additional scale has been added in place of the single datum line. This scale is referred to as a *vernier scale* and its purpose is to determine the intermediate readings more accurately.

Figure 7.10(c) shows how the scale is read in practice. To read the vernier, the first gradation of the vernier to be in line (coincident) with a main scale division is added to the main scale gradation immediately to the left of the vernier zero.

(a) the zero reading of the vernier scale lies between 0·2 and 0·3 on the main scale;

(b) the point of coincidence is 2 on the vernier scale and 0·4 on the main scale.

Thus the reading is 0·2 + (2 × 0·01) = 0·22 units.

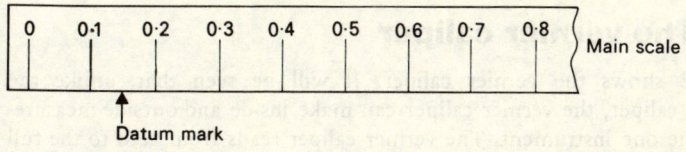

(a) Simple linear scale

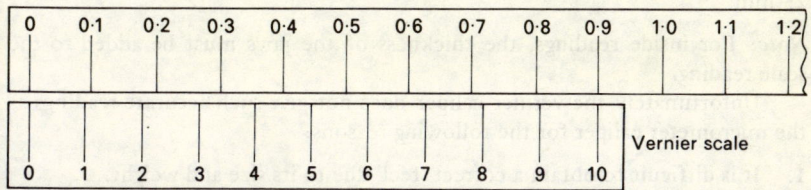

(b) The vernier scale

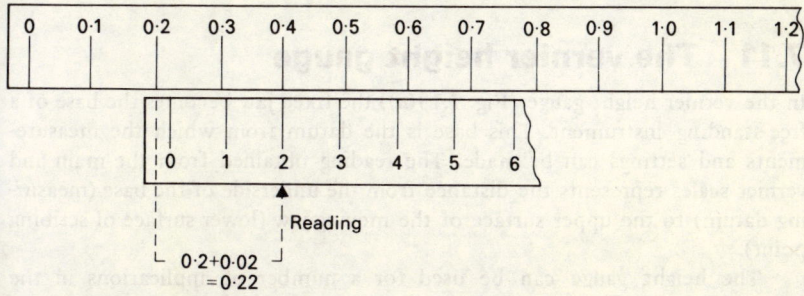

(c) Reading the vernier scale

Fig. 7.10   The vernier scale — principle

Figure 7.11 shows a typical 50-division vernier scale as used on modern metric measuring instruments.

The reading shown equals:

$$32 \text{ 'whole' millimetres} = 32 \cdot 00$$
$$+ 11 \times 0 \cdot 02 \qquad = \underline{ 0 \cdot 22}$$
$$32 \cdot 22 \text{ mm}$$

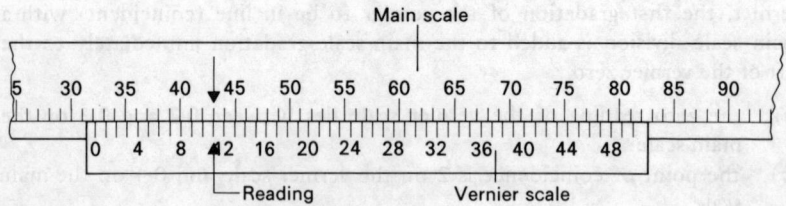

Fig. 7.11   The vernier scale (50 division)

7.10   The vernier caliper

Figure 7.12 shows the vernier caliper. It will be seen that, unlike the micrometer caliper, the vernier caliper can make inside and outside measurements on the one instrument. The vernier caliper reads from zero to the full length of its beam scale, whereas the micrometer only reads over a range of 25 mm.

Note: For inside readings, the thickness of the jaws must be added to the scale reading.

Unfortunately the vernier caliper does not give such accurate readings as the micrometer caliper for the following reasons:

1. It is difficult to obtain a correct 'feel' due to its size and weight.
2. The scales are difficult to read even with the aid of a magnifying glass.
3. The reading accuracy is only 0·02 mm (micrometer is 0·01 mm).

7.11   The vernier height gauge

In the vernier height gauge (Fig. 7.13(*a*)) the fixed jaw becomes the base of a free-standing instrument. This base is the datum from which the measurements and settings can be made. The reading obtained from the main and vernier scales represents the distance from the underside of the base (measuring datum) to the upper surface of the moving jaw (lower surface of scribing point).

The height gauge can be used for a number of applications in the workshop and inspection department. Figure 7.15(*b*) shows the height gauge being used for marking out, whilst Fig. 7.13(*c*) shows how it can be used to check the height of a surface. The accuracy of this latter application is limited

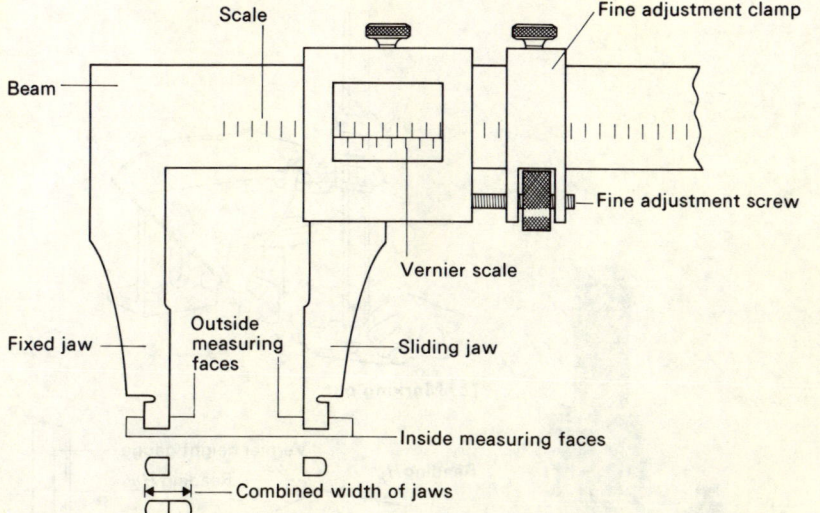

(a) **The vernier caliper**

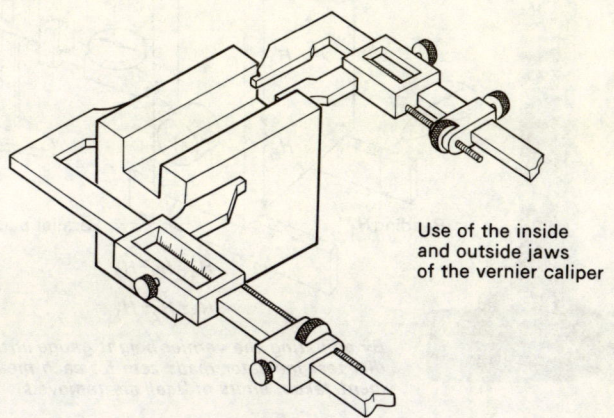

Use of the inside
and outside jaws
of the vernier caliper

(b) **Applications of the vernier caliper**

Fig. 7.12   The vernier caliper

by the skill of the operator. It is very difficult to obtain a satisfactory 'feel'
due to the mass of the instrument and the friction between its base and the
datum surface it is standing upon. As an aid to accuracy the surface being
checked can be smeared with a thin film of engineer's blue. The jaw is slowly
lowered by the fine adjustment screw until it just disturbs the blue. At this
point it can be assumed that contact has been made.

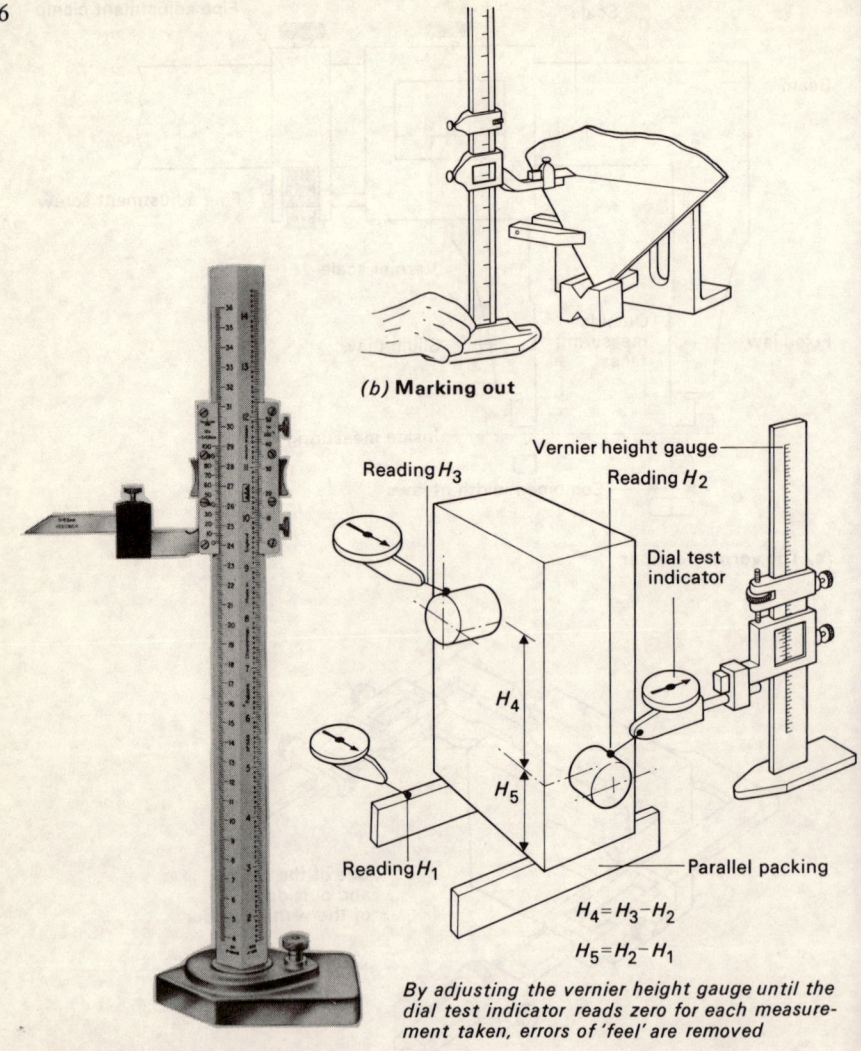

(b) Marking out

Reading H_3

Vernier height gauge

Reading H_2

Dial test indicator

H_4

H_5

Reading H_1

Parallel packing

$H_4 = H_3 - H_2$

$H_5 = H_2 - H_1$

By adjusting the vernier height gauge until the dial test indicator reads zero for each measurement taken, errors of 'feel' are removed

(a) The vernier height gauge   (c)   Measuring the height of a surface

Fig. 7.13   The vernier height gauge

The dial test indicator (section 7.15) can also be used to ensure a constant measuring pressure as shown in Fig. 7.13(c). In this case, the readings obtained from the scales are no longer 'absolute' (correct distances from the datum surface), but are 'difference' readings as shown in the figure. The dial gauge is used in this example as a 'fiducial indicator'; that is, it removes errors due to 'feel' by ensuring a constant measuring pressure.

7.12   The vernier depth gauge

The vernier principle can also be applied to the depth gauge and an example is shown in Fig. 7.14(a). As with the micrometer depth gauge, care has to be taken when reading the scales as these are reversed to those normally found

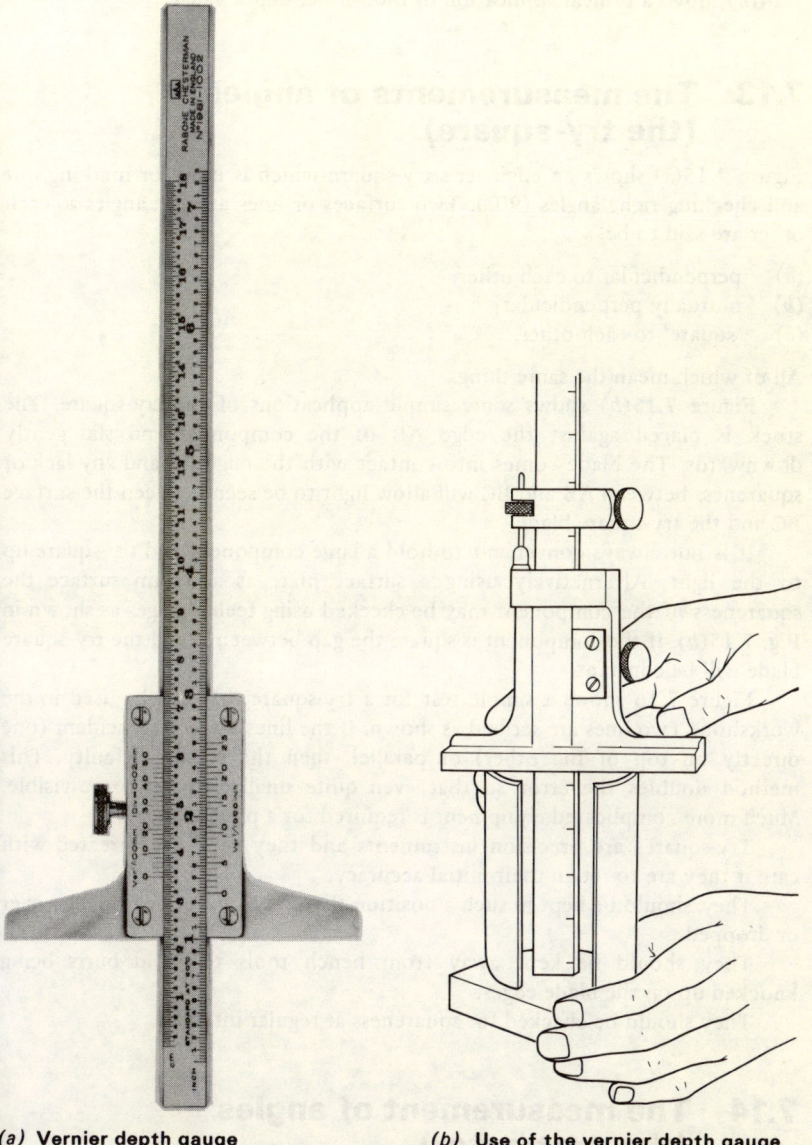

(a) **Vernier depth gauge**   *(b)* **Use of the vernier depth gauge**

Fig. 7.14   The vernier depth gauge

on vernier calipers and vernier height gauges. Again it is difficult to obtain a satisfactory feel with this instrument as it is easily lifted off the datum surface if the adjusting screw is used too vigorously. The vernier depth gauge has the advantage over the micrometer depth gauge of having a large measuring range without having to resort to the use of extension rods. Figure 7.14(*b*) shows a typical application of the vernier depth gauge.

7.13 The measurements of angles (the try-square)

Figure 7.15(*a*) shows an engineer's try-square which is used for marking out and checking right angles (90°). Two surfaces or lines at right angles to each other are said to be:

(*a*) perpendicular to each other;
(*b*) mutually perpendicular;
(*c*) 'square' to each other.

All of which mean the same thing.

Figure 7.15(*b*) shows some simple applications of the try-square. The stock is placed against the edge AB of the component and slid gently downwards. The blade comes into contact with the edge BC and any lack of squareness between AB and BC will allow light to be seen between the surface BC and the try-square blade.

It is not always convenient to hold a large component and try-square up to the light. Alternatively using a surface plate as a datum surface the squareness of the component may be checked using feeler gauges as shown in Fig. 7.15(*b*). If the component is square the gap between it and the try-square blade will be constant.

Figure 7.16 shows a simple test for a try-square that can be used in the workshop. Two lines are scribed as shown. If the lines are not coincident (one directly on top of the other) or parallel, then the square is faulty. This method doubles the error so that even quite small errors become visible. Much more complicated equipment is required for a precise check.

Try-squares are precision instruments and they should be treated with care if they are to retain their initial accuracy.

They should be kept in such a position that they cannot be knocked over or dropped.

They should be kept away from bench tools to avoid burrs being knocked up on the blade edges.

They should be checked for squareness at regular intervals.

7.14 The measurement of angles (the protractor)

Figure 7.17(*a*) shows a simple protractor. It is used for measuring angles other

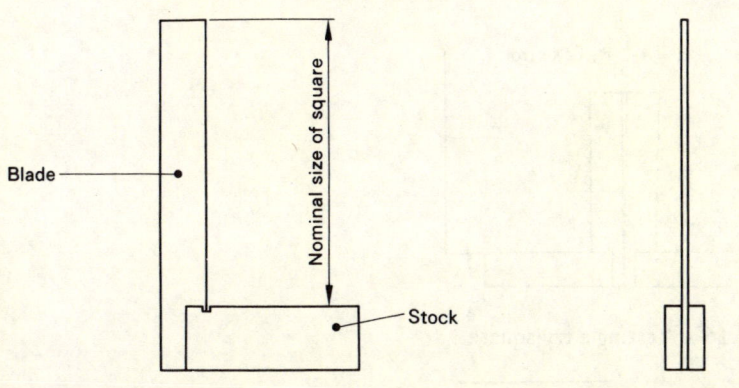

Blade

Nominal size of square

Stock

(a) **The try-square**

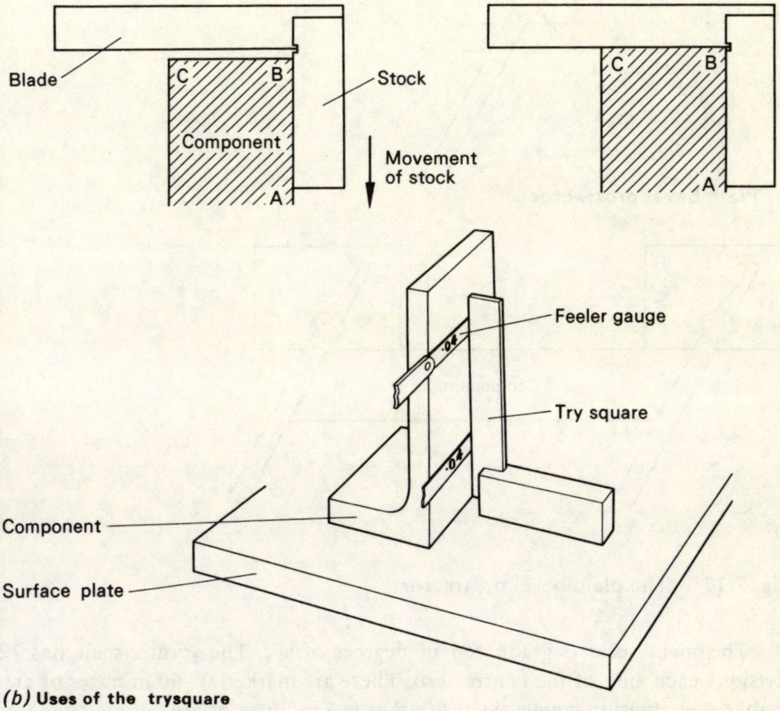

Blade

C B

Component

A

Stock

Movement
of stock

C B

A

Feeler gauge

Try square

Component

Surface plate

(b) **Uses of the trysquare**

Fig. 7.15   The try-square

than 90° but has only a limited accuracy (±½°). Figure 7.17(*b*) shows a typical application.

The simple protractor shown in Fig. 7.17 can be extended in its accuracy by the application of a circular vernier scale as shown in Fig. 7.18.

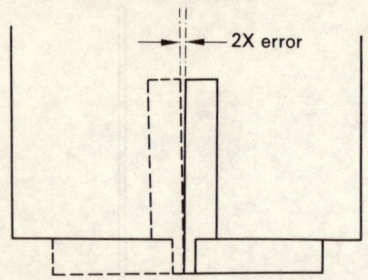

Fig. 7.16 Testing a try-square

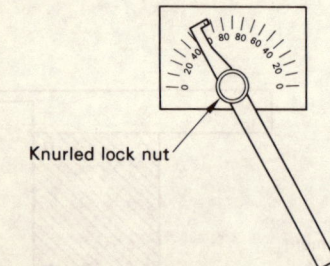

(a) **Plain bevel protractor**

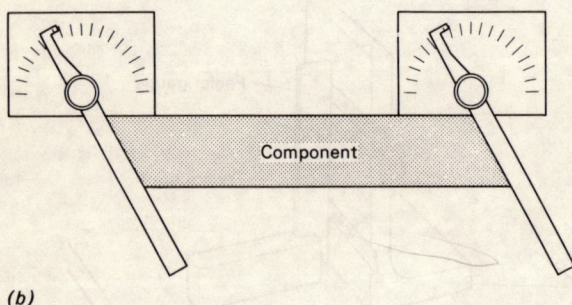

(b)

Fig. 7.17 The plain bevel protractor

The main scale is graduated in degrees of arc. The vernier scale has 12 divisions each side of the centre zero. These are marked 0–60 minutes of arc, so that each division equals $\frac{1}{12}$ of 60, that is 5 minutes of arc.

These 12 divisions occupy the same space as 23 degrees on the main scale. Therefore, each division of the vernier is equal to:

$\frac{1}{12}$ of $23°$ or $1\frac{11}{12}°$

Since two divisions on the main scale equals 2 degrees of arc, the difference between two divisions on the main scale and one division on the vernier scale is:

$2^{\circ} - 1^{11}/_{12}^{\circ} = ^{1}/_{12}^{\circ}$ or 5 minutes of arc.

Thus the reading of the vernier bevel protractor equals:

(a)   the largest 'whole' degree on the main scale indicated by the vernier zero division,

     plus

(b)   the reading on the vernier scale in line with a main scale division. The reading in Fig. 7.18 is:

Vernier zero slightly beyond 17°

$$
\begin{array}{ll}
\text{17 'whole' degrees} & = 17^{\circ} \\
\text{+ Vernier 25 mark in} & \\
\text{line with main scale} & = \underline{00\ 25'} \\
& \quad 17^{\circ}25'
\end{array}
$$

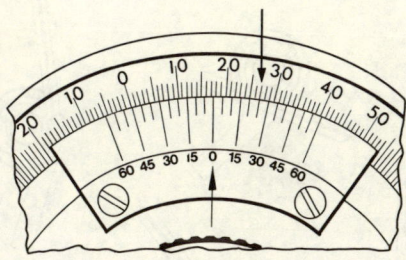

Fig. 7.18   Vernier protractor scales

Providing the vernier protractor has been manufactured to the specifications of BS 1685, it will have an accuracy within 5 minutes of arc. However, as with all precision measuring instruments, its actual performance will largely depend upon the skill of the user.

The scales should be satin chrome finished to prevent corrosion and to improve the ease of reading by reducing glare when working in artificial light.

The ends of the blades are ground off at 45° and 60° for working in corners and to act directly as mitre gauges. The blades and stock should be hardened and tempered to reduce wear. Figure 7.19 shows a typical vernier protractor.

7.15   The dial test indicator

Essentially, the **dial test indicator** (DTI) measures the displacement of its plunger or stylus on a circular dial by means of a rotating pointer. It is from the appearance of this dial that the dial test indicator is familiarly known as a 'clock' in the workshop. Figure 7.20 shows the two most popular types of this instrument normally used.

182

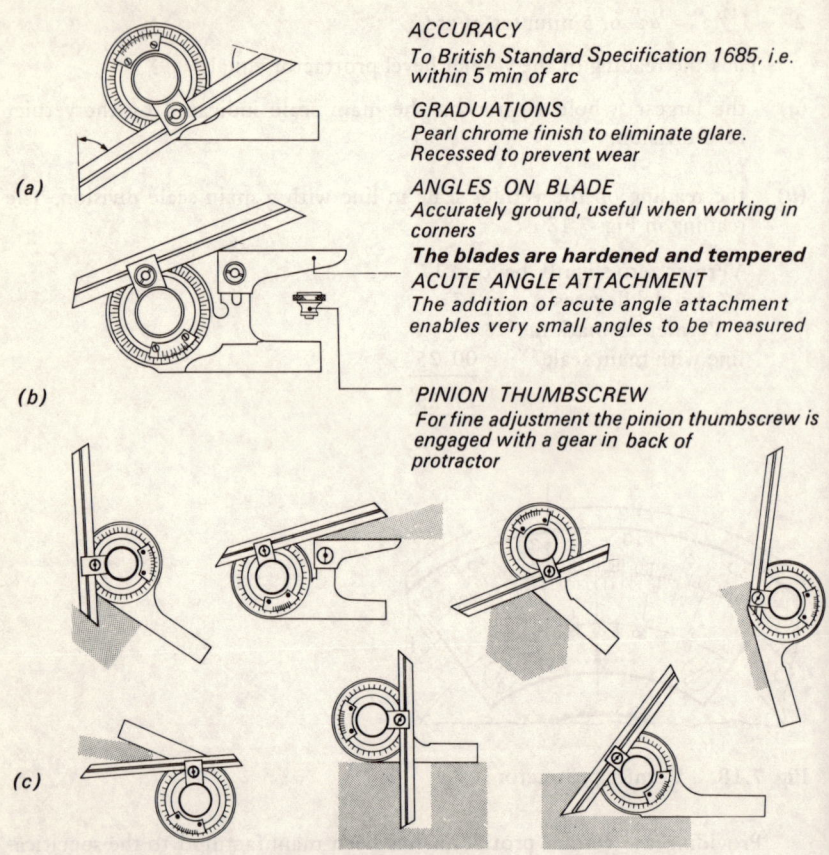

ACCURACY
To British Standard Specification 1685, i.e.
within 5 min of arc

GRADUATIONS
Pearl chrome finish to eliminate glare.
Recessed to prevent wear

ANGLES ON BLADE
Accurately ground, useful when working in
corners
The blades are hardened and tempered
ACUTE ANGLE ATTACHMENT
The addition of acute angle attachment
enables very small angles to be measured

PINION THUMBSCREW
For fine adjustment the pinion thumbscrew is
engaged with a gear in back of
protractor

Fig. 7.19 The vernier protractor

Figure 7.20(*a*) shows a **plunger type** dial test indicator. This type relies upon a rack and pinion followed by a gear-train to magnify the movement of the plunger to the main pointer. This type of instrument has a long plunger movement and is fitted with a secondary scale and pointer for indicating the number of complete revolutions made by the main pointer. Various magnifications and dial markings are available.

Figure 7.20(*b*) shows a **lever type** dial test indicator. This type relies upon a lever and scroll system of magnification. It has only a limited range of stylus movement; little more than one revolution of the pointer. It is more compact than the plunger type and is very popular for both inspection and machine setting.

The dial gauge is widely used in the machine shop for machine setting and examples are given in the later chapters of this book. Figure 7.21 shows the mechanisms of the plunger and lever type dial gauges.

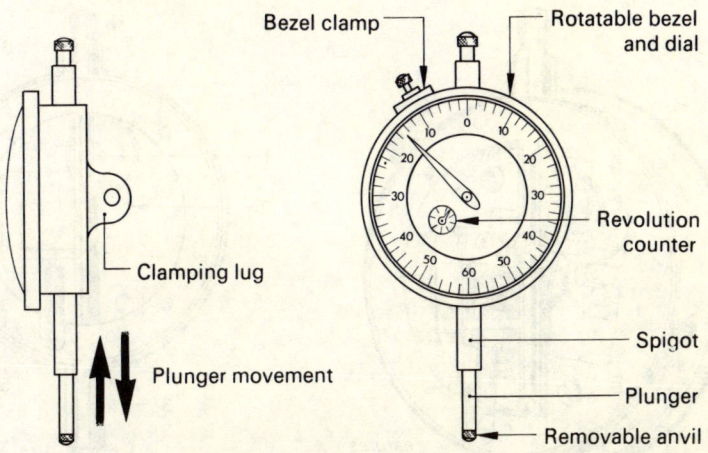

(a) **Plunger type**

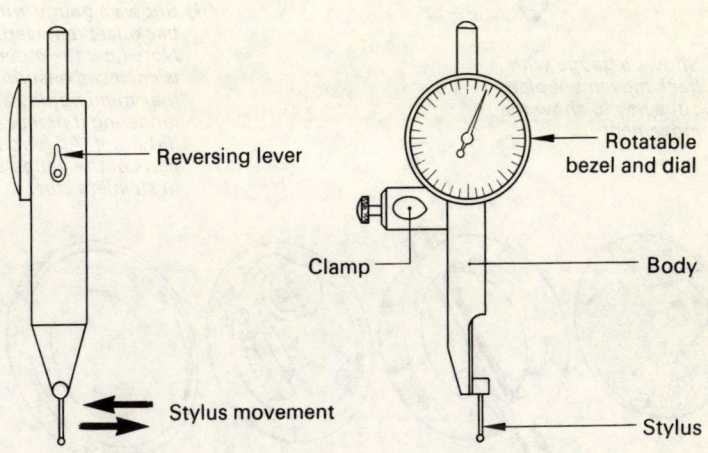

(b) **Lever type**

Fig. 7.20   Types of dial indicator

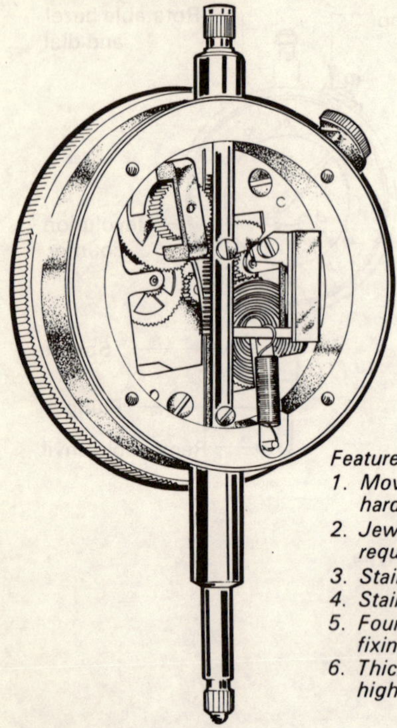

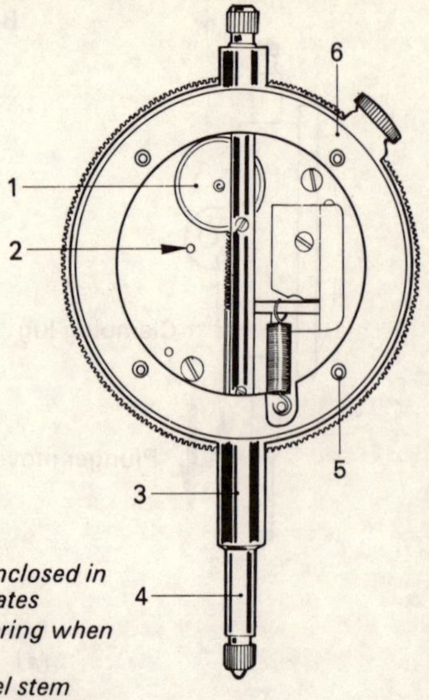

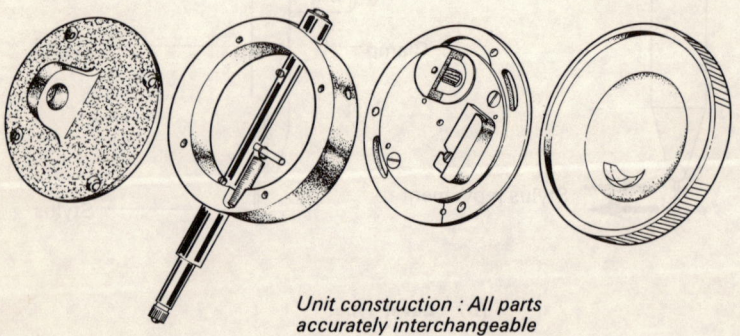

Features :
1. Movement enclosed in hard brass plates
2. Jewelled bearing when required
3. Stainless steel stem
4. Stainless steel spindle
5. Four-hole back plate fixing
6. Thick section case in high duty alloy

(i) Shows a gauge with back movement plate cut away to show the movement.

(ii) Shows a gauge with the backplate removed. Note how the movement is enclosed ensuring maximum rigidity and rendering it practically dirt proof. All ferrous parts of the gauge are in stainless steel.

Unit construction : All parts accurately interchangeable

(a) Mechanism of plunger type dial test indicator

Fig. 7.21 Dial test indicator mechanisms

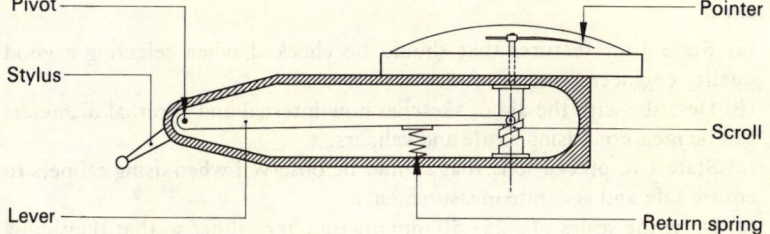

Pivot ——————

Pointer

185

Stylus ——————

Scroll

Lever ——————

Return spring

(b) **Mechanism of the lever type**

Fig. 7.21 *(continued)*

Problems

Section A

1. The reading of the metric micrometer scales shown in Fig. 7.22 is: (*a*) 14·17 mm;   (*b*) 14·67 mm;   (*c*) 14·72 mm;   (*d*) 15·17 mm.

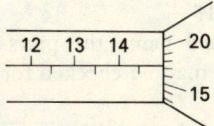

Fig. 7.22

2. The most accurate standard of length normally available in the workshop is the: (*a*) rule;   (*b*) micrometer caliper;   (*c*) vernier caliper; (*d*) slip (block) gauge.

3. To avoid reading errors due to parallax an engineer's rule should be (*a*) as thin as possible;   (*b*) as thick as possible;   (*c*) satin chrome finished;   (*d*) engine engraved.

4. When using a micrometer a constant measuring pressure can be obtained by using the: (*a*) thimble;   (*b*) spindle;   (*c*) ratchet;   (*d*) barrel.

5. The vernier scale shown in Fig. 7.23 has a reading accuracy of ¼₀ mm. What is the reading shown? (*a*) 43·00 mm;   (*b*) 43·15 mm;   (*c*) 44·20 mm;   (*d*) 49·15 mm.

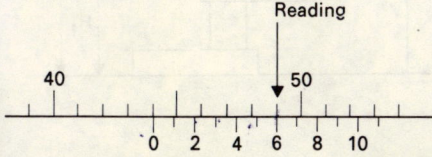

Reading

40   50

0 2 4 6 8 10

Fig. 7.23

6. (a) State four features that should be checked when selecting a good quality engineer's rule.

    (b) Describe with the aid of sketches how internal and external diameters can be measured using a rule and calipers.

    (c) State five precautions that should be observed when using calipers to ensure safe and accurate measurements.

7. (a) Draw the scales of a 25—50 mm micrometer caliper so that they show a reading of 43·74 mm.

    (b) Draw the scales of a 50 division vernier caliper so that they show a reading of 150·28 mm.

    (c) Draw the scales of a 0—25 mm depth micrometer so that they show a reading of 15·63 mm.

8. (a) Describe in detail the principle of operation of the micrometer.

    (b) Explain what precautions should be taken to maintain the high initial accuracy of a micrometer caliper.

    (c) Explain, with the aid of sketches, how a micrometer caliper should be adjusted to remove initial zero error.

9. (a) Sketch a typical vernier protractor and draw, in good proportion, an enlarged view of the scales to show a reading of 13° 15′.

    (b) Sketch a good quality engineer's try-square and name the parts. Explain, with the aid of sketches, how a try-square may be checked for perpendicularity.

10. (a) With reference to Fig. 7.24, calculate the height of the step (H_3) if the reading of $H_1 = 50·82$ mm and the reading of $H_2 = 96·24$ mm.

    (b) Explain the purpose of the dial test indicator in this application of the vernier height gauge and the need for a datum surface.

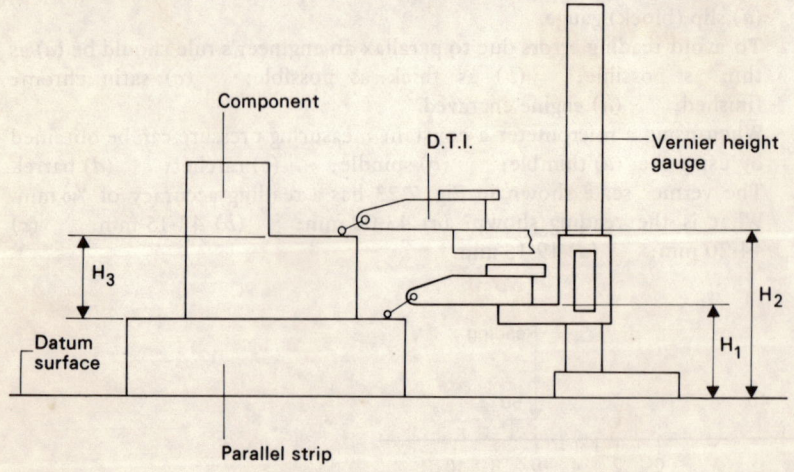

Fig. 7.24   Checking milled component (vernier height gauge)

Chapter 8

Marking out

8.1 The need for marking out

Except for very simple machining operations such as 'squaring up' a blank preparatory to marking it out, it is usually advisable to mark out a component before commencing to work on it by hand or machine. There are three reasons for marking out.

1. To provide guide lines which are worked to, and provide the only control to the size and shape of the finished component.
2. To indicate the outline of a component to the machinist as an aid to setting up and roughing out. The final dimensional control would come, in this instance, from the use of precision measuring instruments in conjunction with the micrometer dials on the machine.
3. To ensure that adequate machining allowances have been left on forgings and castings; that webs, flanges and cores have not been incorrectly positioned or displaced during pouring; that holes will be centrally positioned in their bosses after machining. (See Fig. 8.1.)

In order that scribed lines will show up clearly the metal surface is usually coated in a contrasting colour.

Whitewash

This is usually applied to rough forgings and castings which have a dark,

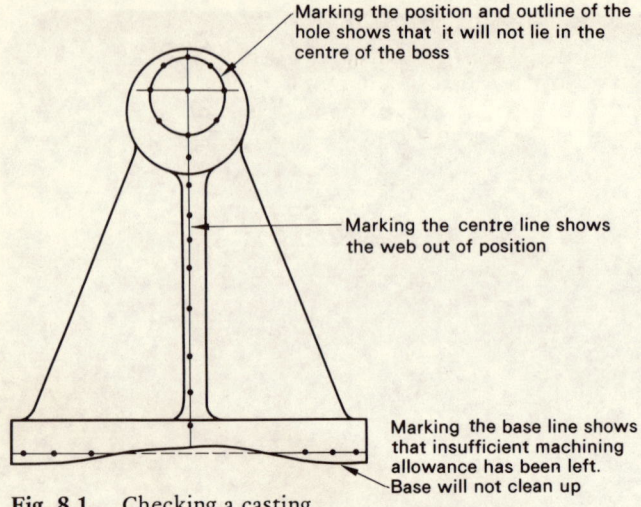

Marking the position and outline of the hole shows that it will not lie in the centre of the boss

Marking the centre line shows the web out of position

Marking the base line shows that insufficient machining allowance has been left. Base will not clean up

Fig. 8.1   Checking a casting

heavily oxidised surface (black scale).

Cellulose lacquer

This is made in a variety of colours and can be applied to any clean, bright surface.

Copper sulphate solution

A solution of copper sulphate in water plus a few drops of nitric acid deposits a thin film of copper on a clean plain carbon steel and low alloy steel surface. This method of surface preparation has been used for many years, but compared with a lacquer it has certain disadvantages:

(*a*)   if it comes into contact with measuring and marking out instruments it corrodes them;

(*b*)   it can only be applied to plain carbon and low alloy steels;

(*c*)   it is difficult to remove the copper coating once it has been deposited.

Note: in the interests of safety all sharp edges and burrs should be removed from sawn blanks before marking out. Further, burrs may disguise the true edge of the blank and cause inaccuracies.

8.2   The scribed line

For accurate work a clean, fine line is required. The correct way to produce such a line is shown in Fig. 8.2(*a*). It is assumed that the scriber point has been correctly sharpened (see section 8.4). Then:

(*a*)   the point, and not the side of the scriber is guided by the rule, straight edge or template to produce the line required (Fig. 8.2(*b*));

(*b*) the scribing point trails the direction of movement so that it does not 189
    'dig' into the surface being marked out.

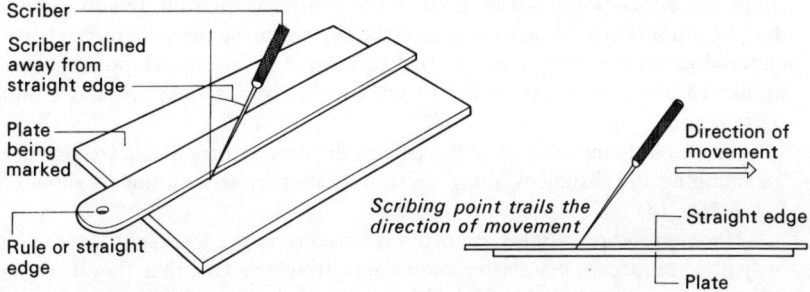

Scriber

Scriber inclined
away from
straight edge

Plate
being
marked

Rule or straight
edge

Direction of
movement

*Scribing point trails the
direction of movement*

Straight edge

Plate

(a) **Use of the scriber**

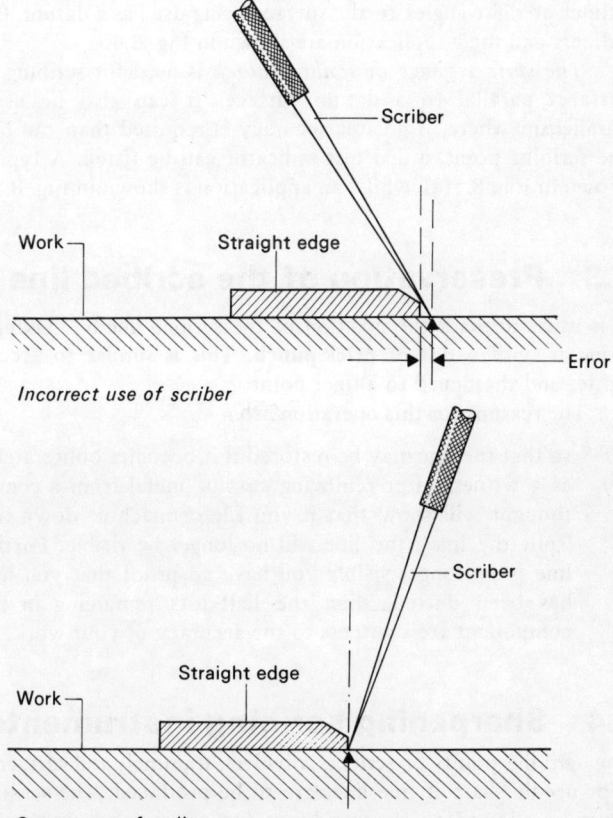

Work

Straight edge

Scriber

Error

Incorrect use of scriber

Work

Straight edge

Scriber

Correct use of scriber

(b) **Possible error when using a scriber**

Fig. 8.2   The scribed line

For scribing circular lines up to about 150 mm diameter *dividers* are used. Above this diameter *trammels* are used. Figure 8.3 shows examples of these instruments. The points are symmetrical, either point can scribe the circle and either point can be used as the centre of rotation. In soft material the pivot point can be depressed into the surface being marked, but in harder materials a small indentation has first to be made using a prick punch. This is similar to a centre punch (used to start drills) but is finer and has a more acute point.

As well as being used to scribe circles, dividers and trammels are also used for stepping off distances along or around another scribed line as shown in Fig. 8.3(*c*), (*d*).

Hermaphrodite calipers are often referred to in the workshop as 'odd-leg' or 'jenny' calipers. They derive their name from the fact that they have one caliper leg and one divider leg. They are used for scribing a line that is equi-distant from an edge or surface throughout its length. To avoid positional error in the scribed line, care must be taken to keep the hermaphrodite caliper at right angles to the surface being used as a datum. Examples of such calipers and their application are shown in Fig. 8.4.

The *surface gauge* or *scribing block* is used for scribing lines and setting surfaces parallel to a datum surface. It can also be used for checking parallelism where, if greater accuracy is required than can be obtained from the scribing point, a dial test indicator can be fitted. A typical instrument is shown in Fig. 8.5(*a*), whilst an application is shown in Fig. 8.5(*b*).

8.3 Preservation of the scribed line

It is usual to mark the position of the scribed line by making a series of dots along it with a **dot** or **prick-punch**. This is similar to a centre-punch, but lighter and sharpened to a finer point.

The reasons for this operation are:

(*a*)   so that the line may be restored if it becomes obliterated;

(*b*)   as a witness after removing surplus metal from a component. A little thought will show that if you file or machine down to a line so as to 'split the line', the line will no longer be visible. Further, because the line is no longer visible you have no proof that you have split it. If it has been dotted, then the half-dots remaining in the edge of the component are a witness to the accuracy of your work.

8.4 Sharpening scribing instruments

The scribing points of scribers, dividers, trammels and surface gauges must be kept needle sharp if fine lines are to be produced. The scribing point should never be allowed to become blunt for, apart from the inaccurate line produced, too much metal must be removed to restore the point and the temptation to use a grinding wheel becomes greater. A fine oilstone or 'slip'

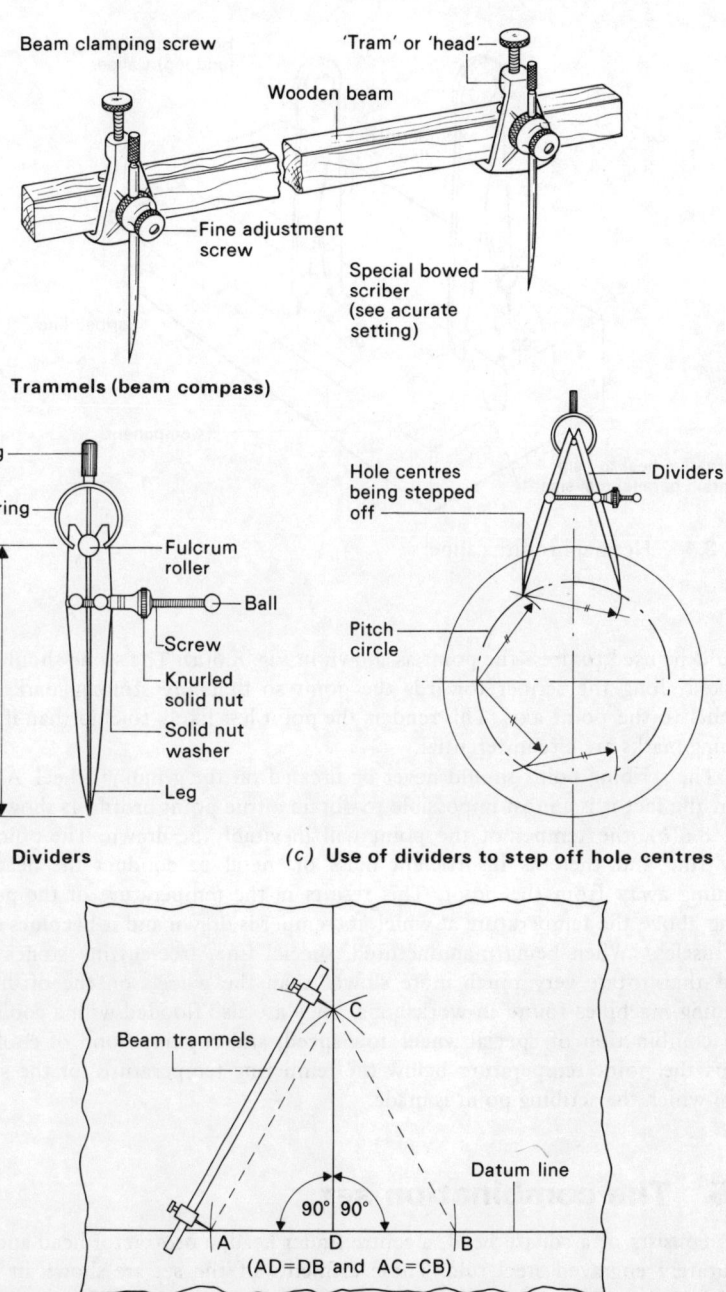

(a) Trammels (beam compass)

(b) Dividers

(c) Use of dividers to step off hole centres

Beam clamping screw

'Tram' or 'head'

Wooden beam

Fine adjustment screw

Special bowed scriber (see acurate setting)

Peg

Spring

Fulcrum roller

Ball

Screw

Knurled solid nut

Solid nut washer

Leg

Nominal size

Hole centres being stepped off

Dividers

Pitch circle

Beam trammels

Datum line

90° 90°

A D B

(AD=DB and AC=CB)

C

(d) Use of trammels to step off distances (to construct a right angle)

Fig. 8.3 Use of dividers and trammels

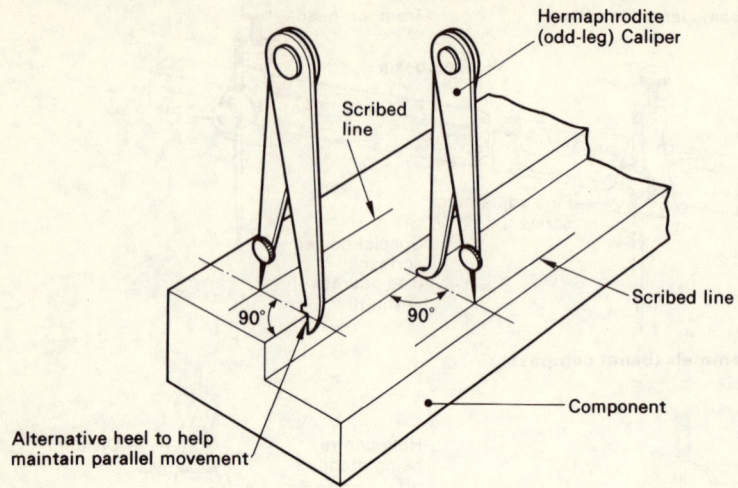

Fig. 8.4   Hermaphrodite calipers

should be used to dress the point as shown in Fig. 8.6(*a*). The stone should be stroked along the scriber towards the point so that any stoning marks are parallel to the point axis. This renders the point less likely to chip than if the stoning marks are circumferential.

The scribing point should never be dressed on the grinding wheel. Apart from the fact it is almost impossible to obtain a true point profile as shown in Fig. 8.6(*b*), the temper of the point will inevitably be drawn. The point is very fine and there is insufficient mass of metal to conduct the heat of grinding away from the point. This results in the temperature of the point rising above the temperature at which its temper is drawn and it becomes soft and useless. When being manufactured, special fine, free-cutting stones are used that rotate very much more slowly than the wheels on the off-hand grinding machines found in workshops. They are also flooded with a coolant. The combination of special wheel, low speed, and copious flood of coolant keeps the point temperature below the tempering temperature for the steel from which the scribing point is made.

8.5   The combination set

This consists of a square head, a centre finder head, a protractor head and an accurately engraved steel rule. These elements of the set are shown in Fig. 8.7(*a*). Only one head at a time is mounted on the rule when it is being used, the three heads being mounted on the rule to prevent loss when not in use.

Examples of the use of the square head and the centre finder head are shown in Fig. 8.7(*b*).

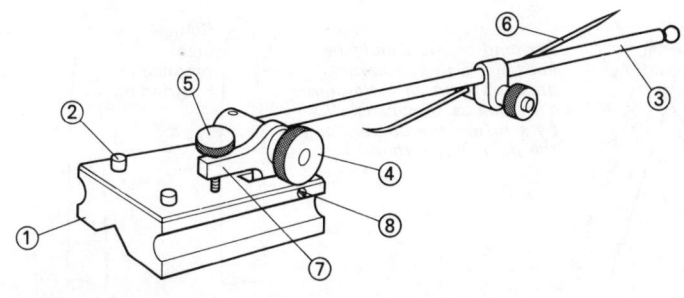

(1) Base
(3) Mast
(5) Fine adjustment
(7) Rocker arm

(2) Edge pins
(4) Clamp nut
(6) Scriber
(8) Fulcrum

(a) Surface gauge

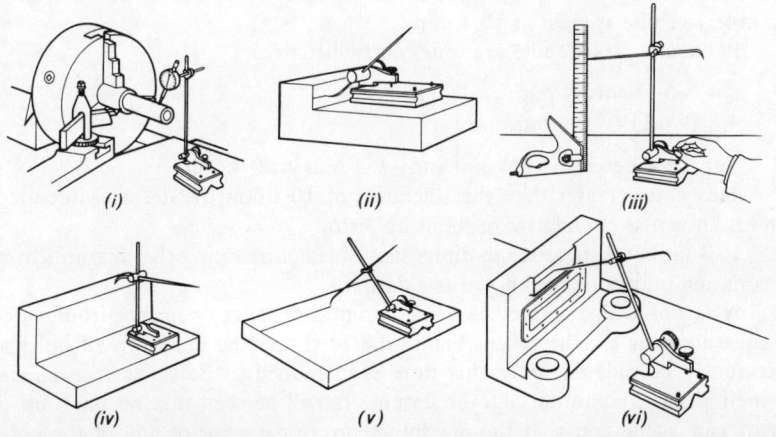

(i) As a dial gauge stand.

(ii) Scribing a line parallel to a surface.

(iii) Setting to a combination square rule.

(iv) Checking a surface for parallelism.

(v) Using the setting pins to scribe. parallel to an edge.

(vi) Marking out a casting.

(b) Typical applications

(Courtesy of Moore & Wright Ltd)

Fig. 8.5   The universal surface gauge

8.6   Datum lines and datum edges

Figure 8.8(a) shows a row of holes whose centre are fixed by 'chain' dimensioning. That is, each hole is dimensioned from the centre of the hole

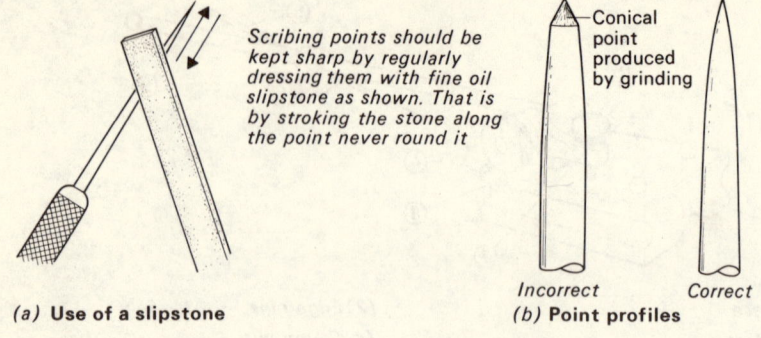

Scribing points should be kept sharp by regularly dressing them with fine oil slipstone as shown. That is by stroking the stone along the point never round it

Conical point produced by grinding

Incorrect   Correct

(a) **Use of a slipstone**   *(b)* **Point profiles**

Fig. 8.6   Sharpening scribing points

next to it. In the example shown the total positional error of hole 5 relative to hole 1 can be as great as ±0·4 mm.

Between holes 1 and 5 are four centre distances

$$4 \times (+0\cdot1) = +0\cdot4 \text{ mm}$$
$$4 \times (-0\cdot1) = -0\cdot4 \text{ mm}$$

The error between +0·4 mm and −0·4 mm is ±0·4 mm.

This is far greater than the tolerance of ±0·1 mm the designer intended, and is known as *cumulative* or 'built-up' error.

It is far better practice to dimension hole centres and other features from a common point or edge known as a *datum.*

A *datum* can be defined as a fixed point, line, edge or surface from which a measurement can be taken. Figure 8.8(*b*) shows the same row of holes as previously considered, only this time each individual hole has been dimensioned from a common edge or datum. It will be seen that no build up of error can occur and that the maximum positional error of any of the holes can only be ±0·1 mm.

The dimensions from the datum to the hole centre or feature (e.g. 75 ± 0·1 mm for hole 5) is known as an *ordinate.* In practice two such dimensions are required to fix the position of a point on a flat surface. These two ordinates are known as *coordinates.* There are two systems of coordinates in common use.

1. Rectangular coordinates;   The point is positioned by a pair of ordinates (coordinates) lying at right angles to each other and at right angles to the axes or datum edges from which they are taken. This system requires the preparation of a pair of mutually perpendicular datum edges before marking out can commence. It will be seen that this is the same system as is used to position a point on a graph. Figure 8.9(*a*) shows an example of the centres of a hole dimensioned by means of rectangular coordinates.

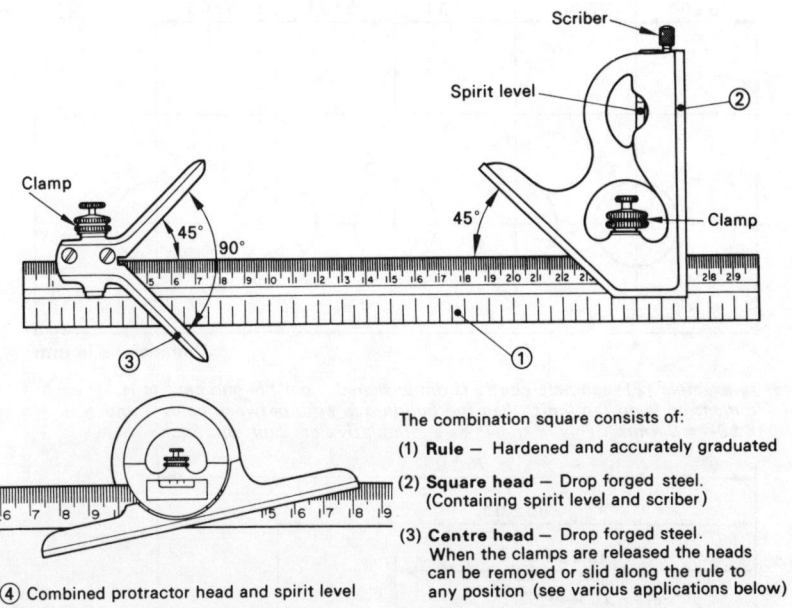

The combination square consists of:

(1) **Rule** — Hardened and accurately graduated

(2) **Square head** — Drop forged steel. (Containing spirit level and scriber)

(3) **Centre head** — Drop forged steel. When the clamps are released the heads can be removed or slid along the rule to any position (see various applications below)

(4) **Protractor head** — With the rule this forms a plain bevel protractor

④ Combined protractor head and spirit level

(When not used with the rule, this head forms a simple clinometer)

(a) **The combination set**

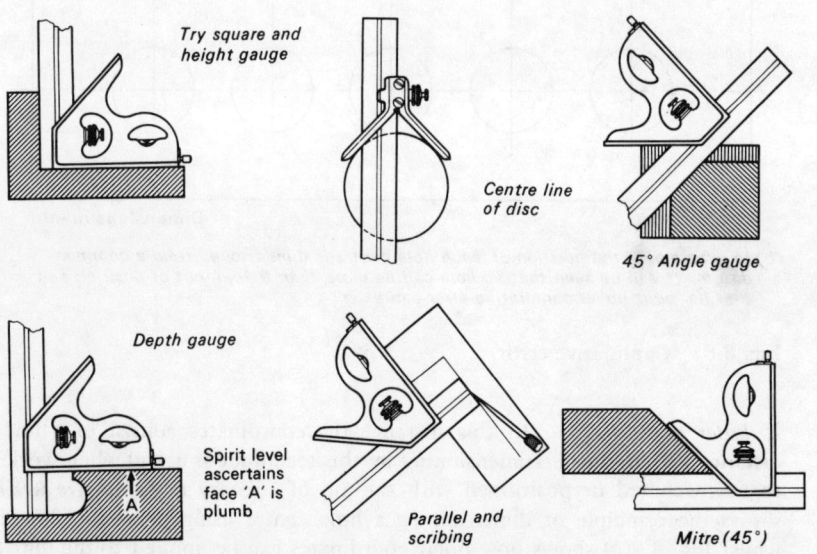

(b) **Uses of the combination set**

(Courtesy of Moore and Wright Ltd)

Fig. 8.7 The combination set

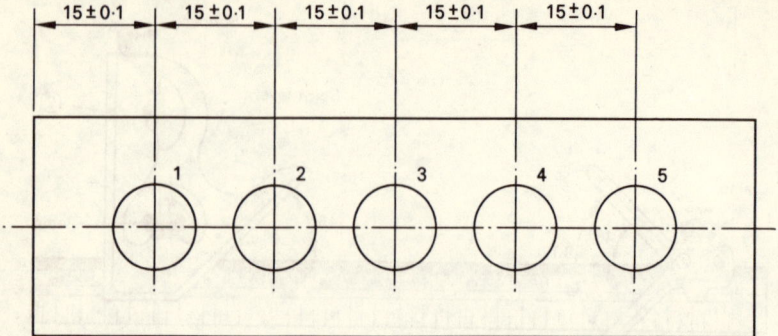

(a) In example (a) each hole centre is dimensioned from the one next to it. If each dimension is on top limit, then the build up in error between holes 1 and 5 is $4 \pm 0.1 = 0.4$mm. This is known as a cumulative or 'built up' error

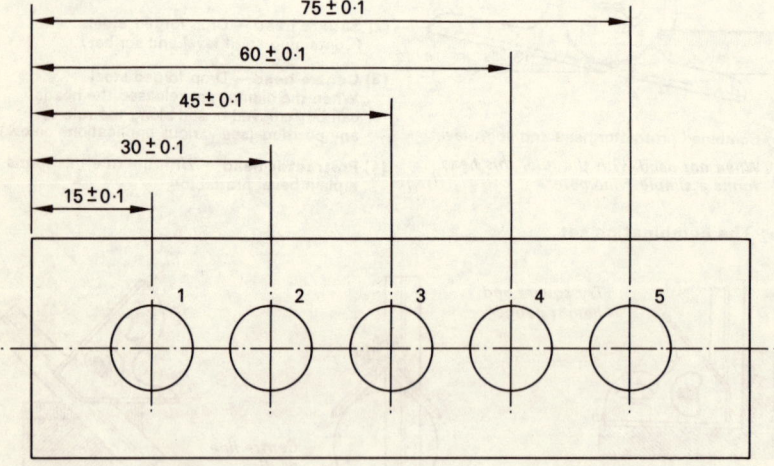

(b) In example (b) the position of each hole centre is dimensioned from a common datum. It will be seen that no hole can be more than 0.1mm out of position and that no 'built up' or cumulative error can exist

Fig. 8.8 Cumulative error

2. *Polar coordinates:* In this instance the coordinates consist of a linear distance and an angle. Dimensioning by this technique is useful when work is to be machined or positioned with the aid of a rotary table. Figure 8.9(*b*) shows the principle of dimensioning a hole centre using polar coordinates, whilst Fig. 8.9(*c*) shows how polar coordinates can be applied to the dimensioning of holes on a pitch circle.

It will be seen that polar coordinates use a point datum instead of edge datums. In practice, this point datum would have to be established from edge

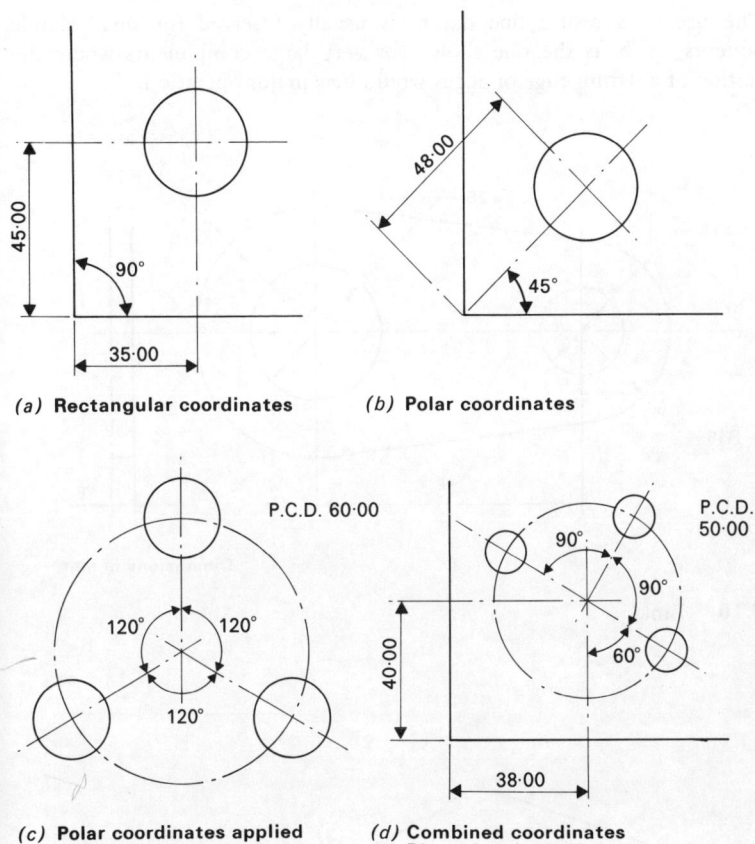

(a) **Rectangular coordinates**

(b) **Polar coordinates**

(c) **Polar coordinates applied to holes on a pitch circle**

(Note: P.C.D. = Pitch circle diameter)

(d) **Combined coordinates**
Dimensions in mm

Fig. 8.9 Coordinates

datums by means of rectangular coordinates as shown in Fig. 8.9(d), thus becoming a secondary datum.

In the subsequent sections of this chapter various examples of marking out techniques are shown. It will be seen that the datum chosen varies with the type of component being considered.

8.7 Marking out (on the bench)

Figure 8.10 shows a simple link. It can be marked out by two methods:

(a) use of a centre line datum;
(b) use of a datum edge.

The use of a centre line datum is usually reserved for small simple components, such as the one shown, or very large components where the production of a datum edge or edges would be a major operation.

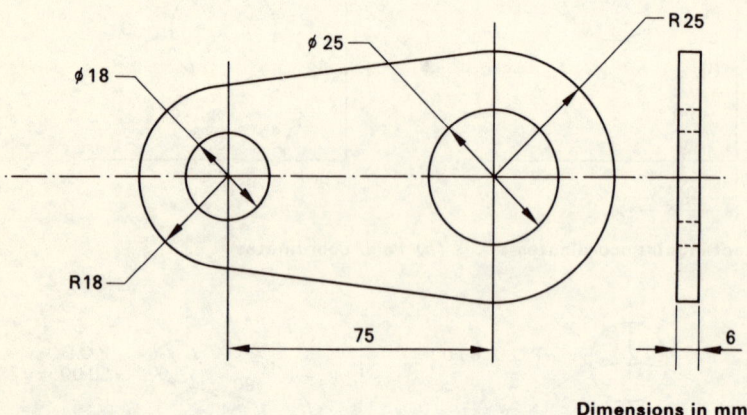

Dimensions in mm

Fig. 8.10 Link

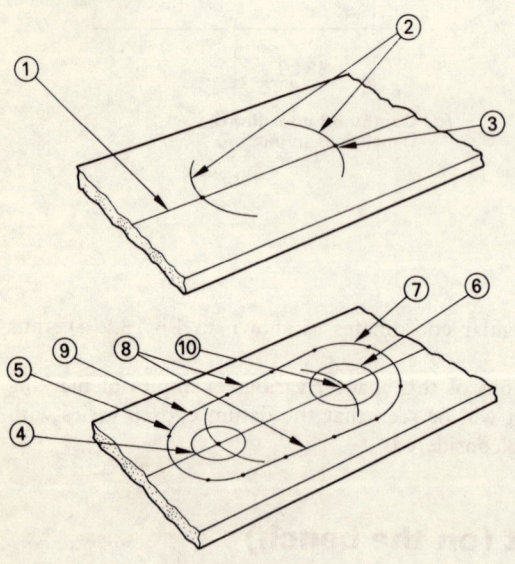

Fig. 8.11 Method of marking out from a centre line datum

Assuming a piece of clean, bright mild steel plate of suitable thickness is available and that it has had one surface treated with lacquer, marking out would be as follows (see Fig. 8.11).

1. Scribe the centre line using a rule and scriber.
2. Step off the hole centres at 75 mm using a rule and dividers.
3. Lightly prick punch the hole centres using a prick punch and very light hammer.
4. Scribe in the 18 mm diameter hole using a rule and dividers.
5. Using the same centre, scribe in the 18 mm radius using a rule and dividers.
6. Scribe the 25 mm diameter hole using a rule and dividers.
7. Using the same centre, scribe in the 25 mm radius using a rule and dividers.
8. Join the 18 mm and 25 mm radii with tangential lines. Use a rule as a straight edge to guide the scriber.
9. To preserve the scribed lines, prick punch them lightly.
10. Enlarge the hole centre marks with a centre punch reading for drilling.

Method 2 (edge datum)

Assuming a piece of clean, bright mild steel plate of suitable thickness is available, that one edge has been machined or filed straight and true, and that it has had one surface treated with marking out lacquer, the operation sequence for marking out would be as follows (see Fig. 8.12).

1. Scribe a centre line parallel to the datum edge with hermaphrodite calipers.
2. Scribe the position of the centre line of the 18 mm diameter hole at right angles to the centre line datum using a try-square and scriber.
3. Mark off the position of the second centre line using a rule and scriber.
4. Scribe the position of the centre line of the 25 mm diameter hole at right angles to the centre line datum using a try-square and scriber at the mark made in (3).
5. Lightly prick punch the hole centres.
6. Having established the hole and radii centres, the profile of the link is marked out as in operations (4) to (10) of method 1.

8.8 Marking out (surface table)

Most components require to be marked out using rectangular coordinates from mutually perpendicular datum edges. Note that in some parts of the country datum lines are called 'foundation lines' and datum edges are called 'service edges'.

When marking out using rectangular coordinates the most accurate method of working is from a datum surface. Such a surface is provided by a marking out table which is made of cast iron and whose upper surface is

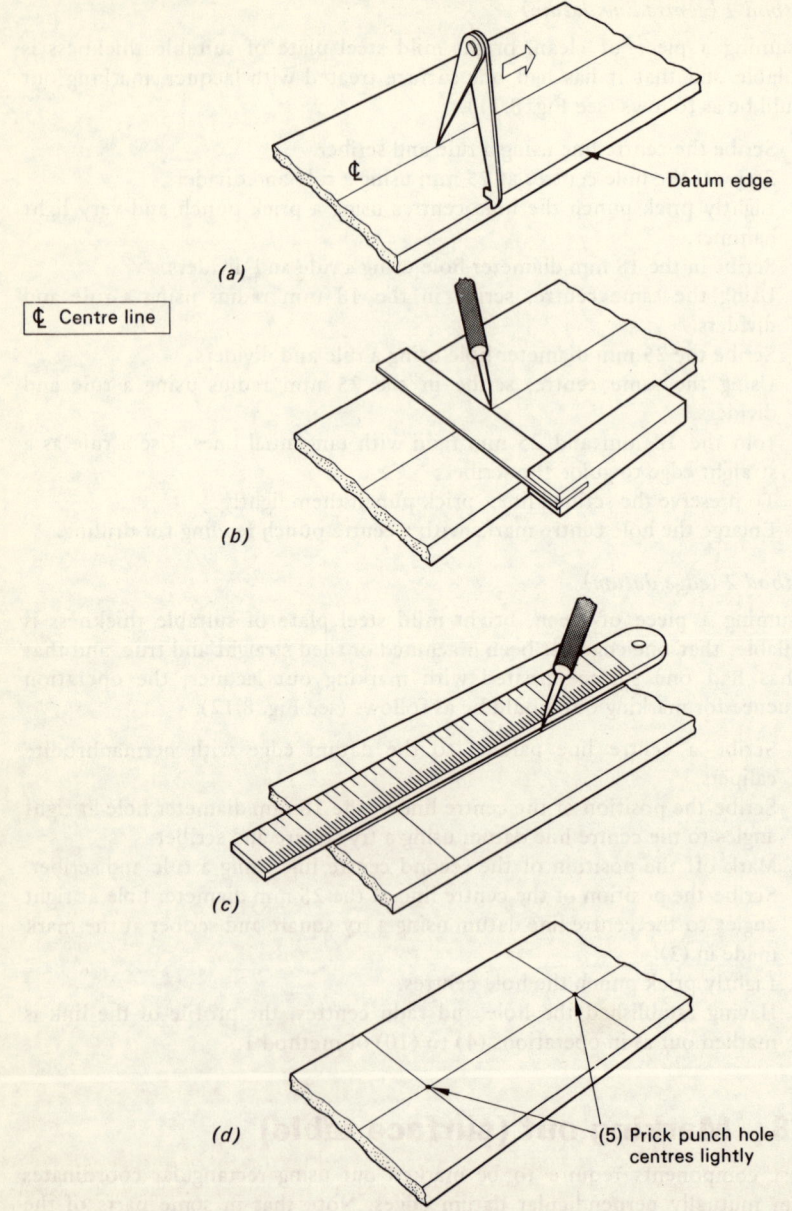

Ɫ Centre line

(a)

(b)

(c)

(d)

Datum edge

(5) Prick punch hole centres lightly

Fig. 8.12 Method of marking out from a datum edge

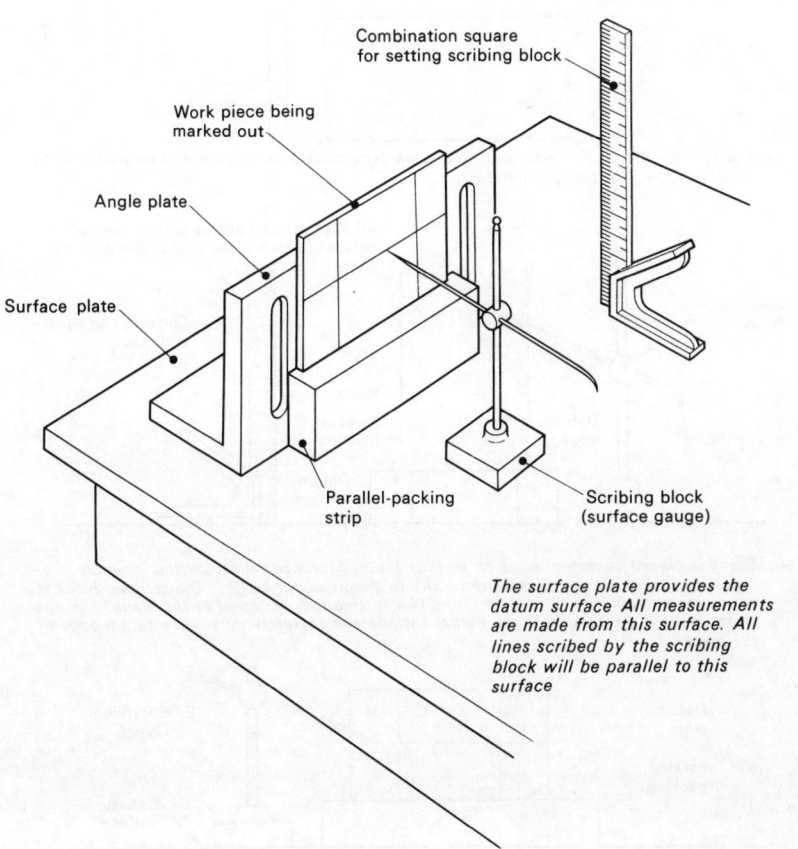

Combination square
for setting scribing block

Work piece being
marked out

Angle plate

Surface plate

Parallel-packing
strip

Scribing block
(surface gauge)

*The surface plate provides the
datum surface All measurements
are made from this surface. All
lines scribed by the scribing
block will be parallel to this
surface*

Fig. 8.13 Marking out from a datum surface

planed or scraped to a high degree of flatness. Figure 8.13 shows such a table together with the equipment set up to mark out the link previously described. The individual steps of marking out are shown in the simplified sketches of Fig. 8.14.

To ensure that a drilled hole has not 'wandered' off centre the hole should be 'boxed' as shown in Fig. 8.15 during marking out. If the hole is accurately positioned and of correct size it will split the dot punch marks at PQRS.

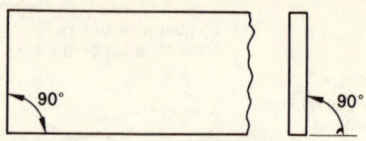

(a) File or machine up two edges at right angles (perpendicular) to each other and at right angles to the face being marked out

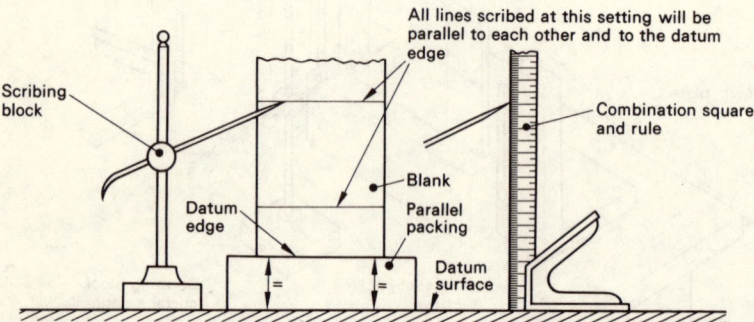

All lines scribed at this setting will be parallel to each other and to the datum edge

Scribing block

Combination square and rule

Blank

Datum edge

Parallel packing

Datum surface

(b) Blank is placed on datum edge on surface plate (datum surface). In this example parallel packing is used to raise the blank to a convenient height. The scribing block is set to the combination square and rule. The setting is transferred to the blank. The line so scribed will be parallel to the datum surface and therefore parallel to datum edge of the blank

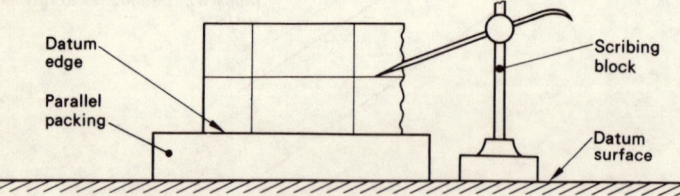

Datum edge

Parallel packing

Scribing block

Datum surface

(c) Blank is turned through 90° so that it rests on the other datum edge. This enables the remaining centre line to be scribed in at right angles to the first two. The marking out of the link is completed as in operations (4)–(10), Fig. 8.11

Fig. 8.14 Marking out procedure when using a datum surface

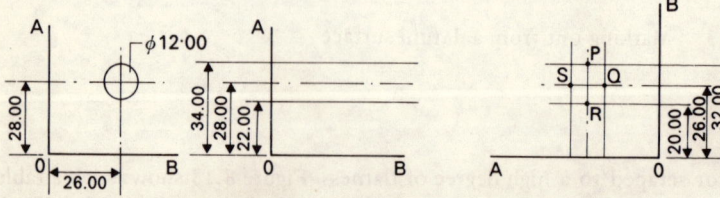

(a) Coordinates of Ø12·00 hole

(b) Hole centre is marked out together with parallel lines half a hole diameter above and below the centre line.

(c) The workpiece is rotated through 90°. The remaining centre line is scribed, together with parallel lines half a hole diameter above and below the centre line.

Fig. 8.15 Boxing a hole

Problems

Section A

1. A cellulose lacquer is painted on to a surface prior to marking out in order to: (a) prevent corrosion; (b) provide a means of identification; (c) provide a contrasting background; (d) protect the scriber point.

2. Hermaphrodite (odd-leg) calipers are used to scribe (a) circular lines; (b) lines parallel to an edge; (c) lines perpendicular to an edge; (d) irregular profiles.

3. Scribing instruments should be kept sharp with the aid of: (a) a grinding wheel; (b) a smooth file; (c) metal polish; (d) an oil stone.

4. The hole shown in Fig. 8.16 has been dimensioned using: (a) polar coordinates; (b) retangular coordinates; (c) a centre line datum; (d) an edge datum.

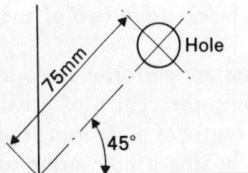

Fig. 8.16

5. A fixed point, line, edge or surface from which a measurement can be taken is known as a: (a) coordinate; (b) centre; (c) datum; (d) dimension line.

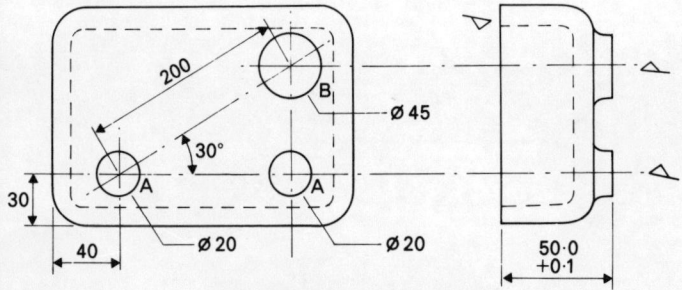

Dimensions in millimetres
Bosses A are of equal diameter
Fig. 8.17   *Material: grey cast iron*

Section B

6. (a) Figure 8.17 shows a casting, the faces of which marked ▽ have been machined. Describe with the aid of sketches:

(i) the method of locating and holding the casting relative to the datum surface;

(ii) the method of preparing the boss surfaces so that the scribed lines will show up clearly;

(iii) the process of marking out the boss centres.

(b) Calculate the rectangular coordinates for the bosses.

7. (a) Explain why it is usual to mark the position of a scribed line with a series of fine dots.

(b) Explain, with the aid of sketches, how dimensioning a row of holes; from an edge datum results in a smaller cumulative error than 'chain' dimensioning.

8. Show, with the aid of sketches, how a universal surface gauge can be used for: (a) scribing lines parallel to a datum surface; (b) scribing lines parallel to an edge; (c) setting surfaces parallel to a datum surface; (d) carrying a dial test indicator to check the parallelism of a surface.

9. Describe **three** reasons for marking out a component before manufacture and describe how a sawn mild steel blank 100 mm × 75 mm × 12 mm should be prepared for marking out on the flat faces, using two of the edges as datums.

10. Describe how the following marking out processes are performed. Your answers should contain clear sketches. (a) Finding the centre of a bar using the 'four arc' technique; (b) finding the centre of a bar using the centre finder from a combination set; (c) boxing a hole prior to drilling; (d) constructing a perpendicular line using trammels; (e) stepping off equi-distant hole centres round a pitch circle.

Chapter 9

Metal cutting

9.1 The wedge in metal cutting

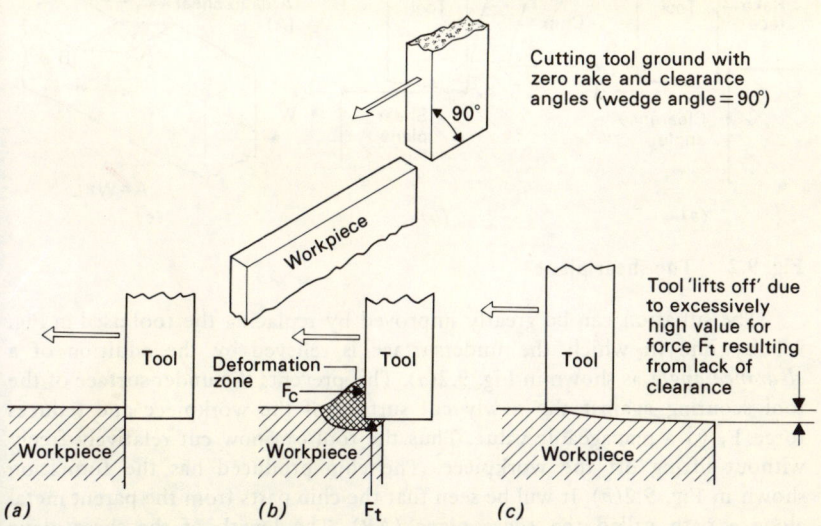

Cutting tool ground with zero rake and clearance angles (wedge angle = 90°)

Workpiece

Tool

Deformation zone F_c

Workpiece

Tool

Workpiece F_t

Tool

Tool 'lifts off' due to excessively high value for force F_t resulting from lack of clearance

Workpiece

(a)　　　　(b)　　　　(c)

Fig. 9.1　The need for clearance

All metal cutting tools require a basic wedge shape at the cutting edge. This shape is no accident, but is fundamental to the needs of metal cutting tools.

Consider the shaping machine set-up shown in Fig. 9.1(a). Here, a cutting tool is being used that has been ground flat on the end so that it is devoid of rake and clearance. The workpiece, which is slightly narrower than the cutting tool, is a low strength ductile material such as mild steel. Figure 9.1(b) shows what happens when the tool starts to cut. The metal just ahead of the tool is compressed until it starts to shear away from the test piece and piles up ahead of the cutting tool. Obviously, this deformation of the metal ahead of the cutting tool sets up reaction forces to the movement of the cutting tool. The most important of these are:

F_c, which is the cutting reaction force.
F_t, which is the thrust reaction force.

Assuming that the machine and tool are strong enough to keep cutting without mechanical failure, the thrust force (F_t) would gradually push the cut off, as shown in Fig. 9.1(c), by springing the tool and workpiece apart. The underside of the tool would be heavily scored; the cutting edge would soon be destroyed, and the newly cut surface of the workpiece would be very rough and uneven.

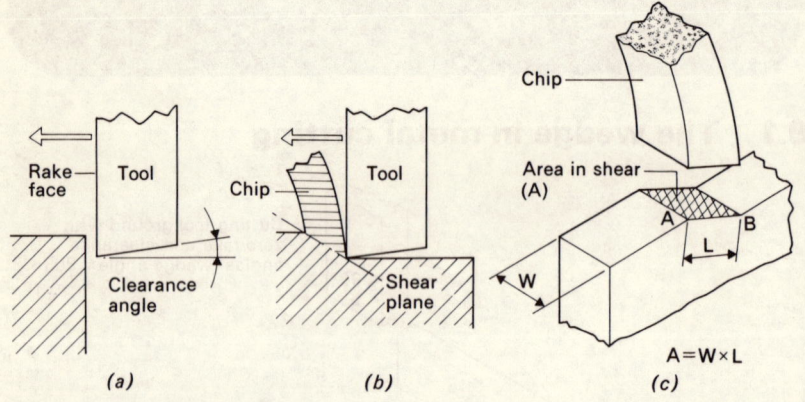

Fig. 9.2 The shear plane

The situation can be greatly improved by replacing the tool used in Fig. 9.1 by one in which the undersurface is relieved by the addition of a *clearance angle* as shown in Fig. 9.2(a). This prevents the under-surface of the tool scouring against the newly cut surface of the workpiece and reduces force F_t to a manageable value. Thus the tool can now cut relatively freely without lifting off the workpiece. The chip produced has the formation shown in Fig. 9.2(b). It will be seen that the chip parts from the parent metal along a path called the *shear plane* (AB). The length of the shear plane multiplied by the width of the cut gives the area in shear for the metal being cut. This is shown diagrammatically in Fig. 9.2(*c*) where the chip has been 'lifted away' from the workpiece to expose the area in shear.

For any given material, the smaller this area can be made, the lower will be the cutting force (F_c), and the greater will be the cutting efficiency. Since any reduction in the width of the cut would cause a reduction in the rate of metal removal, the most effective way of reducing the shear area is to reduce its length AB.

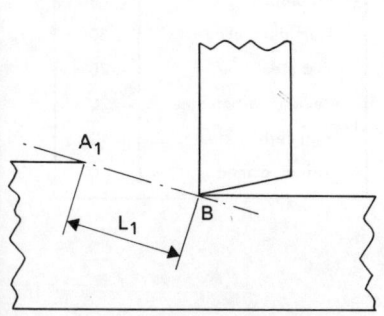

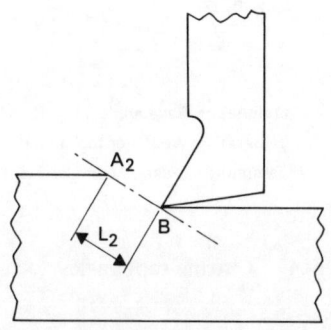

$A_1 B$ = Shear plane with zero rake

$A_2 B$ = Shear plane with positive rake

Comparing L_1 with L_2, it is apparent that the shear plane is shortened by increasing the rake angle from zero to a positive value

Fig. 9.3 The effect of rake on the shear plane

It has been shown by experimentation that if the rake face of the tool is inclined away from the perpendicular the shear plane tends to become normal to the rake face. That is, giving a tool a *rake angle* decreases the length of the shear path. Figure 9.3 shows how the shear path — and therefore the shear area — decreases as the rake angle increases for a ductile material.

Further, an inclined rake face enables the chip to peel away from the parent metal without having to turn through such an acute angle. Thus a high rake angle reduces the cutting force (F_c) by reducing the shear area, and it also reduces the pressure of the chip on the rake face of the tool. Both these factors lead to increased cutting efficiency.

Unfortunately there is a limit to how much the rake angle can be increased. Figure 9.4 shows the metal cutting *wedge.* It will be seen that three angles are involved:

(*a*)   rake angle α,
(*b*)   wedge or tool angle β,
(*c*)   clearance angle γ.

The clearance angle is generally fixed, depending upon the geometry of the surface being cut, at the following values:

(*a*)   external cylindrical surface $5°$ to $7°$
(*b*)   flat surface $6°$ to $8°$
(*c*)   internal cylindrical surface $8°$ to $10°$ plus secondary (heel) clearance.

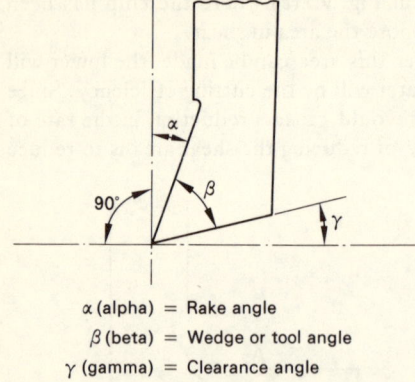

Some typical rake angles for high speed steel tools	
Material being cut	Rake
Cast iron	0°
Free-cutting brass	0°
Ductile brass	14°
Tin bronze	8°
Aluminium alloy	30°
Mild steel	25°
Medium carbon steel	20°
High carbon steel	12°
'Tufnol' plastic	0°

α (alpha) = Rake angle
β (beta) = Wedge or tool angle
γ (gamma) = Clearance angle

Fig. 9.4   Cutting tool angles

Angles less than this cause rubbing.

Angles greater than this cause chatter and a tendency for the tool to 'dig in'. It also reduces the wedge angle and, therefore, the strength of the tool.

With the clearance angle fixed within narrow limits the rake angle and wedge angle have to be balanced to form a compromise.

1. Increasing the *rake angle* increases the cutting efficiency of the tool but decreases the wedge angle.
2. Decreasing the *wedge angle* reduces the mechanical strength of the tool and reduces the mass of metal available at the cutting edge to conduct away the heat generated by the cutting process. This leads to a build up in the temperature of the tool at the cutting edge, softening of the tool and early failure.

Generally, therefore, low strength, ductile materials are cut with high rake tools to take advantage of the increased cutting efficiency. High strength, ductile materials are cut with low rake tools having a large wedge angle to give them adequate strength and heat dissipation capability at the cutting edge. Figure 9.4 gives typical rake angles for high speed tools cutting with positive rake. The values given are for roughing cuts and can be slightly increased, with advantage, for finishing cuts.

9.2   Types of chip

There are three basic types of chip produced when cutting metals. These are:

1. the **discontinuous** chip,
2. the **continuous** chip,
3. the **continuous** chip with **built-up edge.**

1. Discontinuous chip

The shearing of the chip from the parent metal of the workpiece has already been discussed in section 9.1. How this shearing action takes place and how the chip is formed is shown in Fig. 9.5(*a*). In forming the chip the metal is very severely strained and, if it is a brittle material, it may fracture in the primary deformation zone; that is, in the vicinity of the shear plane. This will give rise to the *discontinuous* type of chip illustrated in Fig. 9.5(*b*). Discontinuous chips are associated with brittle materials such as cast iron and free cutting brass. Low cutting speeds and lack of rake can also cause discontinuous chips to be formed when cutting ductile materials such as mild steel.

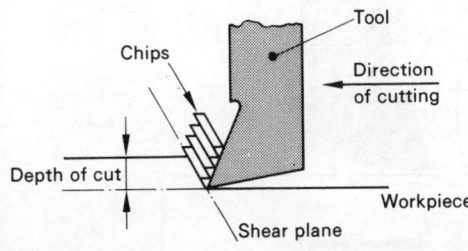

(a) **Chip formations**

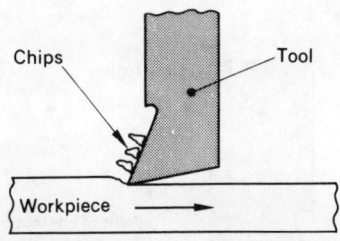

(b) **Discontinuous chip**

Fig. 9.5   Discontinuous chip

Since brittle materials, such as cast iron, form a discontinuous chip with a constant shear plane angle, little advantage is gained from giving the tool a rake angle. Therefore brittle materials are usually cut with a zero rake angle as indicated in Fig. 9.4. This permits the maximum possible wedge angle which results in an appreciable increase in tool life.

2. Continuous chip

This is the long, ribbon-like chip that is produced when machining *ductile* materials such as mild steel, copper and aluminium. The metal behaves as a rigid plastic and, although the chip shears from the parent metal along the shear plane, it remains homogeneous in itself and does not separate up into 'plates', as shown in Fig. 9.5(a). The formation of a continuous chip is shown in Fig. 9.6(a). Some very soft and ductile materials, with a low strength, tend to 'tear' away from the parent metal of the workpiece rather than shear cleanly. This results in a rough surface that has to be cleaned up by a very keen cutting edge as shown in Fig. 9.6(b).

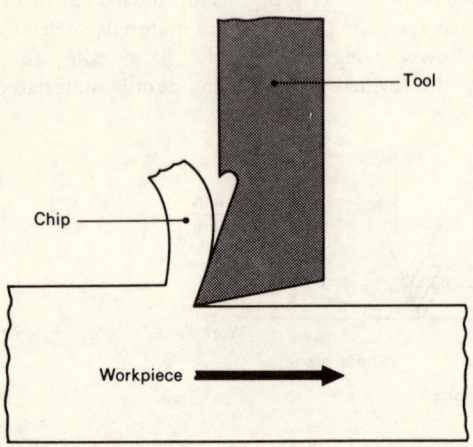

(a) **Continuous chip**

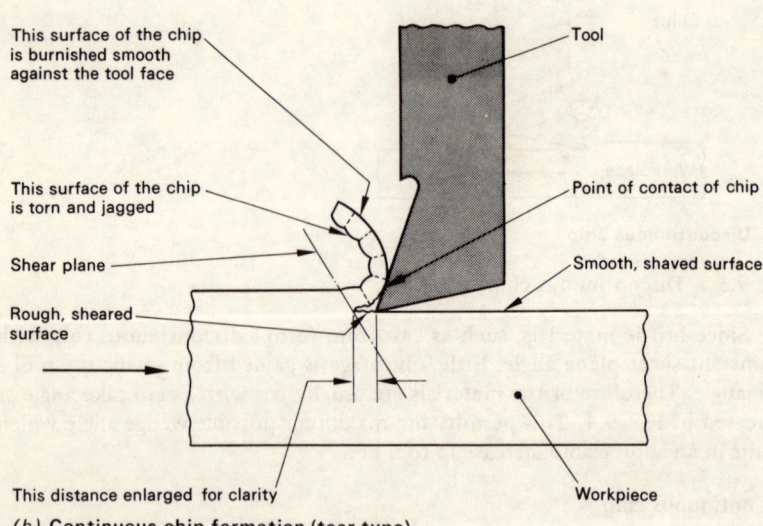

(b) **Continuous chip formation (tear type)
for soft, ductile, low strength metals**

Fig. 9.6 Continuous chip

(a) a blunt tool will cause a chip to shear away from the parent metal, but will leave a rough torn surface;

(b) a soft material tends to tear instead of shearing cleanly. Often the tear will be below the plane of cutting so that the component is undersize and will not clean up even with a sharp tool. This is why it is difficult to produce a good surface when cutting soft, ductile materials such as copper and fully annealed mild steel;

(c) soft materials such as copper and aluminium can be machined to a high finish by cutting at very high speeds using diamond-tipped tools. Under these conditions the metal behaves, at the point of cutting, as though it were very much harder and stiffer.

3. Continuous chip with built-up edge

Under some conditions the friction between the secondary deformation zone of the chip and the rake face of the tool is very great. This results in metal from the chip becoming pressure-welded to the rake face, making it rough. The increased roughness increases the friction and this leads to a building up of layer upon layer of chip material as shown in Fig. 9.7(a). This is referred to as a built-up edge.

Eventually the amount of material grows to such an extent that it tends to become unstable and breaks down. The particles of built-up material that flake away weld themselves to the chip and to the workpiece as shown in Fig. 9.7(b). This produces a dangerously jagged chip and a rough surface on the workpiece. The formation of a built-up edge is called chip welding.

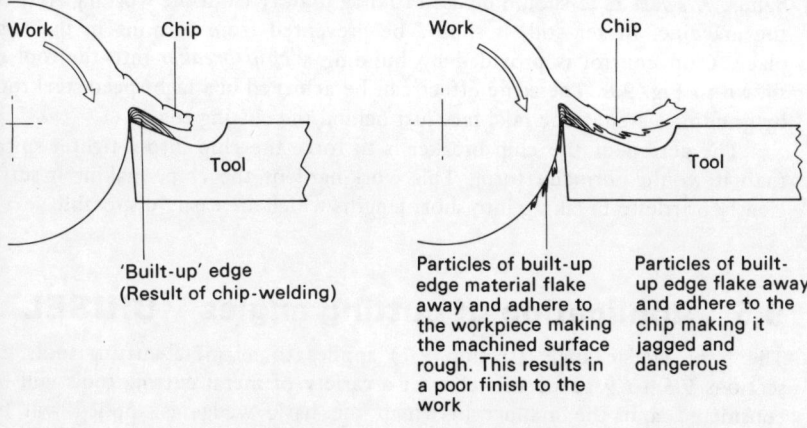

(a) Layering of chip material on rake face of tool during chip-welding

(b) Instability of built up edge if chip-welding becomes excessive

Fig. 9.7 Chip welding (built up edge)

9.3   Prevention of chip welding

Since chip welding has a considerable and adverse influence on tool life, power consumption and surface finish, every attempt must be made to prevent it occurring. This is achieved by reversing one or more of the causes of chip welding as follows:

(a)   *Reducing friction.* This can be achieved by increasing the rake angle; using a lubricant between the rake face and the chip; polishing the rake face.

(b)   *Reducing the temperature.* This can be achieved by reducing friction (see p. 206), and by reducing the cutting speed.

(c)   *Reducing the pressure* between the chip and the tool. This can be achieved by increasing the rake angle; reducing the feed rate; using oblique instead of orthogonal cutting (Fig. 9.17).

(d)   *Preventing metal-to-metal contact.* This can be achieved by use of a high-pressure lubricant between the chip and tool interface. Such lubricants contain chlorine and sulphur additives that build up a non-metallic film on the tool face. Active sulphur compounds attack copper and its alloys and should not be used on such materials. The use of non-metallic cutting tool materials will also prevent chip-welding if carefully selected. These will be considered in a later book.

9.4   The chip breaker

The long and ribbon-like continuous chip has razor-sharp edges and can inflict deep, painful and dangerous cuts. It should never be moved with the bare hands. A swarf rake should be used to drag it away from the working zone of the machine. Better still, it should be prevented from forming in the first place. Chip control is provided by building a *chip breaker* into the tool as shown in Fig. 9.8. The same effect can be achieved in a high-speed steel tool by grinding a step in the rake face just behind the cutting edge.

The action of the chip breaker is to force the chip into a tighter spiral than it would normally form. This work-hardens the chip, making it sufficiently brittle to break up into short lengths which are easily disposable.

9.5   Application of cutting angles – CHISEL

The basic wedge angle (section 9.1) applies to all metal-cutting tools. In sections 9.5 to 9.12 of this chapter a variety of metal cutting tools will be considered and the manner in which the basic wedge is applied will be examined.

Figure 9.9(a) shows how the point of a cold chisel forms a metal cutting wedge with rake and clearance angles. In Fig. 9.9(b) the chisel is inclined to the work at too small an angle. This destroys the clearance angle and prevents the cutting edge biting into the workpiece.

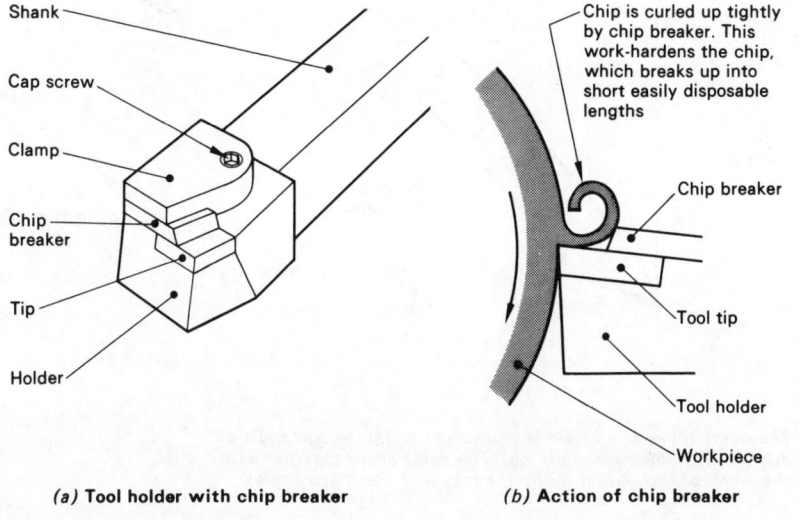

(a) **Tool holder with chip breaker**

(b) **Action of chip breaker**

Fig. 9.8 The chip breaker

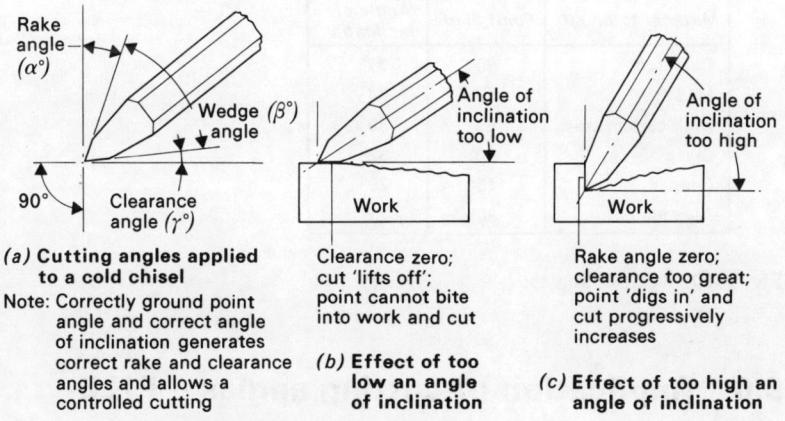

(a) **Cutting angles applied to a cold chisel**

Note: Correctly ground point angle and correct angle of inclination generates correct rake and clearance angles and allows a controlled cutting

(b) **Effect of too low an angle of inclination**

Clearance zero; cut 'lifts off'; point cannot bite into work and cut

(c) **Effect of too high an angle of inclination**

Rake angle zero; clearance too great; point 'digs in' and cut progressively increases

Fig. 9.9 The chisel — correct use

In Fig. 9.9(c) the chisel is inclined at too steep an angle. This destroys the rake angle and also makes the clearance angle excessive. The result is that the chisel digs into the workpiece with loss of control of the cutting processes. Figure 9.10 shows the correct point angles and inclination angles for various materials. It will be seen that these provide appropriate rake and clearance angles for the given materials.

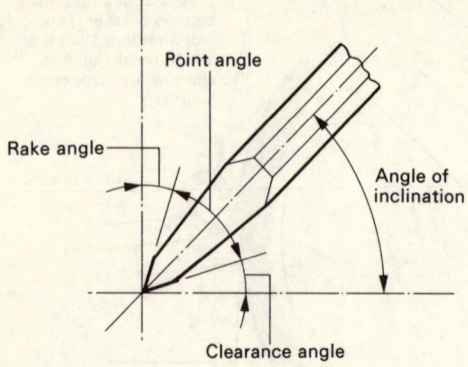

The point angle of a chisel is equivalent to the wedge angle of
a lathe or shaping machine tool. The point angle together with
the angle of inclination forms the rake and clearance angles

$$\text{Rake angle} = 90° - \left\{ \text{angle of inclination} + \tfrac{1}{2}\ \text{point angle} \right\}$$
$$\text{Clearance angle} = 90° - \left\{ \text{rake angle} + \text{point angle} \right\}$$
$$\text{or} = \text{angle of inclination} - \tfrac{1}{2}\ \text{point angle}$$

Material to be cut	Point angle	Angle of inclination
Cast iron	60°	37°
Mild steel	55°	$34\tfrac{1}{2}$°
High carbon steel	65°	$39\tfrac{1}{2}$°
Brass	50°	32°
Copper	45°	$29\tfrac{1}{2}$°
Aluminium	30°	22°

Fig. 9.10 Chisel angles

9.6 Application of cutting angles – FILE

Like any other cutting tool the file tooth must have correctly applied tool
angles in accordance with the principles laid down earlier in this chapter. The
file tooth is formed by a chisel-type cutter hitting the file blank at an angle as
shown in Fig. 9.11(a). This provides a single or 'over-cut'. Single cut files or
'floats' are used to work on hard material where an ordinary file tooth would
chip. They are particularly effective on brittle materials such as cast brass,
free-cutting brass and cast iron.

In order to give easier and smoother filing with greater control a second
or 'up-cut' is required. This produces a definite tooth as distinct from simple
ridges. However, its formation gives the tooth a burr as shown in Fig. 9.11(b).

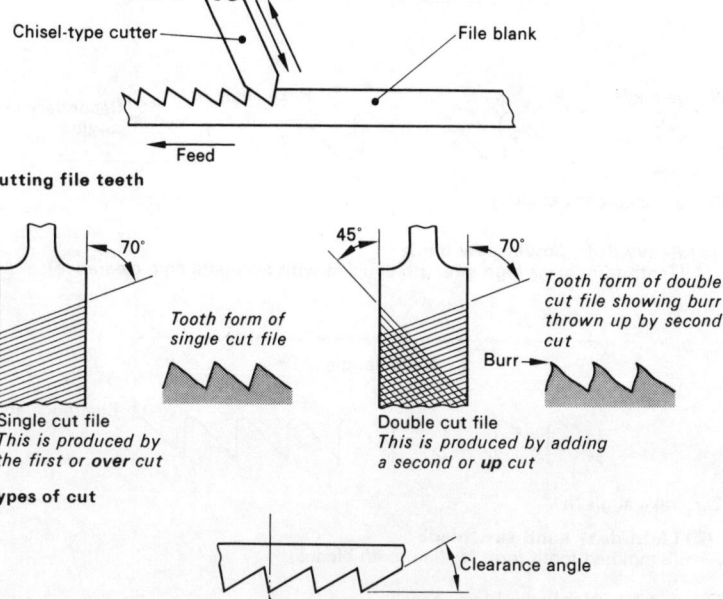

(a) **Cutting file teeth**

(b) **Types of cut**

Single cut file
*This is produced by
the first or over cut*

Double cut file
*This is produced by adding
a second or up cut*

Tooth form of
single cut file

Tooth form of double
cut file showing burr
thrown up by second
cut

(c) **Tooth form**

Fig. 9.11 File teeth

If a brand new file is used on tough material such as medium or high carbon steel or die steel, the burr would chip off, leaving the file blunt and inefficient. A brand new file should always be 'broken in' on softer and weaker materials such as brass and mild steel.

It will be seen from Fig. 9.11(c) that, unlike the chisel, the file tooth has *negative* rake.

9.7 Application of cutting angles – HACKSAW

The teeth of a heavy duty hacksaw blade used on power sawing machines is shown in Fig. 9.12(a). It will be seen that the teeth form a series of metal cutting wedges. Like all multi-tooth cutters designed to work in a slot, the hacksaw blade has to be provided with swarf (secondary) clearance as well as cutting (primary) clearance. This provides room for the chips to be carried out of the slot, without clogging the teeth, whilst maintaining a strong cutting edge.

The finer teeth of a hand saw blade have only a simple wedge shaped tooth as shown in Fig. 9.12(b). Here, swarf clearance is provided by exaggerating the primary clearance, whilst tooth strength is maintained by the use of zero rake angle to increase the available wedge angle to a maximum.

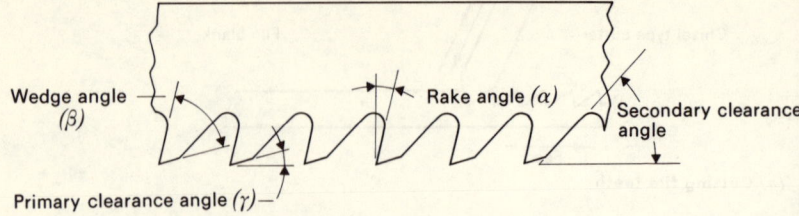

Wedge angle
(β)

Rake angle (α)

Secondary clearance
angle

Primary clearance angle (γ)

(a) Heavy duty power saw blade
(Tooth form gives high strength coupled with adequate chip clearance)

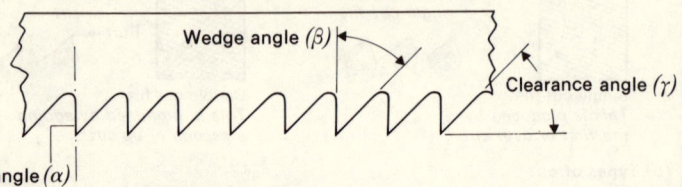

Wedge angle (β)

Clearance angle (γ)

Zero rake angle (α)

(b) Light duty hand saw blade
(Simplified tooth form for fine tooth blades)

Fig. 9.12   Hacksaw blade teeth

9.8   Application of cutting angles – SCRAPER

The scraper is used to remove metal locally from a surface with a high degree of accuracy. The end of the scraper is ground square to the blade so that a *negative rake* angle is formed as shown in Fig. 9.13(a), clearance being provided by the angle of inclination at which the scraper is held. Figure 9.13(b) shows how the vertical component of the cutting reaction force holds the scraper off the work and prevents it digging in. If positive rake were applied, then direction of action of the vertical component force would be reversed. This would tend to pull the scraper into the work and reduce the fitter's control.

9.9   Application of cutting angles – THREAD-CUTTING TAP and DIE

Figure 9.14(a) shows a section through a thread-cutting tap. Since the 'teeth' are form *relieved*, the clearance face is curved, and the clearance angle is formed by the tangent to the thread form at the cutting wedge. The rake angle is formed by the flute, thus the metal cutting wedge is still in evidence. Because the tap is a form cutter with a form-relieved clearance, re-sharpening is more specialised than for the simple single point tools referred to so far.

Figure 9.14(c) shows how the metal cutting wedge is applied to the thread-cutting die.

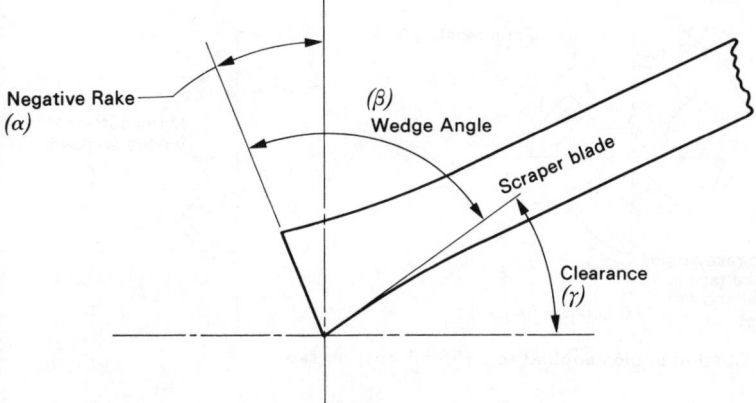

(a) **Scraper cutting angles**

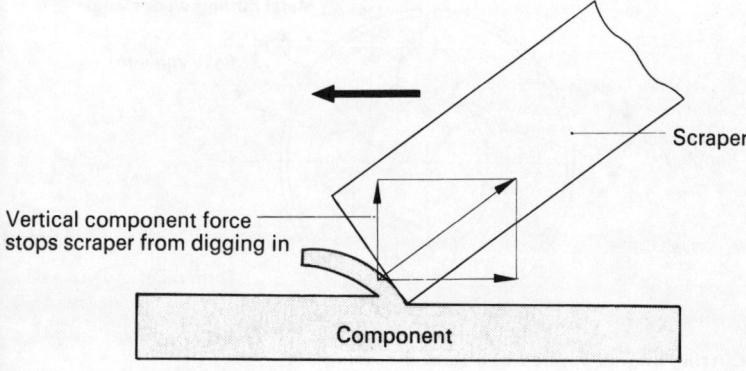

(b) **Forces acting on a scraper**

Fig. 9.13   The scraper

9.10   Application of cutting angles – TWIST DRILL

This is dealt with in greater detail in Chapter 11. However, Fig. 9.15(*a*) shows how the basic metal cutting wedge is applied to this cutting tool. Because the rake angle is formed by a helical groove, the rake angle varies from point to point along the lip of the drill, as shown in Fig. 9.15(*b*), from positive at the outer corner to negative near the centre of rotation. The fact that cutting conditions are poor at the point of the drill does not affect the quality of the hole produced by the outer corner where the cutting conditions are relatively good.

218

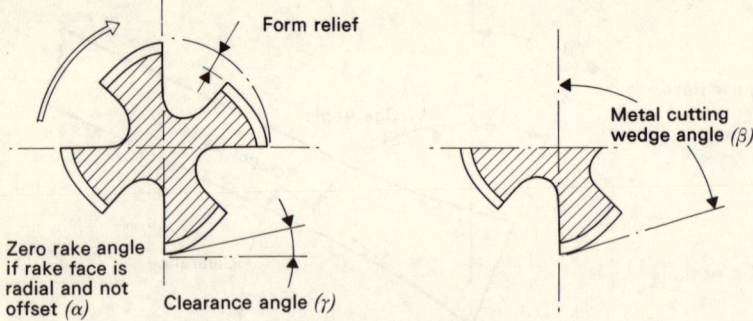

(a) **Cutting angles applied to a thread-cutting tap**

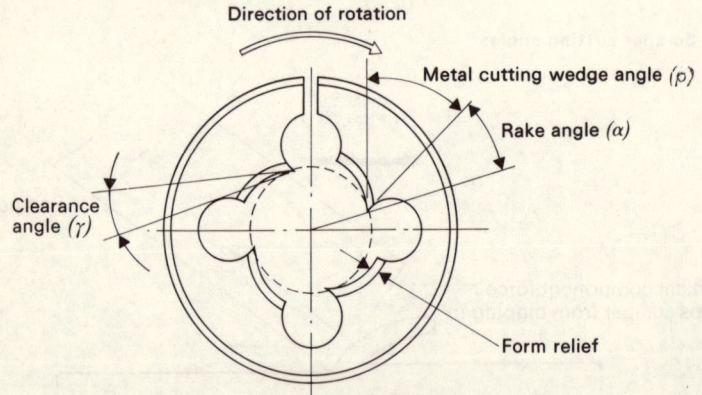

(b) **Cutting angles applied to a thread-cutting die**

Fig. 9.14   The thread cutting tap and die

9.11   Application of cutting angles – REAMERS

Reference to Fig. 9.16 shows that the tooth of a reamer can take several different forms.

Fluted reamer: This term refers to reamers that have a cylindrical land just behind the cutting edge. Such reamers have, therefore, zero clearance and the essential metal cutting wedge is not established. Thus, for the reasons discussed in section 9.1 little or no cutting action can take place peripherally. In fluted reamers, all the cutting action takes place at the bevel lead or the taper lead where the metal-cutting wedge can be established. The terms bevel lead and taper lead are explained in Figs. 10.14 and 11.6.

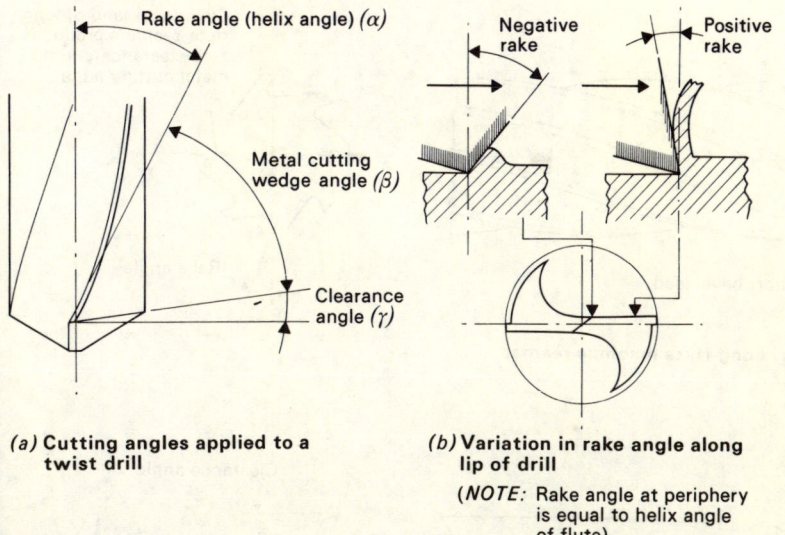

(a) **Cutting angles applied to a twist drill**

(b) **Variation in rake angle along lip of drill**

(*NOTE:* Rake angle at periphery is equal to helix angle of flute)

Fig. 9.15   The twist drill

Rose reamers: This term refers to reamers that have a definite clearance angle behind the cutting edge. Such reamers have a clearly defined metal-cutting wedge action and can cut peripherally. Because of this, rose reamers can remove metal more rapidly but generally produce a less accurate hole than a fluted reamer where the cylindrical land gives close control of the cutting edges. Long fluted reamers seldom have a rose cutting action, this design usually being found on taper pin reamers (section 10.8) and machine 'chucking' reamers (Fig. 11.9). The rose cutting action is useful when cutting materials such as phosphor bronze and reinforced plastics which tend to close and seize on a fluted reamer.

Reamers can only be ground on their rake faces otherwise the diameter that they cut is reduced. Continual regrinding will, in any case, eventually reduce the diameter, but because of the land or of the small primary clearance angle this reduction will usually be within the tolerance of the reamer.

9.12   Application of cutting angles – LATHE and SHAPER TOOLS

These are dealt with in greater detail in Chapters 12 and 13. However, Fig. 9.17(*a*) shows how the metal-cutting wedge is applied to a simple parting off tool. Since this tool has to work in a slot it is given side clearance and body clearance as well as the front clearance angle formed by the basic wedge. Figure 9.17(*b*) shows how rake and clearance angles are applied to the more

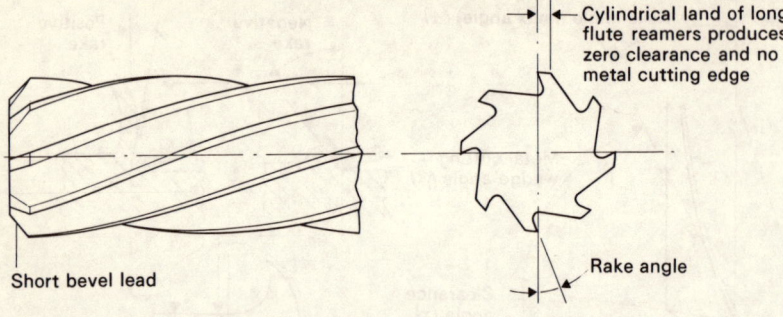

Cylindrical land of long
flute reamers produces
zero clearance and no
metal cutting edge

Rake angle

Short bevel lead

(a) **Long flute machine reamer**

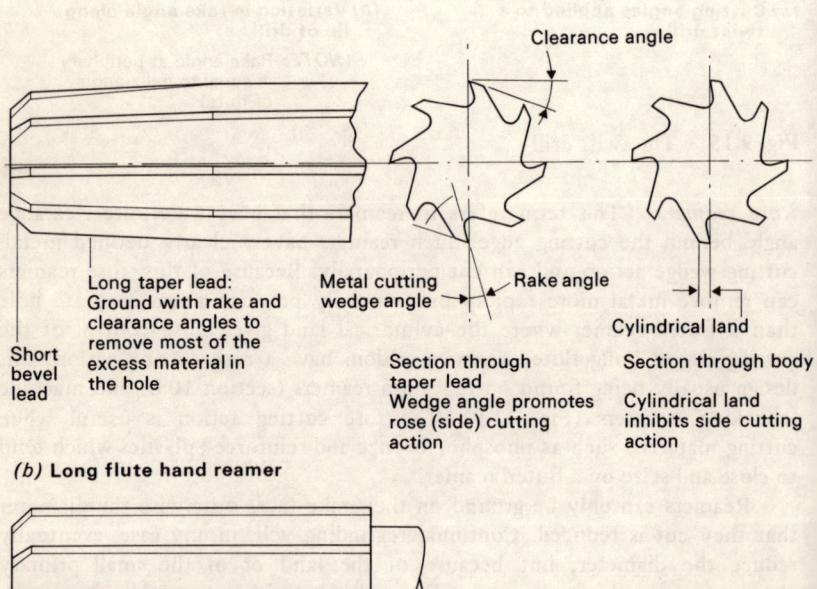

Clearance angle

Long taper lead:
Ground with rake and
clearance angles to
remove most of the
excess material in
the hole

Short
bevel
lead

Metal cutting
wedge angle

Rake angle

Cylindrical land

Section through
taper lead
Wedge angle promotes
rose (side) cutting
action

Section through body

Cylindrical land
inhibits side cutting
action

(b) **Long flute hand reamer**

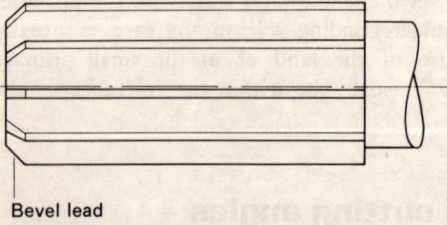

Bevel lead

(c) **Short flute (chucking) machine reamer**

Fig. 9.16 The reamer

complex oblique cutting tool and how they form compound angles. The same
principles apply to shaping machine tools except that the angles are rotated
through 90° as shown in Fig. 9.17(*c*).

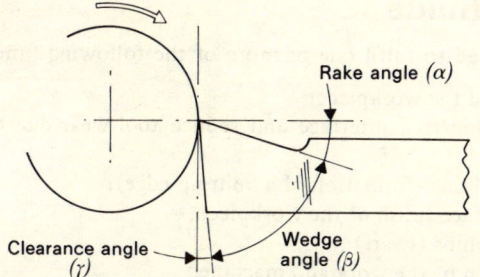

(a) Cutting angles applied to an orthogonal turning tool

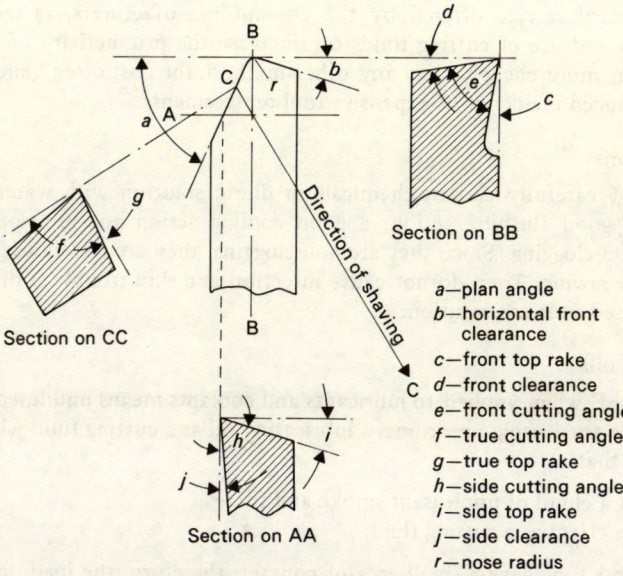

Rake angle *(α)*

Clearance angle *(γ)*

Wedge angle *(β)*

B

C

A

r

b

a

g

f

d

e

c

Section on BB

Section on CC

Direction of shaving

B

C

h

i

j

Section on AA

a—plan angle
b—horizontal front clearance
c—front top rake
d—front clearance
e—front cutting angle
f—true cutting angle
g—true top rake
h—side cutting angle
i—side top rake
j—side clearance
r—nose radius

(b) Cutting angles applied to an oblique turning tool

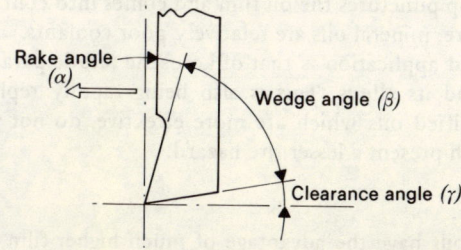

Rake angle *(α)*

Wedge angle *(β)*

Clearance angle *(γ)*

(c) Cutting angles applied to an orthogonal shaping machine tool

Fig. 9.17　Single point cutting tools

9.13 Cutting fluids

Cutting fluids are designed to fulfil one or more of the following functions:

(a) to cool the tool and the workpiece;
(b) to lubricate the chip/tool interface and reduce tool wear due to friction;
(c) to prevent chip welding (formation of a built-up edge);
(d) to improve the surface finish of the workpiece;
(e) to flush away the chips (swarf);
(f) to prevent corrosion of the work and machine.

There are many types of cutting fluid and it is only possible to consider the general principles in the scope of this chapter. It is always best to consult the expert advisory service offered by the coolant manufacturers, as the correct selection and use of cutting fluids can increase the productivity of a machine shop far more cheaply than any other method, the cost often being offset by the reduced incidence of expensive tool replacements.

Chemical solutions

These consist of carefully chosen chemicals in dilute solution with water. They possess a good flushing action, a good cooling action and are non-corrosive and non-clogging. Since they are non-clogging, they are widely used for grinding and sawing. They do not cause infection and skin trouble as did the previously used soda ash solution.

Straight mineral oils

The term 'straight' when applied to lubricants and coolants means undiluted. Anyone who has tried using an ordinary lubricating oil as a cutting fluid will have discovered that:

(a) it gives off a cloud of unpleasant smoke and fumes;
(b) it has little effect as a cutting fluid.

The chip and tool have a small area of contact; therefore, the load per unit area on the lubricant is much higher than in a bearing. The *film strength* of the oil is not sufficient to withstand this load and it ceases to act as a lubricant. That is, the chip punctures the oil film and comes into contact with the tool face. Furthermore, mineral oils are relatively poor coolants.

The only widely-used application is that of kerosene (crude paraffin) for machining aluminium and its alloys. This is also being rapidly replaced by specially designed emulsified oils which are more effective, do not produce noxious fumes, and which present a lesser fire hazard.

Straight fatty oils

Lard oils and vegetable oils have the advantage of much higher film strength (oiliness) than mineral oils. Unfortunately, they are not stable and rapidly lose their lubricating properties when contaminated with impurities. Neither are they satisfactory coolants as they have a high viscosity. They are much

more expensive and less plentiful than mineral oils. Nevertheless, they were
used for heavy duty machining and pressing operations for a long time.
Today, lard oil is mainly used on the bench for thread cutting with taps and
dies.

Compounded or blended oils

These are mixtures of mineral and fatty oils. The film strength of fatty oils is
retained even when diluted with 75 per cent mineral oil. As a result they are
much cheaper and more fluid than the neat fatty oils. They are very versatile
and can be used on most machining operations, especially on light and
medium-duty automatic machines.

Emulsified oil (suds)

When water and oil are added together they refuse to mix, but if an
emulsifier, in the form of a detergent, is added the oil will break up into
droplets and spread through the water as shown in Fig. 9.18. This is what
happens when the so called 'soluble' oils are added to water. The milky
appearance of these emulsions is due to the light being refracted by the oil
droplets. Since the oil is highly diluted with water, these emulsions are very
cheap to use and form the most widely-used group of cutting fluids to be
found in the machine shop.

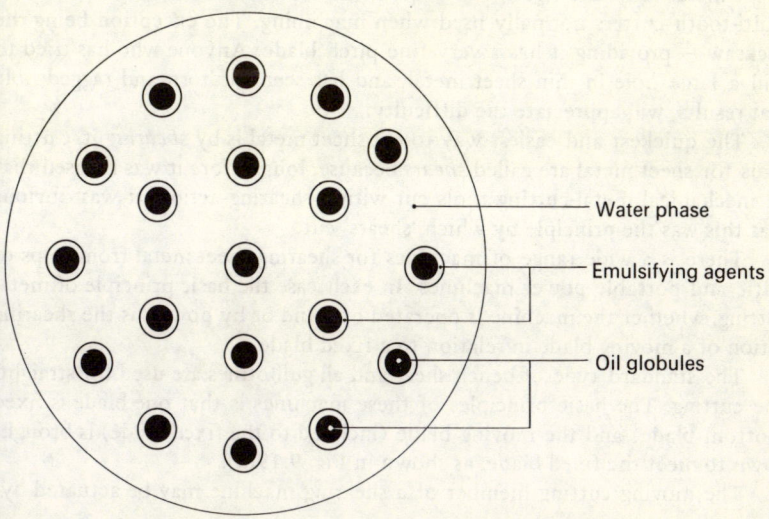

Fig. 9.18   Emulsified oil

Obviously, the dilution with water must reduce the lubricating properties
and as a result soluble oils are not suitable for the very severe conditions
found on many automatic machines or on machines using form cutters such
as broaching and gear cutting machines. However, the high water content
makes them excellent coolants and for the general machine shop they are

ideal, especially for manually operated machines taking fast but light cuts, and where the operator would be affected by the fumes given off by straight oils. The dilution also helps to reduce the tendency for mineral oils to produce skin diseases. The addition of a disinfectant helps to reduce the transmission of disease from one operative to another, and also helps to prevent the emulsion breaking down under bacteriological attack when it is standing.

Emulsified oils must be properly mixed. Modern oils are, generally, very stable. However, incorrect mixing and storing can cause it to separate out:

1. Mixing must be carried on in clean containers, using carefully measured quantities of oil and fresh, clean water.
2. The oil must be added to the water (never the reverse), whilst the mixture is vigorously and continuously stirred with a paddle.
3. If the emulsion is for stock, then it must be kept free from dirt dropping into it, but at the same time be open to fresh air and periodically stirred. This 'oxygenation' of the emulsion prevents it becoming stagnant and subject to bacteriological attack.

9.14 Sheet metal cutting principles

Sheet metal lacks the rigidity to respond successfully to the single point and multi-tooth cutters normally used when machining. The exception being the hacksaw — providing it has a very fine pitch blade. Anyone who has tried to drill a large hole in thin sheet metal, and has seen the torn and ragged hole that results, will appreciate the difficulty.

The quickest and easiest way to cut sheet metal is by *shearing* it. Cutting tools for sheet metal are called *shears* because, long before it was realised that all mechanical metal-cutting tools cut with a shearing action, it was obvious that this was the principle by which 'shears' cut.

There is a wide range of machines for shearing sheet metal from snips to static and portable power machines. In each case the basic principle of metal cutting, whether the machine is operated by hand or by power, is the shearing action of a moving blade in relation to a fixed blade.

The standard type of bench shear and all guillotines are used for straight-line cutting. The basic principles of these machines is that one blade is fixed (bottom blade) and the moving blade (inclined to the fixed blade) is brought down to meet the fixed blade, as shown in Fig. 9.19(*a*).

The moving cutting member of a shearing machine may be actuated by:

1. **Hand-lever** — bench shearing machines.
2. **Foot treadle** — treadle guillotines.
3. **Electric motor** or **hydraulics** — power guillotines.

If the cutting members of a guillotine or shearing machine were arranged parallel to each other, the area under shear would be the **cross-section** of the material to be cut, i.e. 'length × thickness', as shown in Fig. 9.19(*b*).

The top cutting member of a shearing machine is always inclined to the

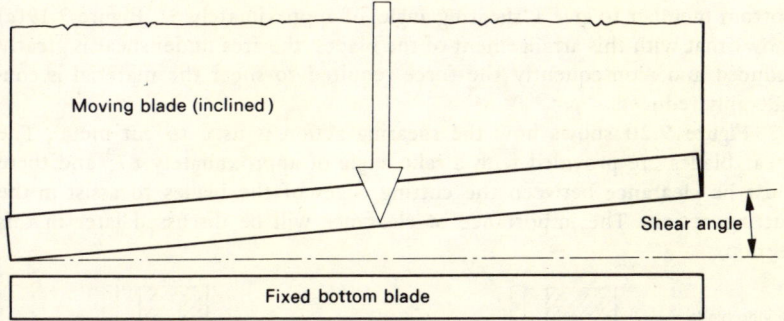

Moving blade (inclined)

Shear angle

Fixed bottom blade

(a) **Shear blade movement**

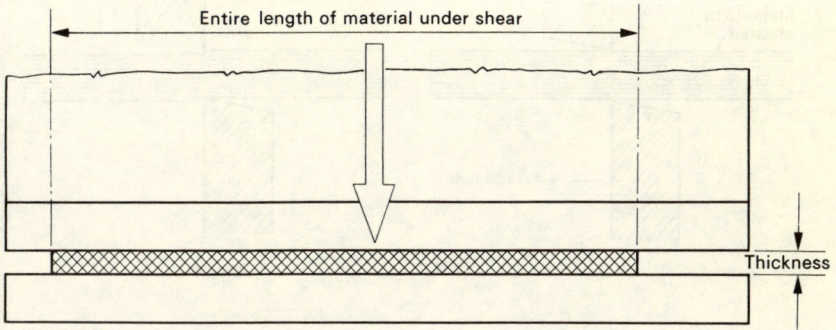

Entire length of material under shear

Thickness

(b) **Parallel cutting blades**

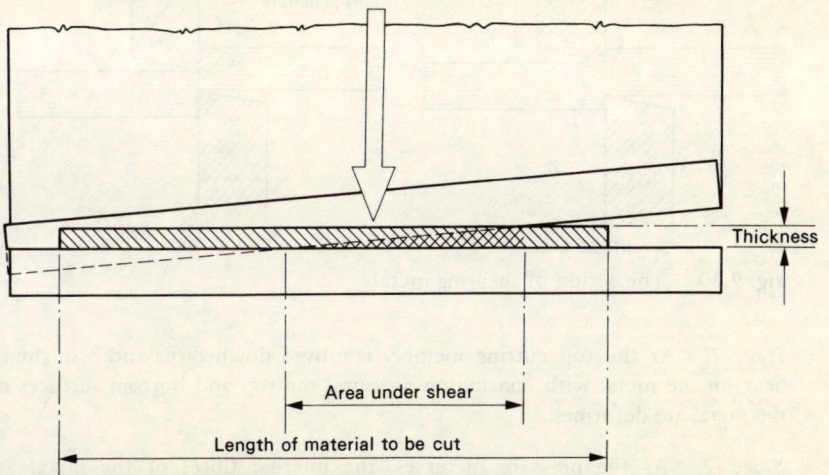

Thickness

Area under shear

Length of material to be cut

The force required for shearing a material is equal to the area under shear multiplied by the 'shear strength' of the material

(c) **Top cutting blade inclined**

Fig. 9.19 The effect of shear angle (shearing machine)

bottom member to give a '**shearing angle**' of approximately 5°. Figure 9.19(c) shows that with this arrangement of the blades, the area under shear is greatly reduced and, consequently the force required to shear the material is considerably reduced.

Figure 9.20 shows how the shearing action is used to cut metal. The shear blades are provided with a **rake angle** of approximately 87° and there must be **clearance** between the cutting edges of the blades to assist in the cutting action. The importance of clearance will be discussed later in this section.

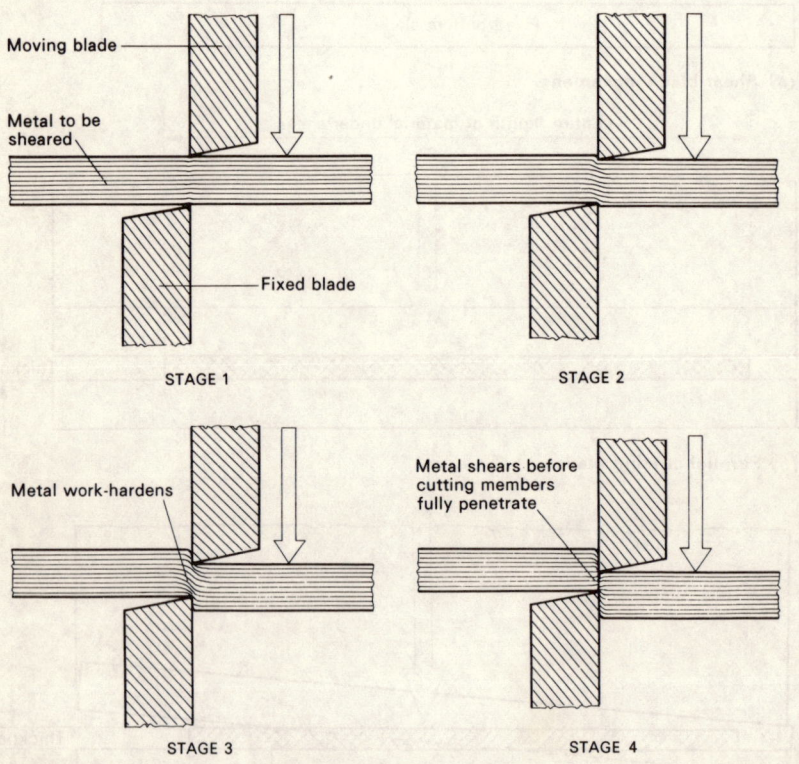

Fig. 9.20   The action of shearing metal

Stage 1.   As the top cutting member is moved downwards and brought to bear on the metal with continuing pressure, the top and bottom surfaces of the metal are deformed.

Stage 2.   As the pressure increases, the internal fibres of the metal are subject to deformation. This is '**plastic deformation**' prior to '**shearing**'.

Stage 3.   After a certain amount of plastic deformation the cutting members begin to penetrate. The uncut metal '**work-hardens**' at the edges.

Stage 4. Fractures begin to run into the work-hardened metal from the points of contact of the cutting members. When these fractures meet, the cutting members penetrate the whole of the metal thickness.

Figure 9.21 shows details of the cutting blades of hand shears. It will be

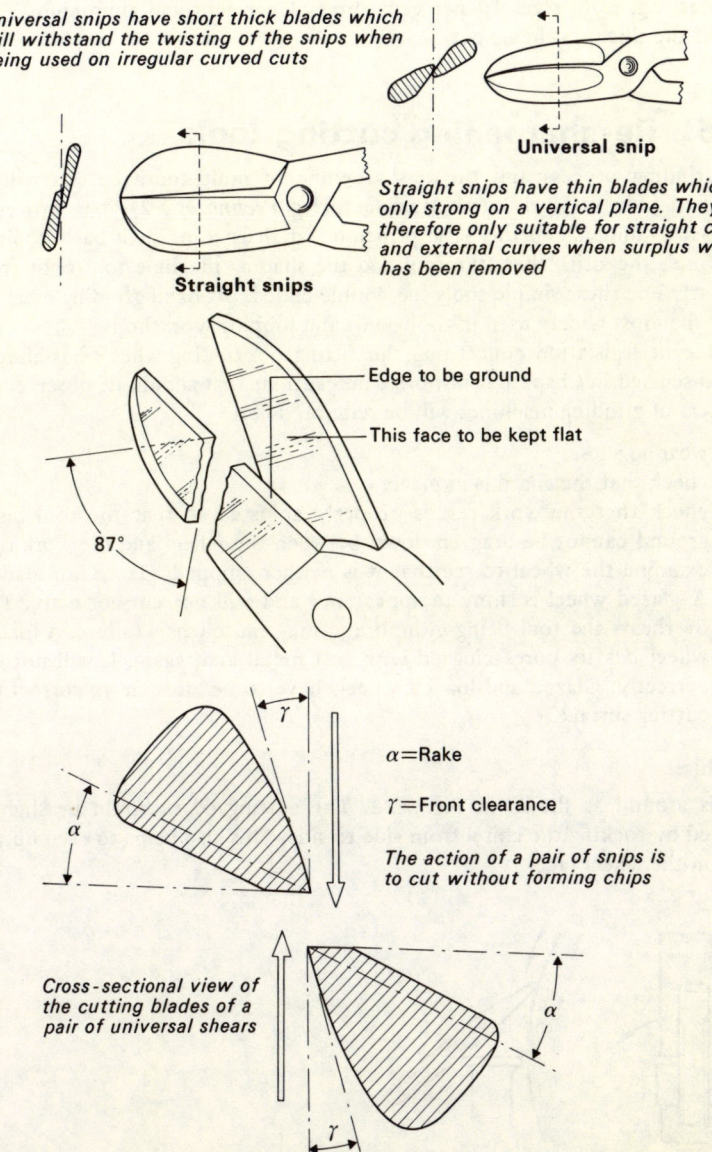

Universal snips have short thick blades which will withstand the twisting of the snips when being used on irregular curved cuts

Universal snip

Straight snips have thin blades which are only strong on a vertical plane. They are therefore only suitable for straight cuts and external curves when surplus waste has been removed

Straight snips

Edge to be ground

This face to be kept flat

87°

α = Rake

γ = Front clearance

The action of a pair of snips is to cut without forming chips

Cross-sectional view of the cutting blades of a pair of universal shears

Fig. 9.21   Details of the cutting blades of hand shears

noticed that the design principles are the same as those employed in respect of guillotine shear blades.

Blade clearances are very important and should be set to suit the material being cut. An approximate rule is that the clearance should not exceed 10 per cent of the thickness to be cut and must be varied to suit the particular material, e.g. mild steel 10 per cent; brass 4 per cent and aluminium 3 per cent of the thickness being cut.

9.15 Re-sharpening cutting tools

The grinding process and the re-sharpening of multi-tooth cutters will be described fully at level 3 (see *Manufacturing Technology* 2); it is, however, essential to understand the basic principles of sharpening such basic tools as the chisel, the drill, the lathe tool and the shaping machine tool right from the start. For these simple tools the double-ended, off-hand grinding machine is still the most widely used in toolrooms and jobbing workshops.

Recent legislation concerning the fitting of grinding wheels has already been discussed in Chapter 1, but basic precautions that should be observed by the users of grinding machines will be reiterated here:

(*a*) wear goggles;
(*b*) check that the guard is in place;
(*c*) check that the work rest is properly adjusted so that the tool being ground cannot be dragged down between the wheel and the work rest;
(*d*) examine the wheel to see that it is neither chipped, glazed nor loaded. A glazed wheel is shiny in appearance and will not cut correctly. This overheats the tool being ground and may cause wheel failure. A loaded wheel has its pores clogged with soft metal and, again, it will not cut correctly. Glazed and loaded wheels have to be 'dressed' to correct the cutting surface.

The chisel

This is ground as shown in Fig. 9.22. The cutting edge should be slightly radiused by rocking the chisel from side to side. This also helps to even up the wear on the wheel face.

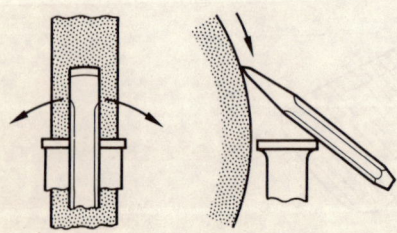

Fig. 9.22   Sharpening a chisel

This is most easily ground on the side of the wheel as shown in Fig. 9.23(a). The flat surface that this provides makes it much easier to generate a true point. The straight cutting lip should lie vertically against the side of the wheel, and the drill should be rocked against the wheel about the axis XX. When the drill has been ground it should be checked on a point gauge as shown in Fig. 9.23(b). This ensures that the angles are equal and the scale checks that the lip length are equal. It will be explained in section 11.4 that equal angles and lip length are required to ensure accurately sized holes.

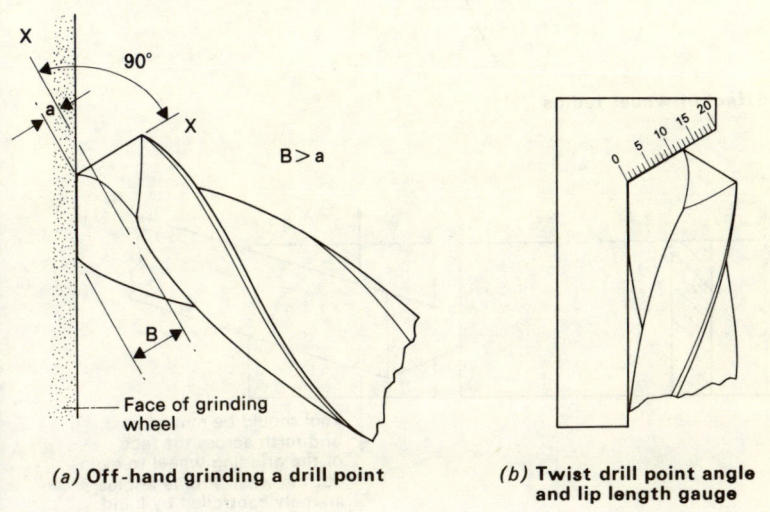

(a) Off-hand grinding a drill point

(b) **Twist drill point angle and lip length gauge**

Fig. 9.23   Twist drill grinding

Single point tools

Lathe and shaping machine tools can also be ground on the off-hand grinding machine. Unfortunately the radius of the wheel prevents flat surfaces being generated and this is particularly noticeable when grinding the front clearance angle as shown in Fig. 9.24(a). The tool being ground in Fig. 9.24 is a straight-nosed roughing tool, but the same principles apply to all single point tools. After grinding, the angles should be checked on a tool protractor as shown in Fig. 9.25.

9.16   Restraints and locations when cutting

When a workpiece is being cut, both the tool and the workpiece are subject to forces which tend to distort and displace them. Thus it is necessary to support and clamp both the tool and the workpiece in such a manner that distortion and displacement cannot occur. In order to locate and restrain the

230

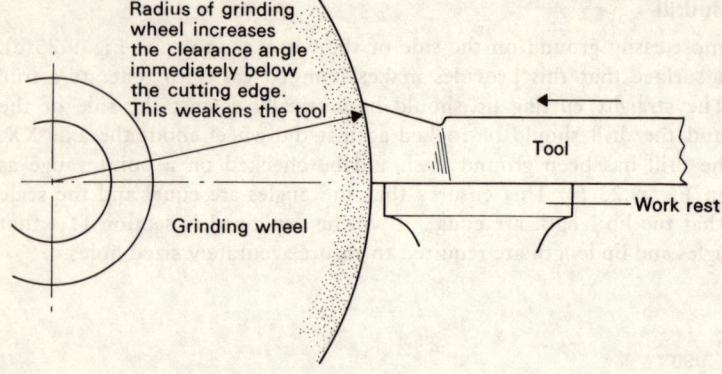

(a) **Effect of wheel radius**

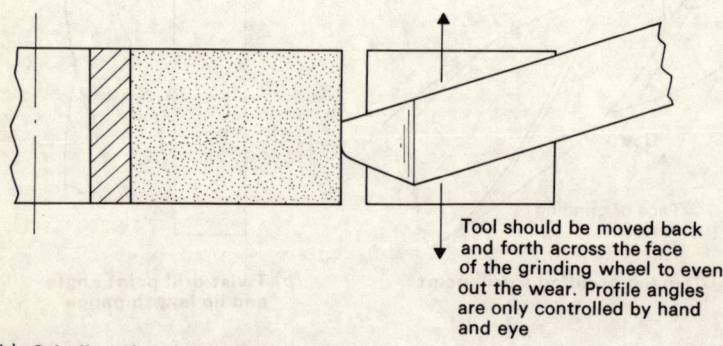

Tool should be moved back
and forth across the face
of the grinding wheel to even
out the wear. Profile angles
are only controlled by hand
and eye

(b) **Grinding the plan profile**

Fig. 9.24   Grinding single point tools

tool and the workpiece effectively it is necessary to understand certain basic principles.

A body in space, free of all restraints, is able to:

(a)   move back and forth along the 'Z' axis;
(b)   move from side to side along the 'X' axis;
(c)   move up and down along the 'Y' axis;
(d)   rotate in either direction about the 'Z' axis;
(e)   rotate in either direction about the 'X' axis;
(f)   rotate in either direction about the 'Y' axis.

That is, the metal block shown in Fig. 9.26 has **six degrees of freedom**.

In order that a body may be worked upon by hand or by machine it must be *located* in a given position by *restraining* its freedom of movement.

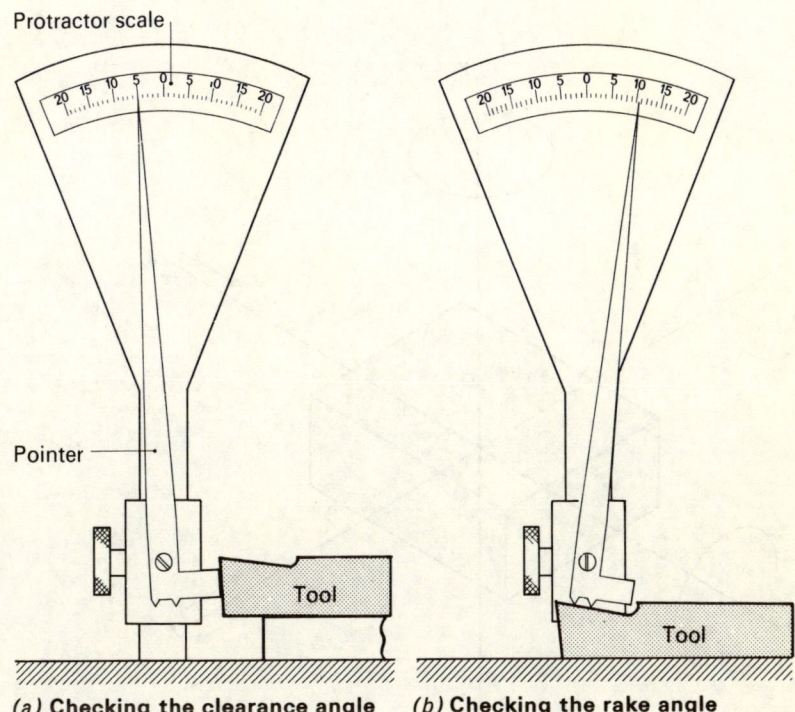

Protractor scale

Pointer

Tool

Tool

(a) **Checking the clearance angle**   *(b)* **Checking the rake angle**

Fig. 9.25   Tool protractor

Figure 9.27 shows how a block of metal can be *located* in a given position by the application of suitable *restraints.* The base plate supports the block and locates it in the vertical plane by restraining its downward movement. At the same time it restrains rotation about the X and Z axes of the block.

The addition of three location pegs adds restraint along the X and Z axes and positions the block on the plate.

Finally screw clamps are provided to complete the restraint of the block by ensuring its contact with the plate and the location pegs at all times. Since the block is restrained by contact with solid metal abutments in every direction it is said to be subjected to *positive* restraint.

Restraint may be *positive* or *frictional* and the difference is explained in Fig. 9.28. Wherever possible cutting forces should be resisted by positive restraints (solid abutments) and not by frictional restraint alone. For example, a component should be positioned in a vice so that the main cutting force is resisted by the fixed jaw. That is, the cutting force should be perpendicular to the fixed jaw and **not** parallel to it.

The application of the principles of restraint and location to work holding and tool holding on the bench and on machine tools will be considered in the remaining chapters on this book.

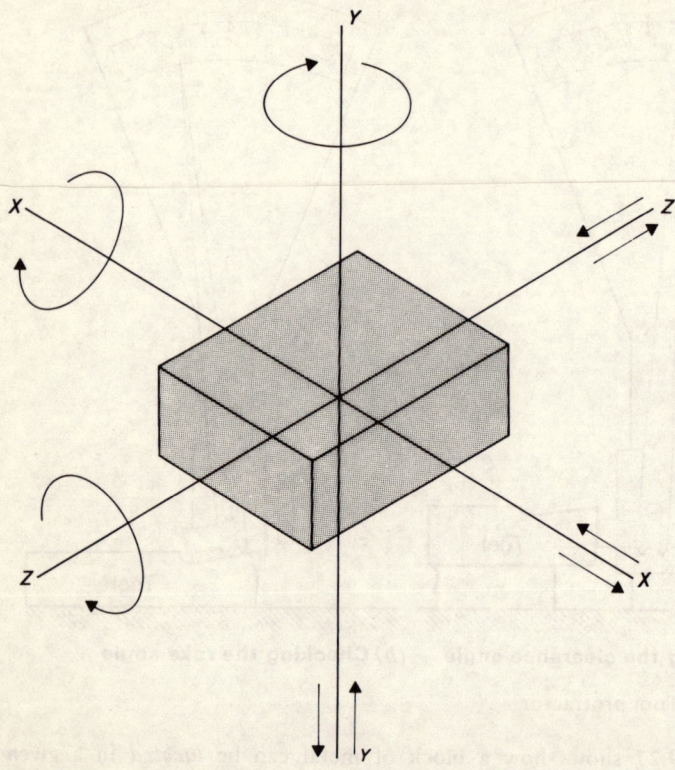

Fig. 9.26   Six degrees of freedom

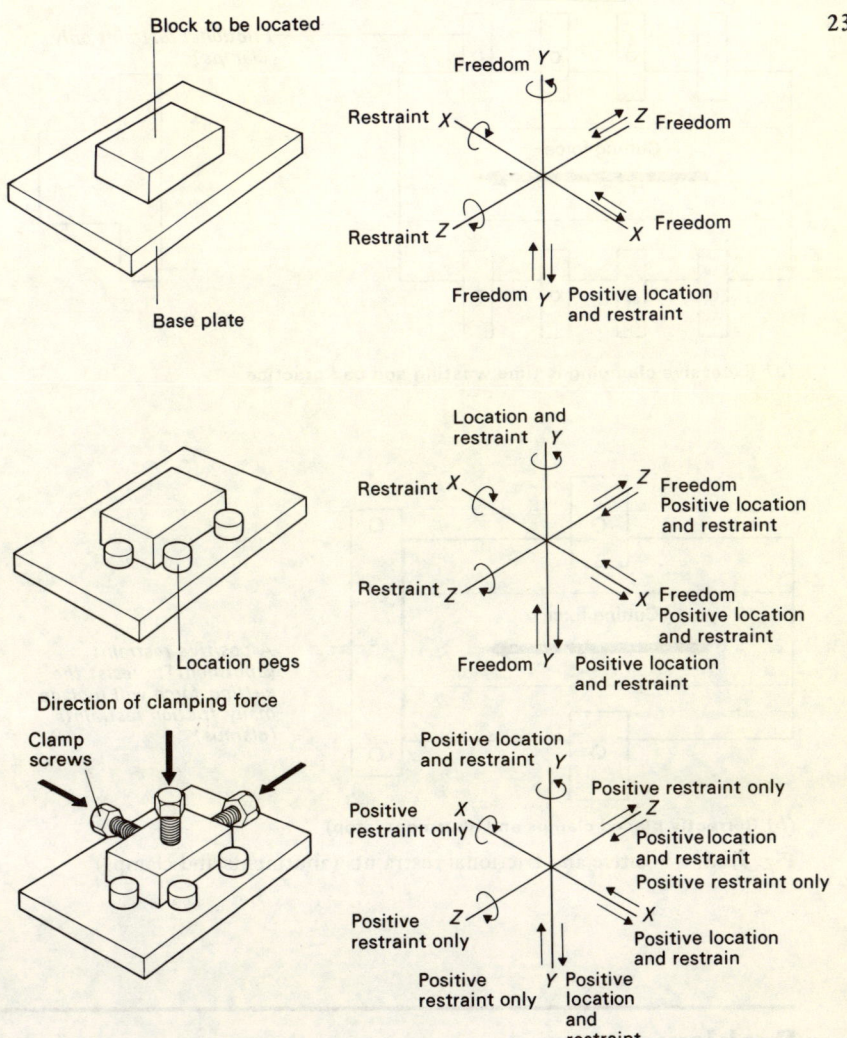

Fig. 9.27 Location and restraint

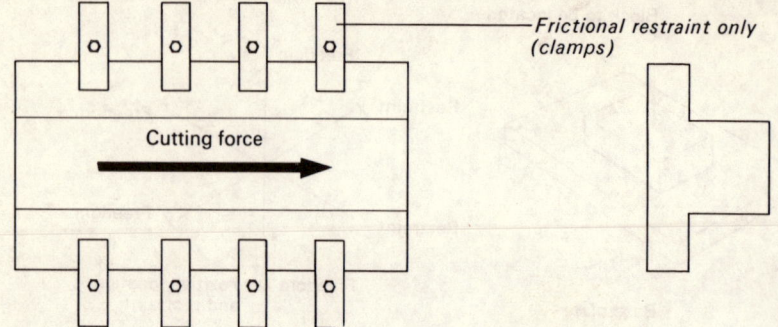

(a) **Excessive clamping is time wasting and bad practice**

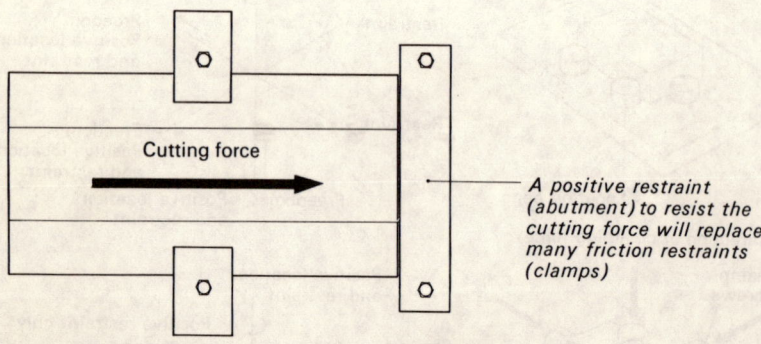

(b) **Correctly placed clamps and abutment (stop)**

Fig. 9.28   Positive and frictional restraints (abutments and clamps)

Problems

Section A

1. The metal cutting wedge is fundamental to the geometry of: (a) hand tools only;   (b) power-driven tools only;   (c) sheet metal cutting tools only;   (d) all cutting tools.

2. Continuous chips are formed when cutting: (a) ductile materials;   (b) brittle materials;   (c) amorphous plastic materials;   (d) free-cutting non-ferrous alloys.

3. When clamping a work piece, ready for machining, the main cutting force should be resisted by: (a) frictional restraint;   (b) a clamp;   (c) a solid abutment;   (d) a spring-loaded abutment.

4. The rake angle of a cutting tool: (*a*) prevents rubbing; (*b*) controls
   the chip formation; (*c*) determines the profile of the tool; (*d*)
   determines whether the cutting action is oblique or orthogonal.
5. The 'suds' used as a coolant in general machine shops consists of: (*a*) a
   solution of detergent and water; (*b*) a straight mineral oil; (*c*) an
   emulsion of oil and water; (*d*) a chemical solution.

Section B

6. (*a*) Draw a simple single point metal cutting tool and indicate: (i) the
   rake angle; (ii) the clearance angle; (iii) the wedge angle.
   (*b*) With the aid of a diagram show how varying the rake angle affects the
   length of the shear plane when cutting.
7. (*a*) Show, with the aid of sketches, what is meant by the terms **oblique**
   cutting and **orthogonal** cutting and the effect they have on the cutting
   action of the tool.
   (*b*) Describe the two basic types of chip formed when cutting metals.
   Name two metals in each case which will form the type of chip discussed.
8. With the aid of clear sketches show how rake and clearance angles are
   applied to the following metal cutting tools; (*a*) hacksaw; (*b*) chisel;
   (*c*) thread cutting tap; (*d*) twist drill; (*e*) lathe parting-off tool;
   (*f*) scraper.
9. (*a*) Describe the main function of a cutting fluid and compare the
   advantages and limitations of: (i) straight fatty oils; (ii) compounded
   (blended) oils; (iii) emulsified oils (suds).
   (*b*) Describe how emulsified oils should be mixed and stored to ensure
   stability and effective use.
10. Describe in detail the precautions that should be taken when re-
    sharpening cutting tools on an off-hand grinding machine.

Chapter 10

Hand processes

10.1 The vice

The work held in a vice must be given adequate support against the cutting force and also against the clamping force of the vice itself. It must also be held securely so that it does not move whilst being worked upon. Figure 10.1 shows the restraints acting on the workpiece and it will be seen that some of these are positive whilst others are frictional. For heavy cutting the work should be positioned so that the cutting forces are resisted by positive restraints.

As well as holding the work securely, the work must be positioned in the vice so that it is not distorted by the cutting force. Figure 10.2 shows how the work should be positioned. In Fig. 10.2(a) the cutting force is applied too far from the vice jaws — the work is said to have excessive *over-hang* — and it will have sufficient 'leverage' to bend the workpiece. Even when the force is too small to permanently bend the workpiece, it will make it vibrate and give off an irritating squealing noise. When the workpiece is held with the least possible over-hang, the cutting force has insufficient 'leverage' to bend the workpiece or make it vibrate. Correct positioning of the workpiece in the vice is shown in Fig. 10.2(b).

Cylindrical work is often a problem to hold securely in a parallel-jaw vice since there is only line contact at two points. Introducing a vee block

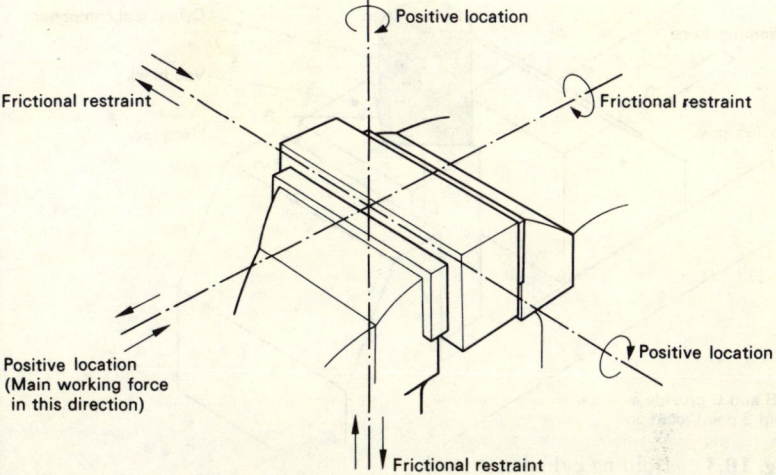

Fig. 10.1 Location in the vice

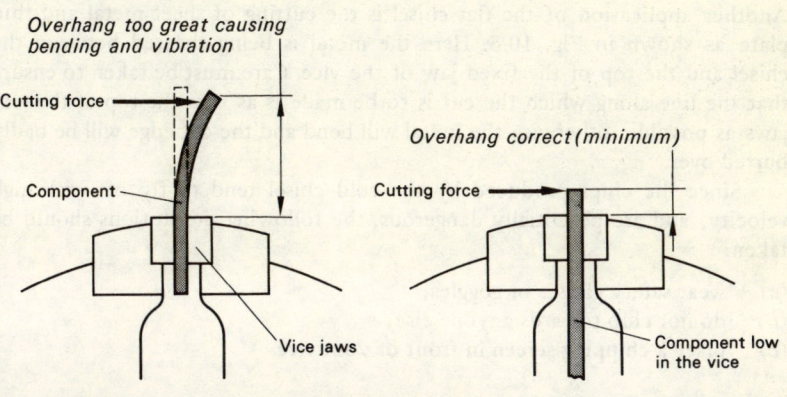

Overhang too great causing bending and vibration

Cutting force

Component

Vice jaws

(a) Incorrect

Overhang correct (minimum)

Cutting force

Component low in the vice

(b) Correct

Fig. 10.2 Positioning the work in the vice

between the workpiece and the fixed jaw provides three-point support and greater security as shown in Fig. 10.3.

10.2 The cold chisel

Cold chisels are used for rapidly breaking down a surface. It is the quickest way to remove metal by hand, but the accuracy is low and the finish is poor. Despite the fact that the chisel is only a simple roughing tool it is still important that the principles regarding tool angles are correctly applied. The application of the metal cutting wedge as applied to the geometry of the cutting edge of the cold chisel has already been discussed in section 9.5.

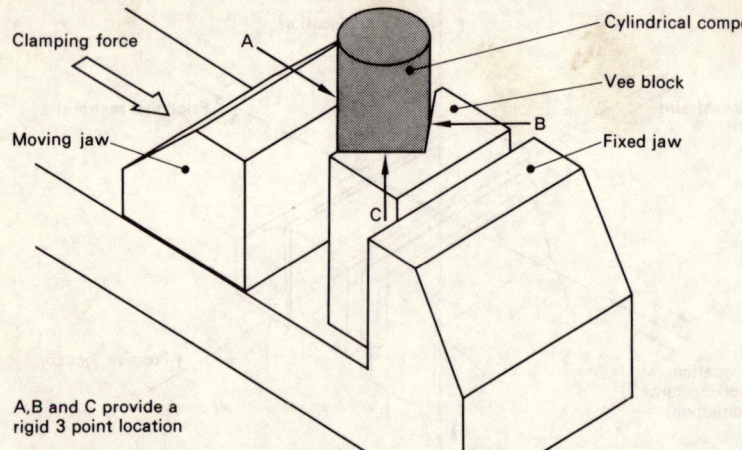

Clamping force

A

Moving jaw

Cylindrical component

Vee block

B

Fixed jaw

C

A,B and C provide a
rigid 3 point location

Fig. 10.3   Holding cylindrical work

Figure 10.4 shows a selection of cold chisels and their applications. Another application of the flat chisel is the cutting of sheet metal and thin plate as shown in Fig. 10.5. Here the metal is being sheared between the chisel and the top of the fixed jaw of the vice. Care must be taken to ensure that the line along which the cut is to be made is as near the top of the vice jaws as possible, otherwise the metal will bend and the cut edge will be badly burred over.

Since the chips produced by the cold chisel tend to fly off with high velocity, and are potentially dangerous, the following precautions should be taken:

(a)   wear safety glasses or goggles;
(b)   do not chip towards anyone else;
(c)   place a chipping screen in front of your vice.

10.3   The file

Filing operations range from roughing down and deburring blanks to finishing operations on flat and curved surfaces of considerable complexity and surprisingly high accuracy. Like any other cutting tool, the file must have correctly applied cutting angles and these have been discussed in section 9.6.

To generate a plane surface with a file it must be moved parallel to the plane of the required surface. Unfortunately, there are no slideways available to guide the file, only the muscular coordination of the user, which largely comes with practice.

There are many different types of file and to specify a particular type the following information must be given (see BS 498 : 1960):

(a)   length;
(b)   grade of cut;
(c)   shape.

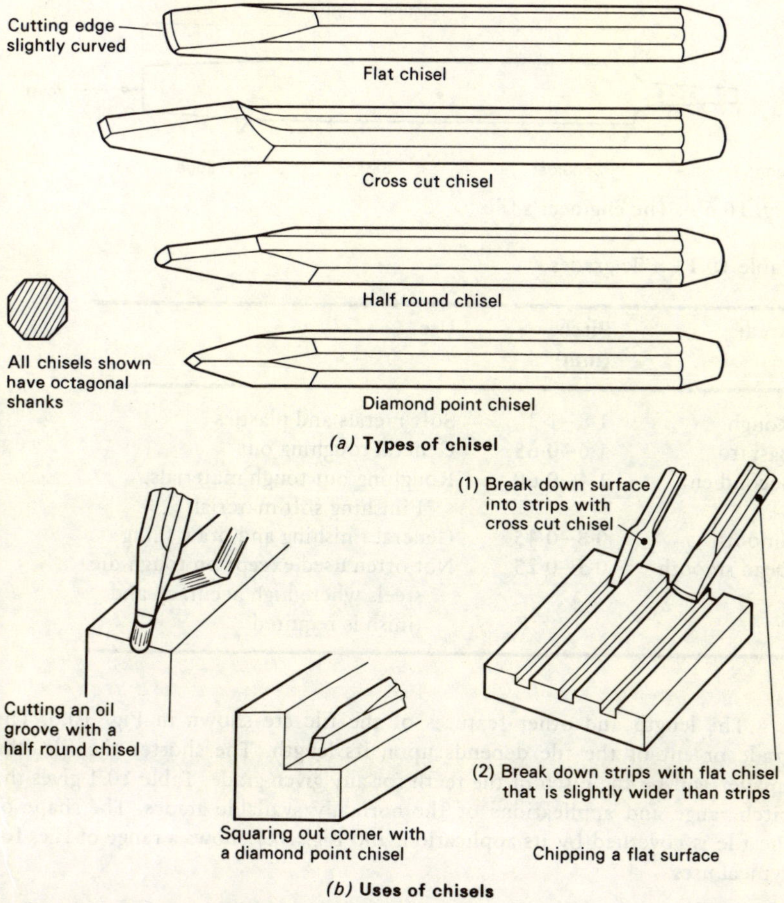

Cutting edge slightly curved

Flat chisel

Cross cut chisel

Half round chisel

All chisels shown have octagonal shanks

Diamond point chisel

(a) **Types of chisel**

Cutting an oil groove with a half round chisel

(1) Break down surface into strips with cross cut chisel

(2) Break down strips with flat chisel that is slightly wider than strips

Squaring out corner with a diamond point chisel

Chipping a flat surface

(b) **Uses of chisels**

Fig. 10.4 Cold chisels

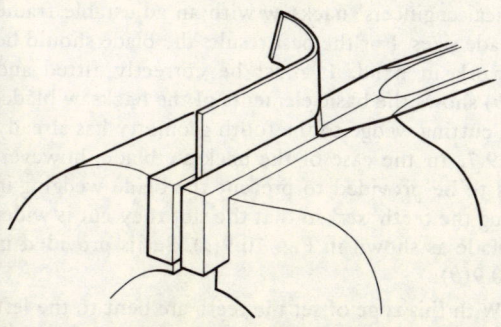

Fig. 10.5 Cutting thin sheet metal using a cold chisel

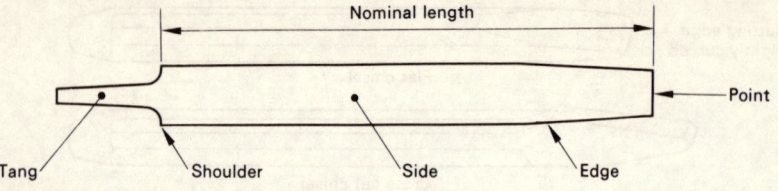

Fig. 10.6 The engineer's file

Table 10.1 File grades

Grade	Pitch (mm)	Use
Rough	1·8—1·3	Soft metals and plastics
Bastard	1·6—0·65	General roughing out
Second cut	1·4—0·60	Roughing out tough materials. Finishing soft materials
Smooth	0·8—0·45	General finishing and draw filing
Dead smooth	0·5—0·25	Not often used except on tough die steels where high accuracy and finish is required

The length and other features of the file are shown in Fig. 10.6. The grade or cut of the file depends upon its length. The shorter the file, the smaller will be the pitch of the teeth for any given grade. Table 10.1 gives the pitch range and applications of the normally available grades. The shape of the file is governed by its application, and Fig. 10.7 shows a range of files for typical uses.

10.4 The hacksaw

Figure 10.8(*a*) shows a typical engineers' hacksaw with an adjustable frame that will accept a range of blade sizes. For the best results the blade should be carefully selected for the work in hand. It must be correctly fitted and correctly used. Figure 10.8(*b*) shows the basic elements of the hacksaw blade. The application of the metal cutting wedge to the tooth geometry has already been considered in section 9.7. In the case of the hacksaw blade, however, additional side clearance has to be provided to prevent the blade wedging in the slot. This is done by giving the teeth '*set*' so that the slot they cut is wider than the thickness of the blade as shown in Fig. 10.9(*a*). Set is provided in two ways as shown in Fig. 10.9(*b*):

(*a*) *Staggered tooth set.* With this type of set the teeth are bent to the left and right of the centre line of the blade. Every third or fifth tooth is left straight as a 'clearing tooth'. This type of set can only be applied to coarse

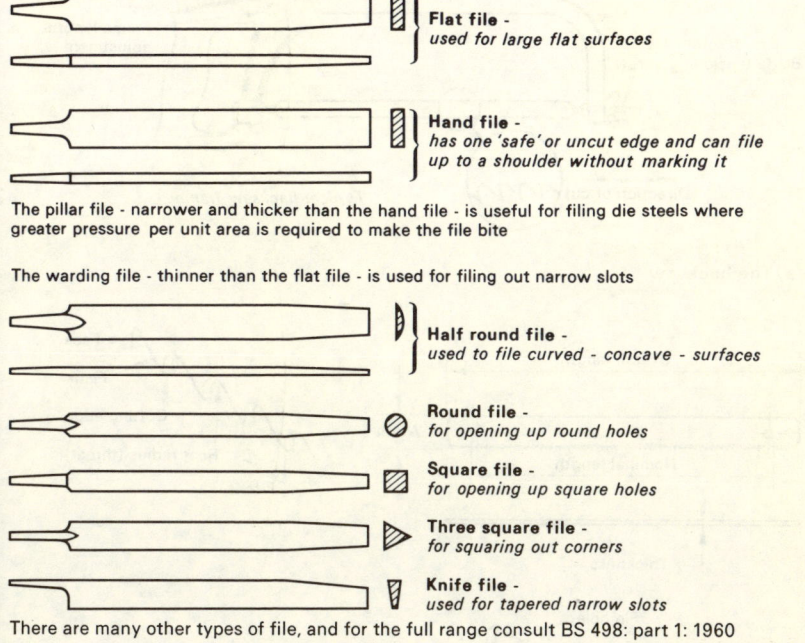

The pillar file - narrower and thicker than the hand file - is useful for filing die steels where greater pressure per unit area is required to make the file bite

The warding file - thinner than the flat file - is used for filing out narrow slots

There are many other types of file, and for the full range consult BS 498: part 1: 1960

Fig. 10.7   Sections and uses of files

pitch blades where the teeth are large enough to be bent.

(*b*) *Wave set.* Where the pitch is fine — as for tube and sheet cutting blades — the teeth are too small to be given an individual set. Here, the cutting edge of the blade is waved so that set is applied to groups of teeth.

10.5   The scraper

Scraping is a hand finishing operation by which very small amounts of metal can be removed locally from a surface. It is not convenient to remove metal locally from a surface with a file. Only the scraper provides sufficient control to produce the accuracy and finish required. Hand scraping is a highly skilled process requiring years of practice. Figure 10.10 shows three basic types of scraper.

Before scraping can commence the high spots on the surface must be found. This is done by rubbing the component on a reference surface (surface plate) that has been lightly smeared with prussian blue. Figure 10.11 shows the sequence of operations for producing a flat surface.

Scraping a flat surface

1.   Remove all burrs and sharp edges from the surface to be scraped by

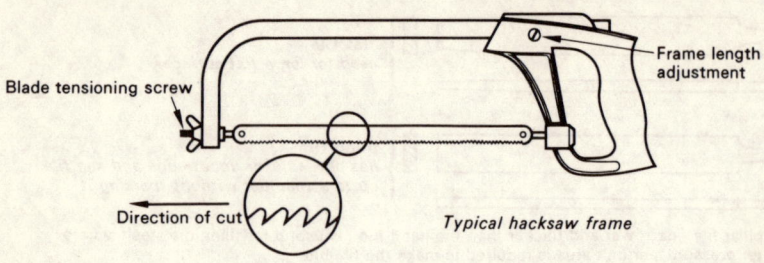

(a) **The hacksaw**

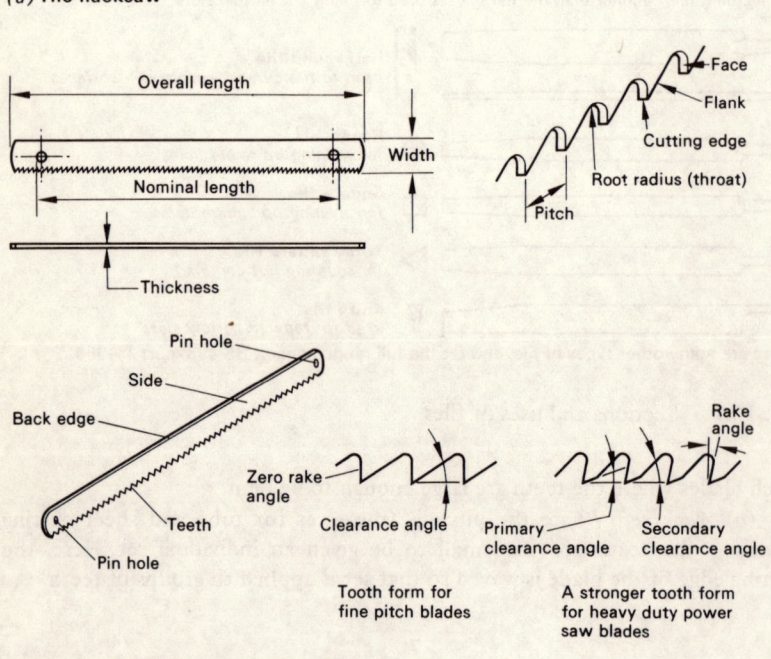

(b) **The hacksaw blade - elements**

Fig. 10.8 The hacksaw and its blade

rubbing a fine, flat oil stone all over the surface as shown in Fig. 10.11(*a*).

2. Wipe surface clean and lightly smear prussian blue on to a surface plate.

3. Place the work upside down on the surface plate and move with a rotary motion. Pressure must not be applied, the weight of the component being sufficient. The appearance of the work upon separation of the surfaces will be as shown in Fig. 10.11(*b*). The large, dark areas (in practice smeared blue) are the high spots.

4. Holding the scraper as shown in Fig. 10.11(*c*) break down the high spots into a number of smaller areas.

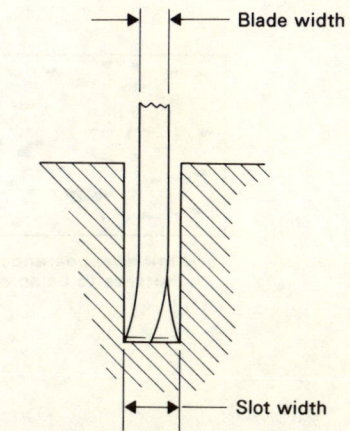

(a) The effect of set

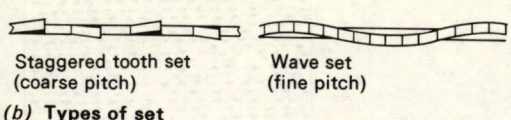

Staggered tooth set
(coarse pitch)

Wave set
(fine pitch)

(b) Types of set

Fig. 10.9 Tooth set

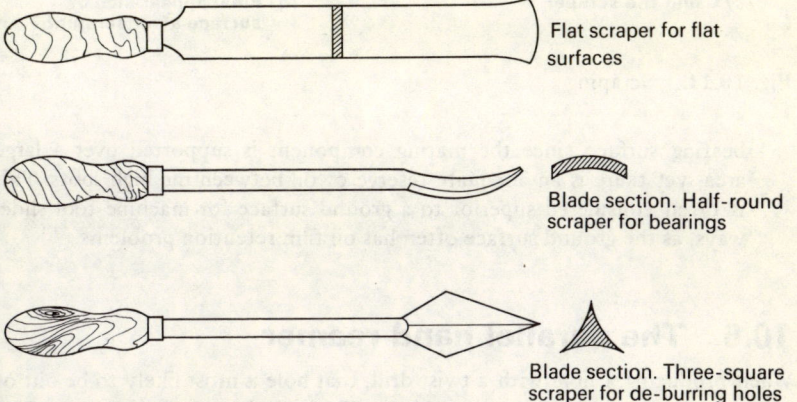

Flat scraper for flat
surfaces

Blade section. Half-round
scraper for bearings

Blade section. Three-square
scraper for de-burring holes

Fig. 10.10 Types of scraper

5. De-burr with a fine oil stone (Fig. 10.11(a)) and wipe clean. Check again on the surface plate for high spots.
6. Repeat operations 1 and 5 until the appearance of the work resembles Fig. 10.11(d). That is, there are about 8 to 10 high spots of uniform size, uniformly distributed in each 25 mm² of the surface. This is an ideal

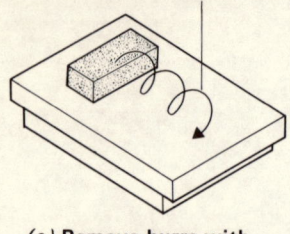

(a) **Remove burrs with fine oil stone**

Move oil stone with a rotary motion

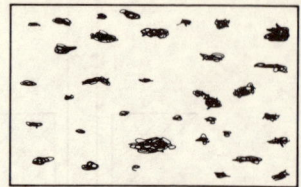

(b) **Initial appearance of surface to be scraped**

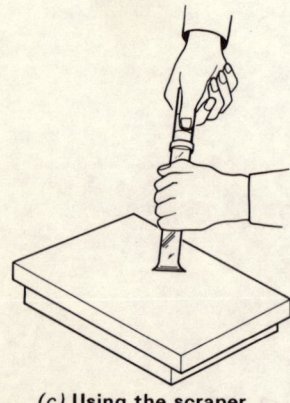

(c) **Using the scraper**

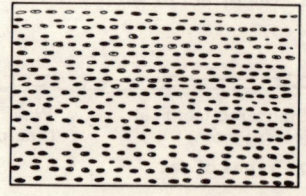

(d) **Final appearance of surface after scraping**

Fig. 10.11   Scraping

bearing surface since the mating component is supported over a large area, yet there is an adequate reserve of oil between the high spots. This is often considered superior to a ground surface for machine tool slideways, as the ground surface often has oil film retention problems.

10.6   The parallel hand reamer

When producing a hole with a twist drill, that hole is most likely to be out of round and oversize (see section 11.6). These faults can be overcome by drilling the hole undersize and correcting it for size, shape and finish by *reaming*. This can be done by hand at the bench or in the drilling machine. Machine reaming will be considered in section 11.7.

Figure 10.12 shows a component restrained by screwed fastenings and located by dowels. Such a component may be removed and replaced without loss of positional accuracy until the holes wear oversize.

Figure 10.13 shows the steps in dowelling a component.

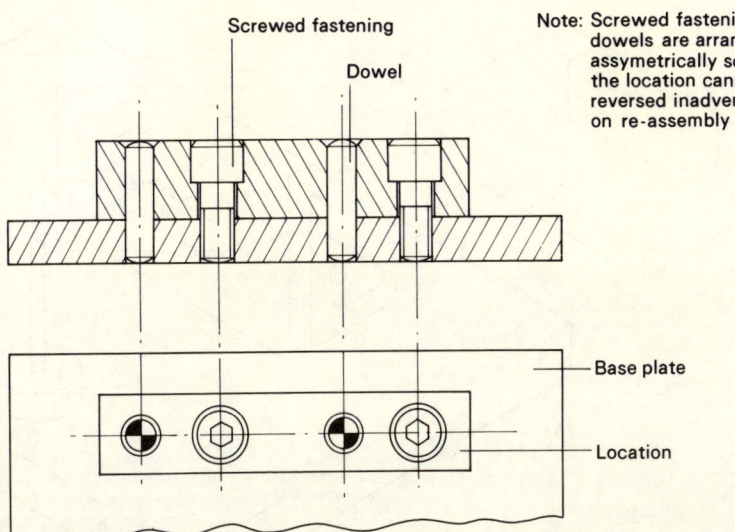

Screwed fastening

Dowel

Note: Screwed fastenings and dowels are arranged assymetrically so that the location cannot be reversed inadvertently on re-assembly

Base plate

Location

Fig. 10.12　Use of dowels

1. The component is lightly nipped into position using the screwed fastenings. These should be in clearance size holes to allow for slight, final adjustment (Fig. 10.13(*a*)).
2. The component should then be set in position. In this case it is being set parallel to a datum edge with a dial test indicator as shown in Fig. 10.13(*b*).
3. The screwed fastenings are tightened so that the component cannot move whilst the dowels are being fitted. After tightening the position should be re-checked and corrected if necessary.
4. The dowel holes are now drilled and reamed *in situ* as shown in Fig. 10.13(*c*).
5. The dowels are now driven home using a soft drift to avoid mushrooming or burring their heads as shown in Fig. 10.13(*d*). A final check is now made of the positional accuracy of the components.

Hand reamers may have straight or spiral (left-hand) flutes. Unlike the machine reamer which has a morse taper shank to fit a machine spindle, the hand reamer has a square on the end of the shank so that it can be rotated by a tap wrench. It is always rotated in a clockwise direction and is *not reversed* when being withdrawn from the hole. A lubricant should always be used when reaming. Figure 10.14 shows the essential features of the hand reamer. Comparison with Fig. 11.9 shows that a hand reamer has a taper lead as well as a bevel lead, whereas a machine reamer only has a bevel lead. The additional taper lead is required to give guidance to the hand reamer and ensure its alignment with the original drilled hole. The parallel portion of the reamer body has radial land and has, therefore, the characteristics of a 'fluted'

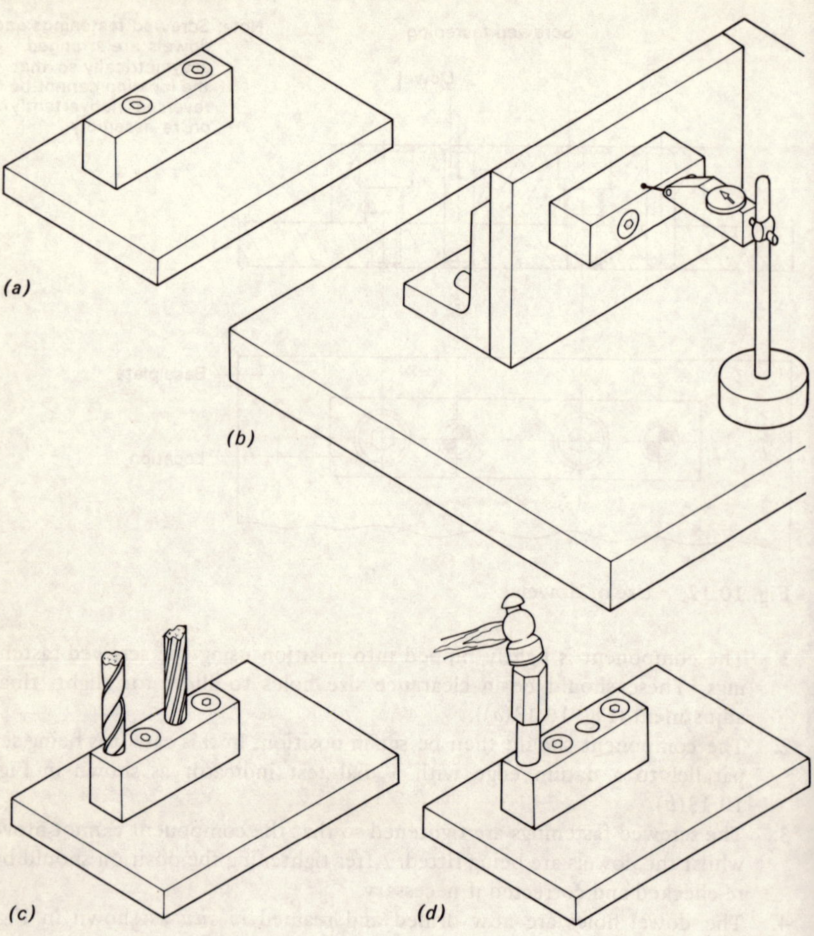

(a)

(b)

(c)

(d)

Fig. 10.13   Dowelling sequence

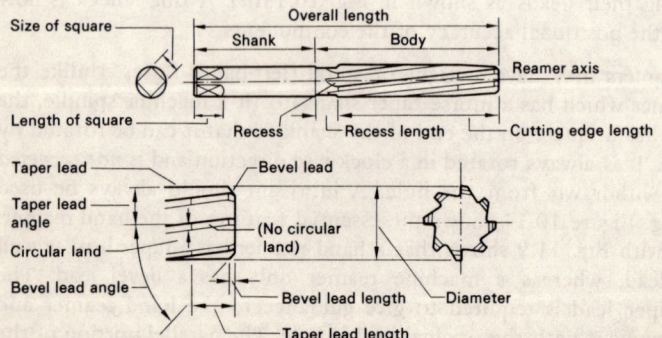

Fig. 10.14   The hand reamer

reamer, whilst the taper lead cuts peripherally and has, therefore, the characteristics of a 'rose' reamer (see section 9.11).

10.7 The taper pin reamer

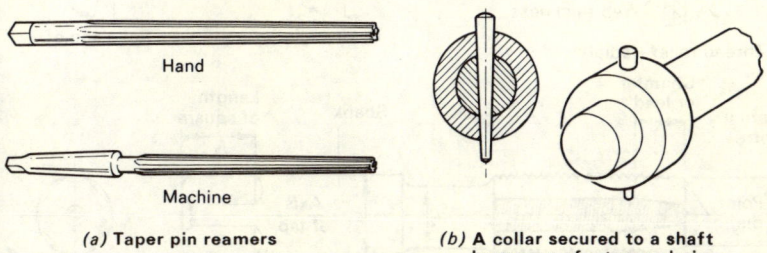

Hand

Machine

(a) **Taper pin reamers**

(b) **A collar secured to a shaft by means of a tapered pin**

Fig. 10.15 Taper pin reamers and their use

Figure 10.15(*a*) shows typical taper pin reamers. These are used for producing tapered holes in components required to be locked in place by a tapered pin as shown in Fig. 10.15(*b*). The tapered pin has obvious advantages over a parallel pin for such purposes. It locks up tight and is retained by a wedging action as it is driven home, but is immediately released when driven back in the reverse direction. Further, any wear in the hole is compensated for by the pin merely having to be driven a little deeper.

10.8 Screwed fastenings – use of taps

Screw thread taps are used to cut internal threads in engineering components. Figure 10.16 shows the essential features of the tap and the three taps that make up a set. In a cheap set, the difference between the taper, second and plug tap is only the length of the taper lead. In a good set of ground thread taps, the thread form is also reduced on the taper tap. This allows the second and plug taps to clean up the thread initially cut and produce a thread with a good surface finish. The cutting action of the tap has already been discussed in section 9.9.

The tap is rotated by means of a tap wrench. This should be selected to suit the size of the tap. Too small a wrench results in excessive force being used to turn the tap. Too large a wrench results in lack of 'feel'. In either case there is lack of proper control that will result in a broken tap. Figure 10.17 shows a selection of typical tap wrenches.

It is essential that the tap is aligned with the tapping size hole both axially and radially. Radial alignment is controlled by the taper and, if present, the pilot. Axial alignment is under the control of the fitter. If the tap is started with lack of alignment, not only will a defective thread be cut but, in a thick component, the forces on the tap will become unbalanced and the tap will break in the hole.

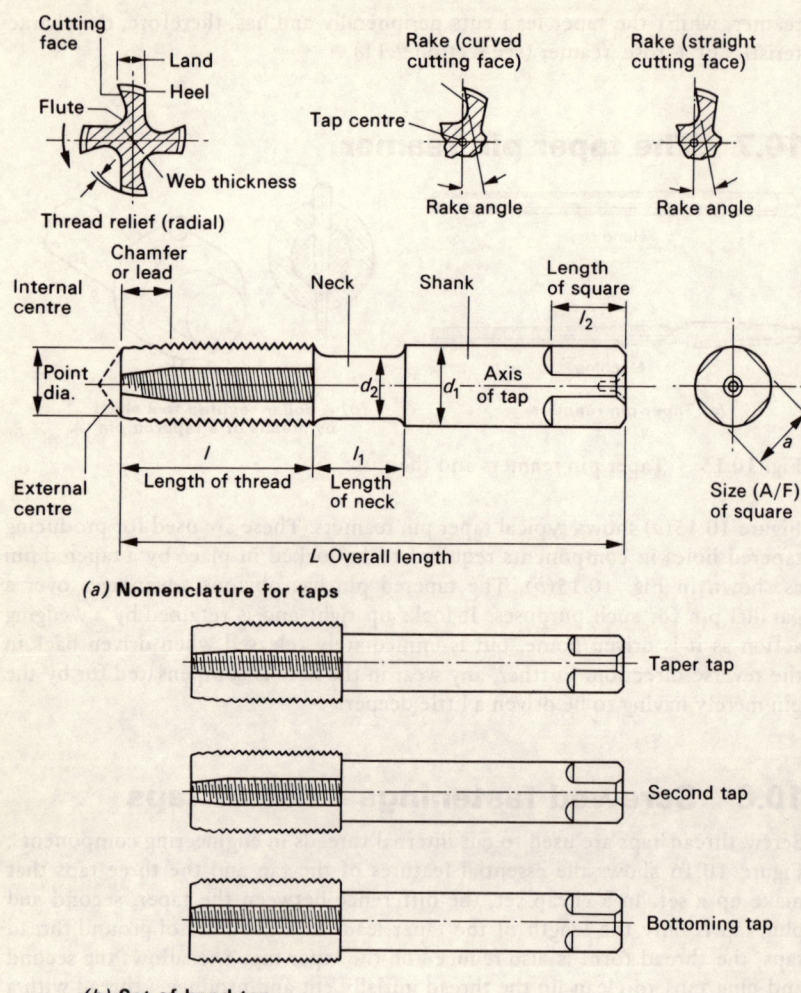

Cutting face
Land
Heel
Flute
Web thickness
Thread relief (radial)

Rake (curved cutting face)
Tap centre
Rake angle

Rake (straight cutting face)
Rake angle

Chamfer or lead
Internal centre
Neck
Shank
Length of square
l_2

Point dia.
d_2
d_1
Axis of tap

External centre
l
Length of thread
l_1
Length of neck

Size (A/F) of square
a

L Overall length

(a) Nomenclature for taps

Taper tap

Second tap

Bottoming tap

(b) Set of hand tap

Fig. 10.16   Screw thread taps

10.9   Screwed fastenings – use of dies

Screw thread dies are used to cut external threads on engineering components. The split button die is most widely used by the fitter and a typical example together with the die stock is shown in Fig. 10.18. This type of die can also be used in the tailstock die holder of the centre lathe (section 12.18). It will be seen that the die has three adjusting screws. The centre screw spreads the die and reduces the depth of cut, whilst the outer screws close the die and increase the depth of cut.

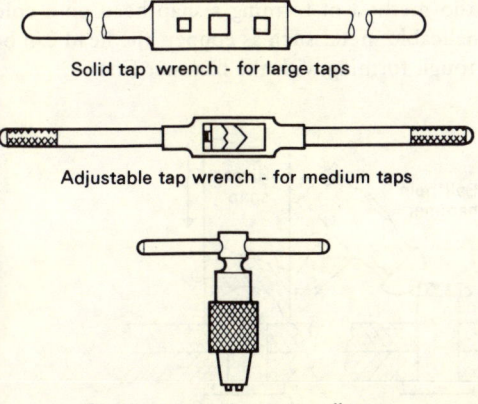

Solid tap wrench - for large taps

Adjustable tap wrench - for medium taps

Chuck wrench - for very small taps

Fig. 10.17   Tap wrenches

It is difficult to control a screw threading die and any attempt to cut a full thread in one pass of the die usually results in a 'drunken' thread with a poor finish. A first, shallow cut should be taken with the die spread open by the centre screw. This is followed by one or more finishing cuts with the outer screws gradually closed down until the thread is a correct fit in the nut.

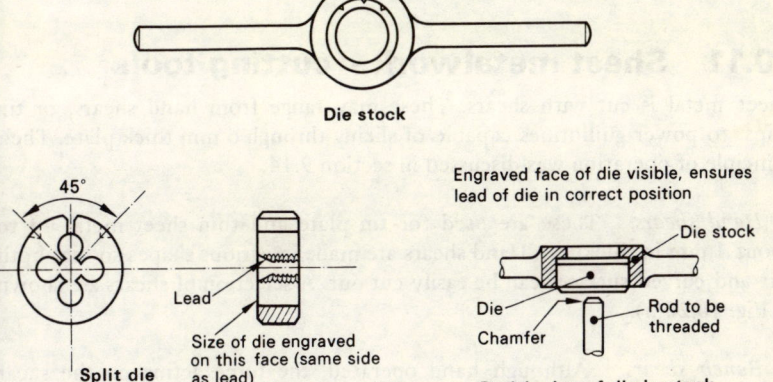

Die adjusting screws

Die stock

45°

Split die

Lead

Size of die engraved on this face (same side as lead)

Engraved face of die visible, ensures lead of die in correct position

Die stock

Die

Chamfer

Rod to be threaded

Positioning of die in stock

Fig. 10.18   Dies and die-stock

10.10   Riveted joints

The type of riveted joints used by engineers have already been discussed in section 5.4 together with the faults most likely to occur with this method of jointing. Figure 10.19 shows the method of forming a snap head on a cold rivet. With a small rivet of a malleable metal such as copper, the head can be closed with rivet snap without rough forming with the hammer first.

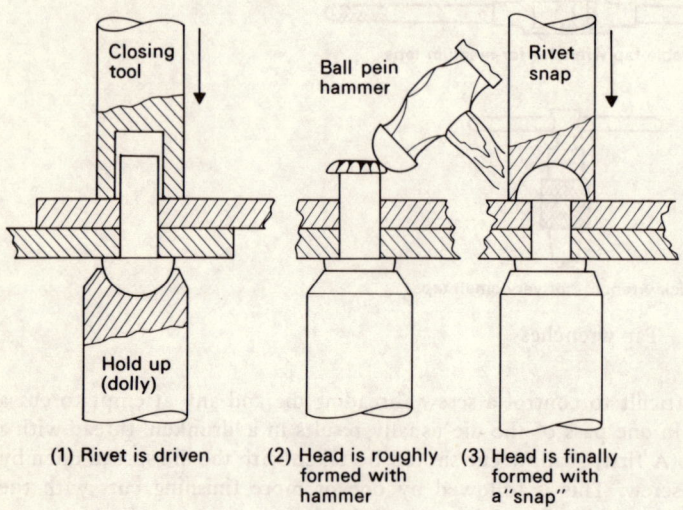

(1) Rivet is driven   (2) Head is roughly formed with hammer   (3) Head is finally formed with a "snap"

Fig. 10.19   Heading a solid rivet

10.11 Sheet metalwork – cutting tools

Sheet metal is cut with shears. These may range from hand shears, or tin snips, to power guillotines capable of slicing through 6 mm thick plate. Their principle of operation was discussed in section 9.14.

1. Hand shears: These are used for tin plate and thin sheet metal up to about 1 mm in thickness. Hand shears are made in various shapes so that both flat and curved surfaces can be easily cut out. A selection of shears are shown in Fig. 10.20(*a*).

2. Bench shears: Although hand operated, the force acting on the shear blades is magnified by a system of levers. This enables thicker sheets up to 3 mm thick to be cut. Figure 10.20(*b*) shows typical bench shears.

3. Guillotine shears: These are used for rapidly breaking down large sheets into strips and rectangular blanks. They cannot be used for cutting curved profiles. Treadle-operated shears can cut sheets up to 1·5 mm thick by 1 m in length at each stroke. Figure 10.20(*c*) shows a typical treadle-operated guillotine shear.

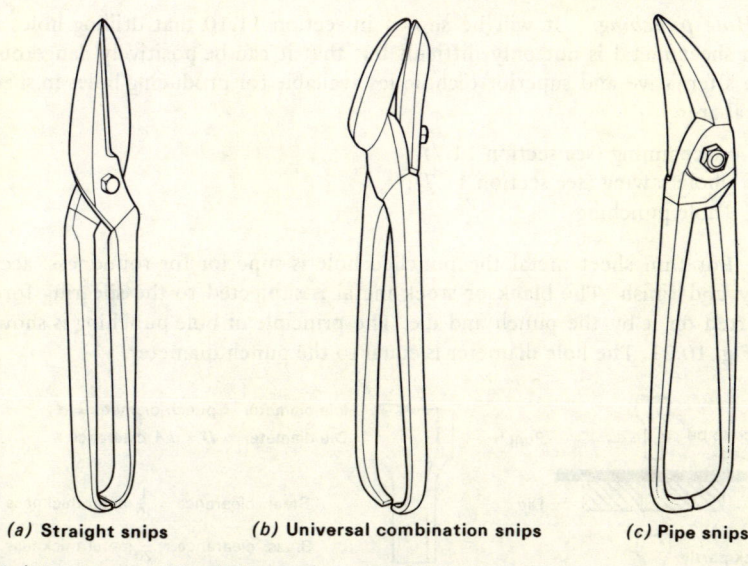

(a) Straight snips   (b) Universal combination snips   (c) Pipe snips

(a) Basic types of hand shears

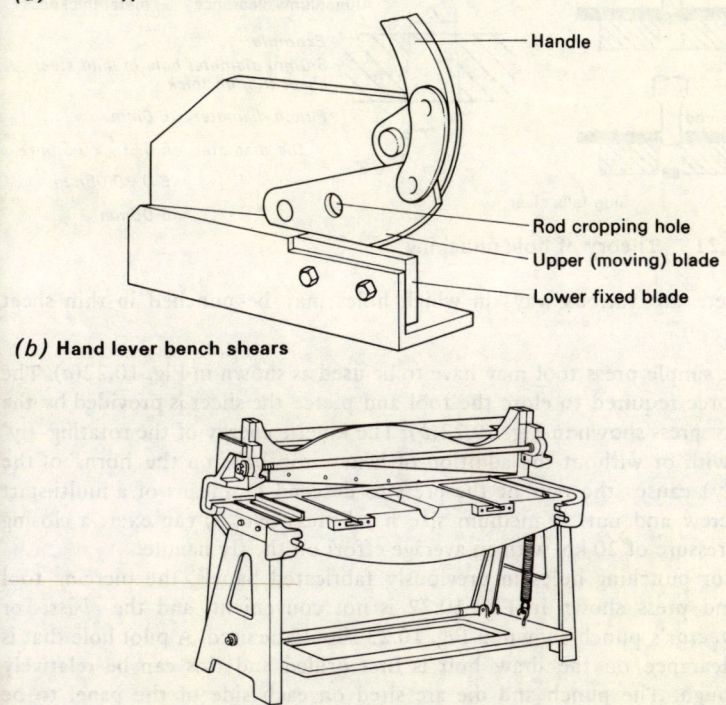

Handle

Rod cropping hole

Upper (moving) blade

Lower fixed blade

(b) Hand lever bench shears

(c) Guillotine shears

Fig. 10.20   Sheet metal cutting devices

4. Hole punching: It will be shown in section 11.10 that drilling holes in thin sheet metal is not only difficult but that it can be positively dangerous. The alternative and superior techniques available for producing holes in sheet metal are:

(a) trepanning (see section 11.7);
(b) hole sawing (see section 11.7);
(c) hole punching.

For thin sheet metal the punched hole is superior for roundness, accuracy and finish. The blank or stock metal is subjected to the shearing force exerted on it by the punch and die. The principle of hole punching is shown in Fig. 10.21. The hole diameter is equal to the punch diameter.

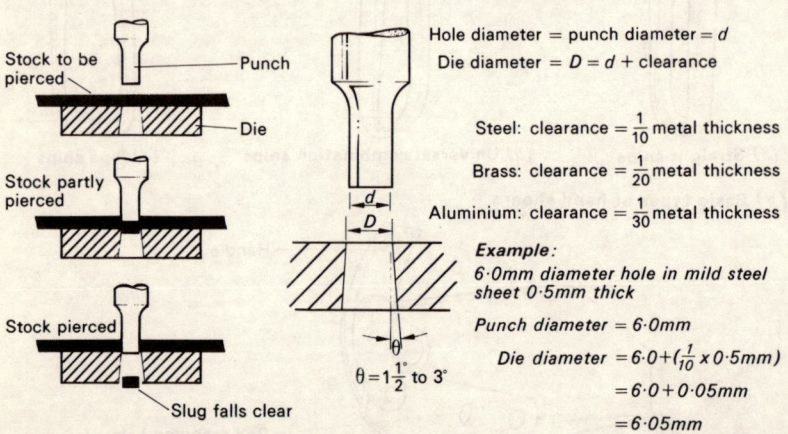

Fig. 10.21   Theory of hole punching

There are various ways in which holes may be punched in thin sheet metal:

(a) A simple press tool may have to be used as shown in Fig. 10.22(a). The force required to close the tool and pierce the sheet is provided by the fly press shown in Fig. 10.22(b). The kinetic energy of the rotating 'fly' (with or without the addition of heavy iron balls on the 'horns' of the fly) causes the ram of the press to descend by means of a multi-start screw and nut. A medium size hand press (No. 3) can exert a closing pressure of 20 kN with an average effort on the fly handle.

(b) For punching holes in previously fabricated panels, the piercing tool and press shown in Fig. 10.22 is not convenient, and the *chassis* or *erector's* punch shown in Fig. 10.23 should be used. A pilot hole that is clearance on the draw bolt is first drilled and this can be relatively rough. The punch and die are sited on each side of the panel to be pierced; the draw bolt is passed through the die and screwed into the punch. As the draw bolt is tightened up it closes the punch and die and provides the shearing force to pierce the hole.

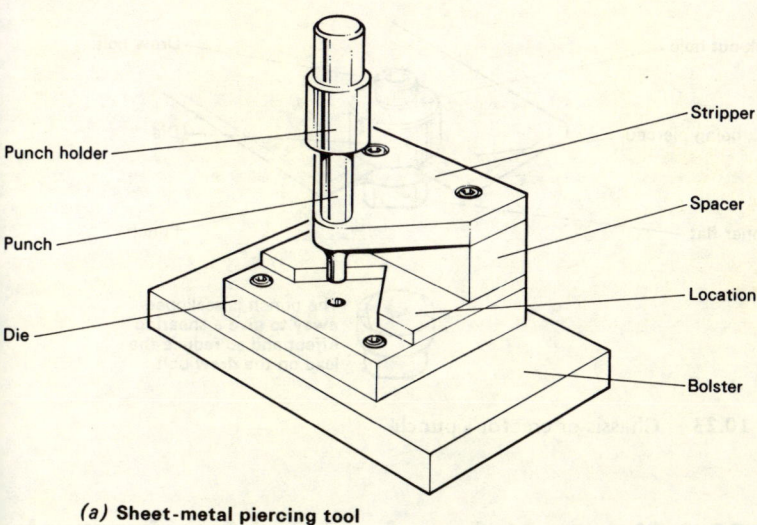

(a) **Sheet-metal piercing tool**

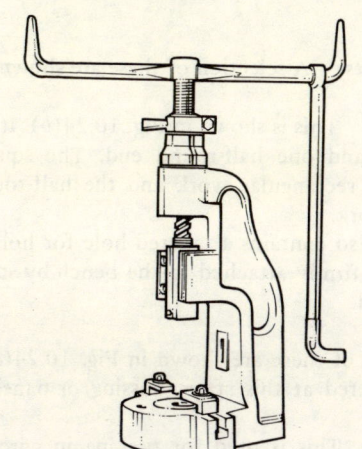

(b) **Fly press (The piercing tool could be fitted into this type of press)**

Fig. 10.22   Piercing tool and press

● *Safety:*   Bench and guillotine shears are designed to slice through sheet steel with ease. They will also slice through flesh and bone with greater ease. Never use shearing machines or hole punching tools unless:

a)   you have been fully instructed in their use;

b)   you have been given permission to use them;

c)   any appropriate guards are in place.

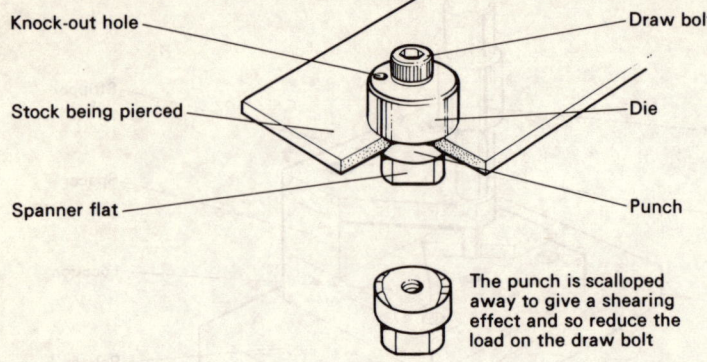

Knock-out hole

Stock being pierced

Spanner flat

Draw bolt

Die

Punch

The punch is scalloped away to give a shearing effect and so reduce the load on the draw bolt

Fig. 10.23 Chassis or erector's punch

10.12   Sheetmetal work – hand forming tools

The full range of tools will not be described; only those necessary for the production of the operations described in the next section of this chapter will be considered.

1. Hammers and mallets: A selection of these are shown in Fig. 10.24(*a*).

2. Tinman's mandrell: This is shown in Fig. 10.24(*b*). It is made of steel and has one square end and one half-round end. The square end is used for forming and riveting rectangular work and the half-round end is used for circular and curved work.

The square end also contains a tapered hole for holding *stakes* and *bick irons*. The mandrel is firmly attached to the bench by strap clamps so that it can be readily reversed.

3. Stakes: A variety of these are shown in Fig. 10.24(*c*). Since only folded work is being considered at this stage, bossing or panel beating heads have been omitted.

   (*a*) *Hatchet stake.* This is used for tucking in wired joints, sharpening bends and dressing edges.

   (*b*) *Half-moon stake.* This is used for the same operations as the hatchet stake when they occur on curved surfaces.

   (*c*) *Bick iron.* This is used to support the hammer or mallet blow while performing such operations as shaping hollow and tapered work, knocking up, wiring and dressing edges.

   (*d*) *Funnel and side stake.* This is used for supporting conical and cylindrically shaped components.

   (*e*) *Creasing iron.* This is used for swaging small work, and hand wiring flat sheets. It is also used for forming hinges.

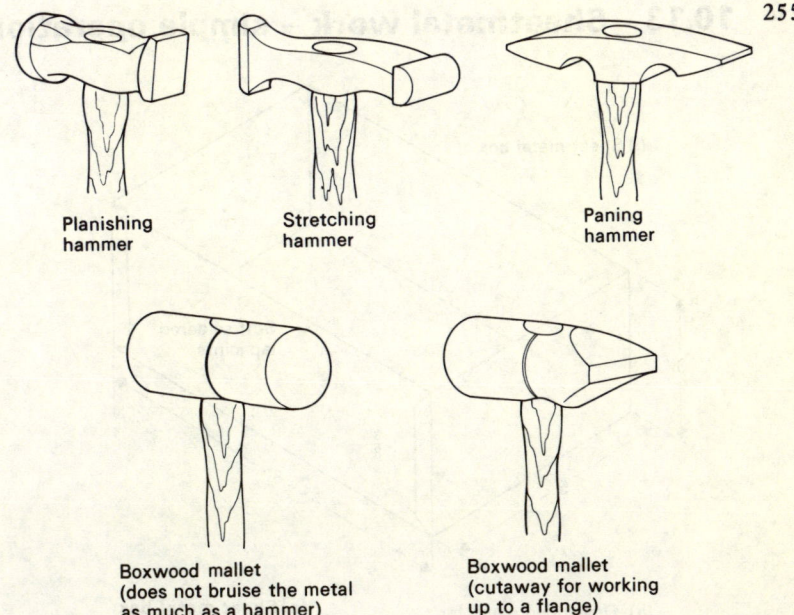

Planishing hammer

Stretching hammer

Paning hammer

Boxwood mallet
(does not bruise the metal
as much as a hammer)

Boxwood mallet
(cutaway for working
up to a flange)

(a) Hammers and mallets

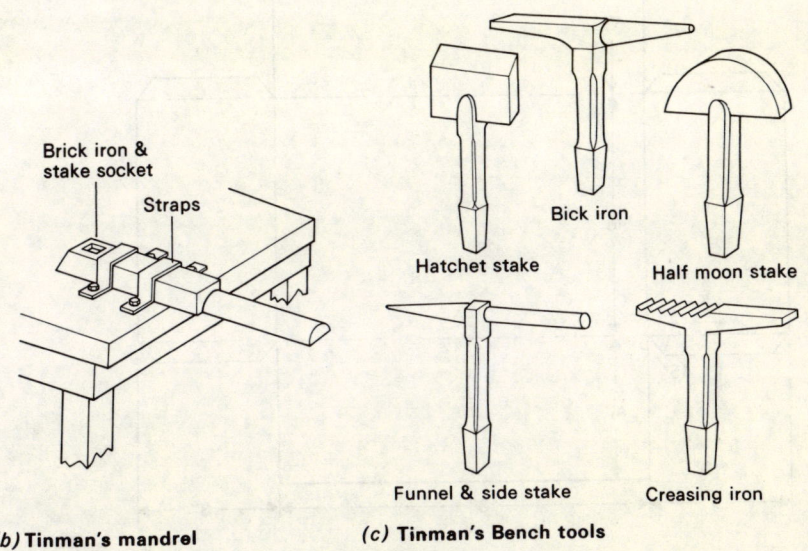

Brick iron &
stake socket

Straps

Hatchet stake

Bick iron

Half moon stake

Funnel & side stake

Creasing iron

(b) Tinman's mandrel

(c) Tinman's Bench tools

Fig. 10.24   Sheet metal bending equipment – bench

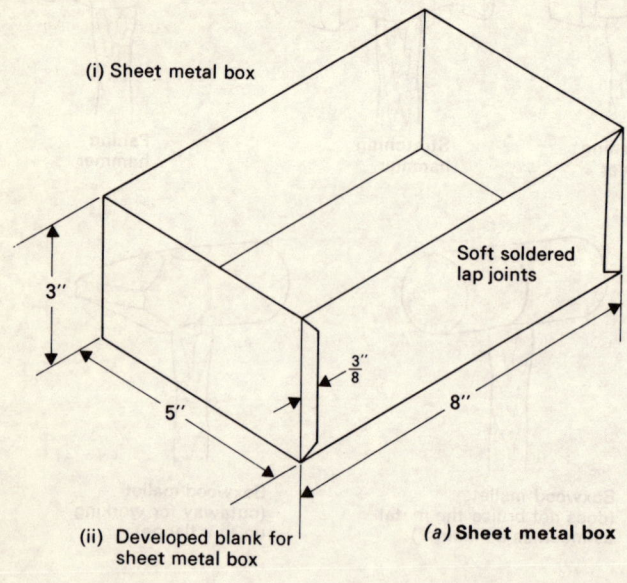

(i) Sheet metal box

Soft soldered
lap joints

3″

3″
8

5″

8″

(ii) Developed blank for
sheet metal box

(a) Sheet metal box

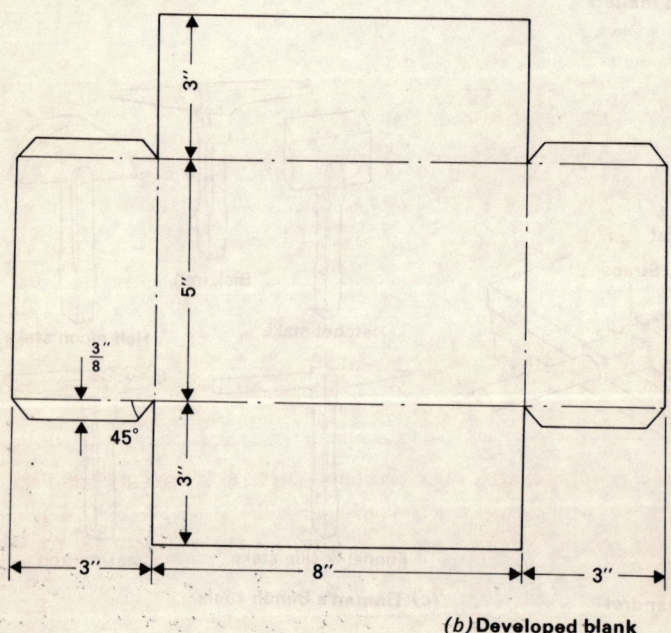

3″

5″

3″
8

45°

3″

3″

8″

3″

(b) Developed blank

Fig. 10.25 Layout for forming a box

1. Development: Consider the template box shown in Fig. 10.25(*a*). Before this box can be made, its shape must be cut from a flat sheet of metal as shown in Fig. 10.25(*b*). This flat shape is called the **developed blank**. If a number of boxes were to be produced, it would be wasteful to draw out the blank from first principles each time. In this case a **pattern** or **template** is first produced. This may be in stiff sheet metal, hardboard or thin plywood. It is laid on the sheet from which the blank is to be cut and a scriber is run round the edge of the pattern.

2. Folded edge (beading): The tin box shown in Fig. 10.25(*a*) would not be very pleasant to use. It would lack rigidity, and the edges would be sharp and dangerous. One way of stiffening up the edges and rendering them safe is shown in Fig. 10.26(*a*).

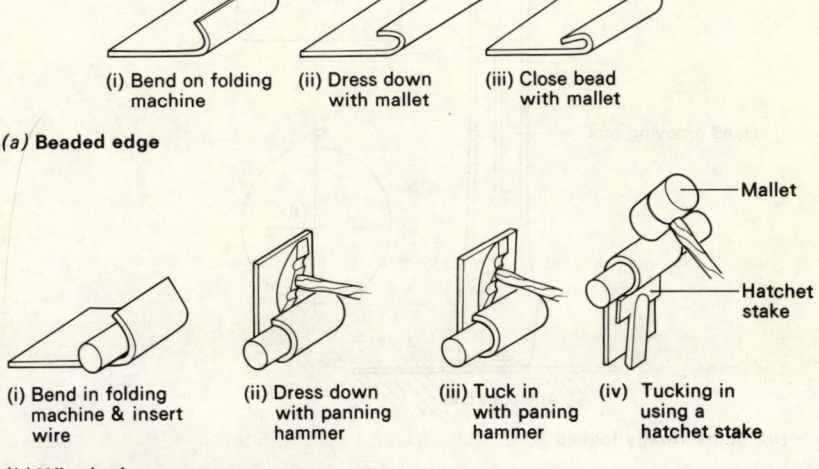

(i) Bend on folding machine   (ii) Dress down with mallet   (iii) Close bead with mallet

(a) **Beaded edge**

Mallet

Hatchet stake

(i) Bend in folding machine & insert wire   (ii) Dress down with panning hammer   (iii) Tuck in with paning hammer   (iv) Tucking in using a hatchet stake

(b) **Wired edge**

Fig. 10.26   Beading and wiring edges

3. Wired edge: This is a superior method of finishing an edge, but requires more skill to produce than the folded edge previously described. The stages in wiring an edge are shown in Fig. 10.26(*b*).

4. Folded joints (self-secured): The various types of self-secured joints together with their advantages, limitations and applications have already been discussed in section 5.11. The technique of forming a **grooved seam**, as shown in Fig. 10.27 will now be considered.

(a)   The edges are folded to form 'locks'. There are two methods of doing this. The joint edges may be folded over and slightly closed (the gap being a little wider than the thickness of metal that it has to accommo-

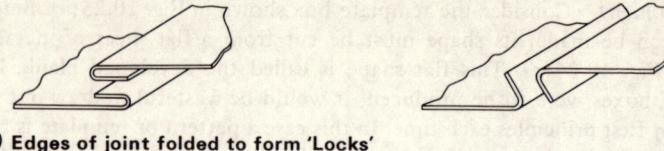

(a) Edges of joint folded to form 'Locks'

(b) Folded edges interlocked (G represents the width of Groove)

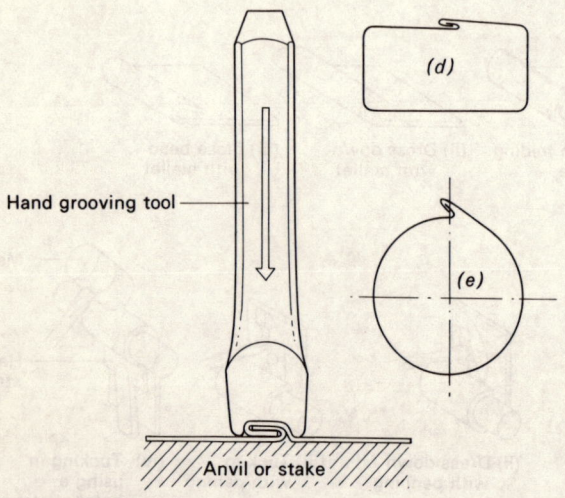

Hand grooving tool

Anvil or stake

(c) Joint finally locked

Fig. 10.27   The grooved seam

date). Edges prepared in this manner present no interlocking problems when making seams on 'flat' surfaces, as shown at Fig. 10.27(*d*). Alternatively, the joint edges are folded over to an approximate angle of 60°. Edges prepared in this manner are more convenient when interlocking seams on cylindrical articles of small diameter as shown at Fig. 10.27(*e*).

(*b*) The two 'locks' (one up and one down) are hooked together, as shown in Fig. 10.27(*b*).

(*c*) The joint is finally secured (locked together) using a grooving tool as shown in Fig. 10.27(*c*). A different width of grooving tool is required for each width of joint being secured.

The **paned down** joint, used for joining two cylindrical components, is shown in Fig. 10.28(a). The **knocked up** joint, used for sealing the ends of cylindrical components, is shown in Fig. 10.28(b).

The cylindrical joints shown in Fig. 10.28(a) and (b) require a knowledge of flanging as a preliminary operation. A method of flanging a component is shown in Fig. 10.28(c).

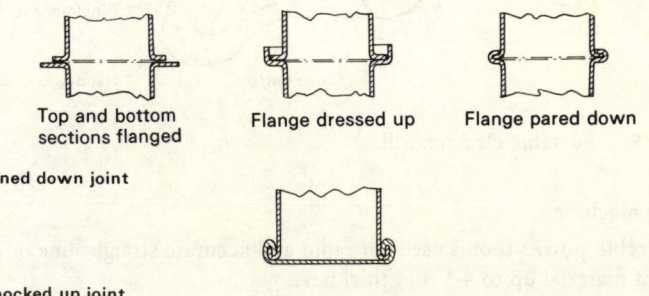

Top and bottom   Flange dressed up   Flange pared down
sections flanged

(a) **Paned down joint**

(b) **Knocked up joint**

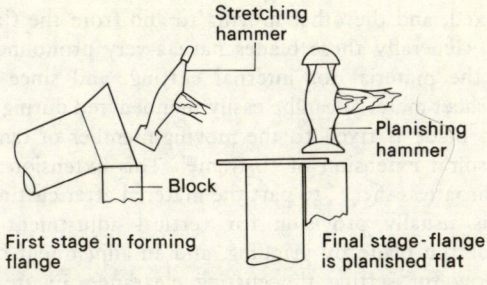

Stretching
hammer

Planishing
hammer

Block

First stage in forming   Final stage - flange
flange   is planished flat

(c) **Flanging**

Fig. 10.28   Paned down and knocked up joints

10.14   Powered hand tools

The fitter has many portable power tools at his disposal. For example:

(a)   rotary drilling machines;
(b)   shearing machines;
(c)   nibbling machines;
(d)   grinding machines;
(e)   riveting machines.

Electric rotary drilling machine

An example of one of these machines is shown in Fig. 10.29. It can be used for drilling metal with a suitable steel twist drill. Some machines have two- or three-speed gearboxes, but for industrial use it is usual to use the more robust single-speed machine and match the size and speed of the machine to the diameter of drill being used.

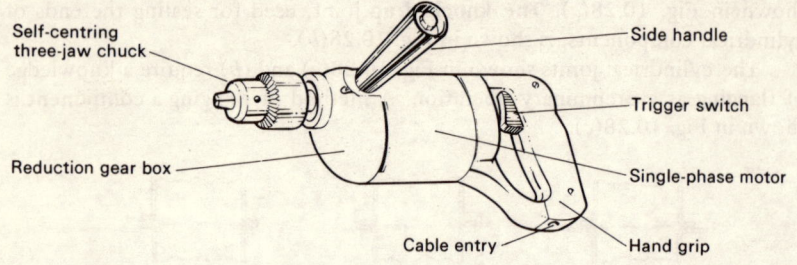

Fig. 10.29   Portable electric drill

Shearing machine

This portable power tool is used for rapid and accurate straight-line or curved cutting of material up to 4·5 mm thickness.

The machine is fitted with a pair of very narrow flat blades, one of which is usually fixed, and the other moving to and from the fixed blade at fairly high speeds. Generally these blades have a very pronounced **rake** to permit piercing of the material for internal cutting, and since the blades are so narrow, the sheet material can be easily manoeuvred during cutting.

The top blade is fixed to the moving member or ram, and the bottom blade on a spiral extension of 'U-frame'. This extension is shaped like the body of a 'throatless shear', to part the material after cutting.

There is usually provision for vertical adjustment to allow for re-sharpening of the blade by grinding, and an adjustment behind the bottom blade to allow for setting the cutting clearance. Figure 10.30(*a*) shows a typical portable, power hand shearing machine, whilst Fig. 10.30(*b*) shows details of the cutting blades.

Nibbling machine

The portable nibbling machine does not operate on the same principle as the shear. A punch and die is employed instead of shearing blades, and the nibbling principle is a special application of punching.

These machines will effect certain operations that cannot be accomplished on other shearing machines. For example, they may be used to cut apertures which could only otherwise be produced by means of specially designed punches and dies set up in a powerful press.

Like the shearing machine the top cutting tool (a punch) reciprocates at fast short strokes. Punch type nibblers are available in various sizes and the punch reciprocates at a rate of 350 to 1400 strokes per minute over a die, nibbling out the material by the simple principle of overlapping punching, and only slight finishing is necessary to produce a smooth clean edge.

One main advantage of nibbling over shearing is that there is less distortion of the work.

Figure 10.31(*a*) shows the punch type nibbling machine, whilst Fig. 10.31(*b*) shows its principle of operation.

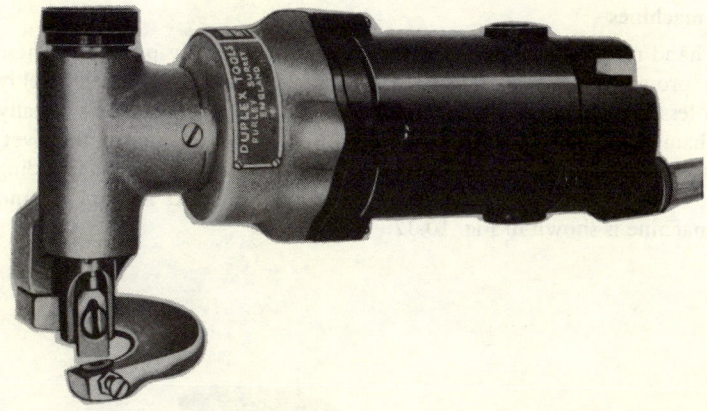

(a) Shearing machine

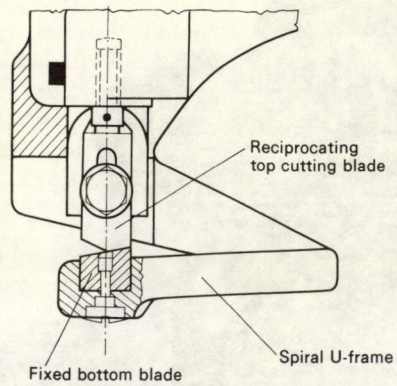

Reciprocating
top cutting blade

The line diagram opposite shows a cross-section of the cutting head of a typical portable vibrating shears driven electrically or pneumatically. The spiral U-frame is designed to assist in parting the metal after it has been sheared

Basic details of cutting blades are given in the diagram below

Fixed bottom blade

Spiral U-frame

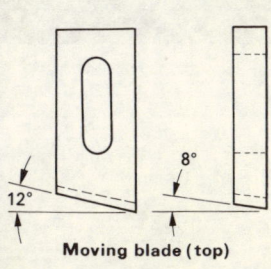

12°   8°

Moving blade (top)

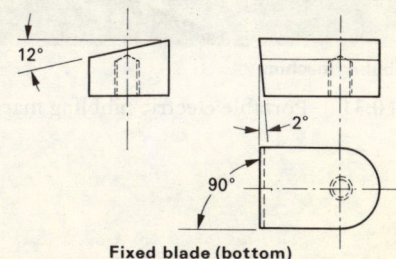

12°

2°

90°

Fixed blade (bottom)

(b) Details of blades

Fig. 10.30   Portable electric shearing machine

(b)   do not use excessive measuring pressure (two 'clicks' of the ratchet are sufficient);

(c)   do not leave the anvil faces in contact when not in use;

(d)   stop the machine before measuring. Attempting to take measurements with the machine working can lead to serious accidents and irreparable damage to the micrometer. This rule applies to all measuring instruments.

In the box with the micrometer will be found a small double-ended spanner. This is for making two basic adjustments.

1.   Looseness in the screw. This is taken up by a slight turn of the screw adjustment nut — item 6, Fig. 7.6.

2.   Zero error. Periodically the anvil faces should be cleaned and closed using the ratchet to give the correct measuring pressure. If the zero line of the thimble does not coincide with the datum line on the barrel, turn the barrel — item 4, Fig. 7.6 — with the 'C' spanner until the zero line coincides with the datum.

7.7   The internal micrometer

The internal micrometer is shown in Fig. 7.8(a). Its range is 50 mm to 210 mm and for any one extension rod its range is 20 mm. It suffers from severe limitations. First, it cannot be used to measure small diameter holes, and second, it cannot be adjusted readily once it is in the hole and this affects the 'feel' that can be obtained.

An alternative and more satisfactory instrument is the 'cylinder gauge' which employs a micrometer-controlled wedge to expand three equispaced anvils until they touch the walls of the bore. This instrument is shown diagrammatically in Fig. 7.8(b).

7.8   The depth micrometer

The depth micrometer is shown in Fig. 7.9. It will be seen that it consists of a micrometer-measuring head together with a number of extension rods. The desired rod can easily be inserted by removing the thimble cap. When the cap is replaced, the rod is held firmly against a positive datum face. The rods are marked with their respective capacity and are square to the base in any position. The measuring faces of the base and rods are hardened. Note that the scales of a depth micrometer give *reverse reading* to the micrometer caliper and the inside micrometer.

7.9   The vernier principle

In Fig. 7.10(a) it will be seen that the datum mark has been displaced from

Portable hand riveting machines are normally pneumatically powered. These machines provide light, rapid blows that form the head on the rivet quickly and with less distortion to the surrounding metal than when using a manually wielded hammer. Further, a lighter hold-up is required to support the rivet. Different types and size of snap can be fitted to the hand riveter depending upon the type of head that is to be formed on the rivet. A typical hand riveting machine is shown in Fig. 10.32.

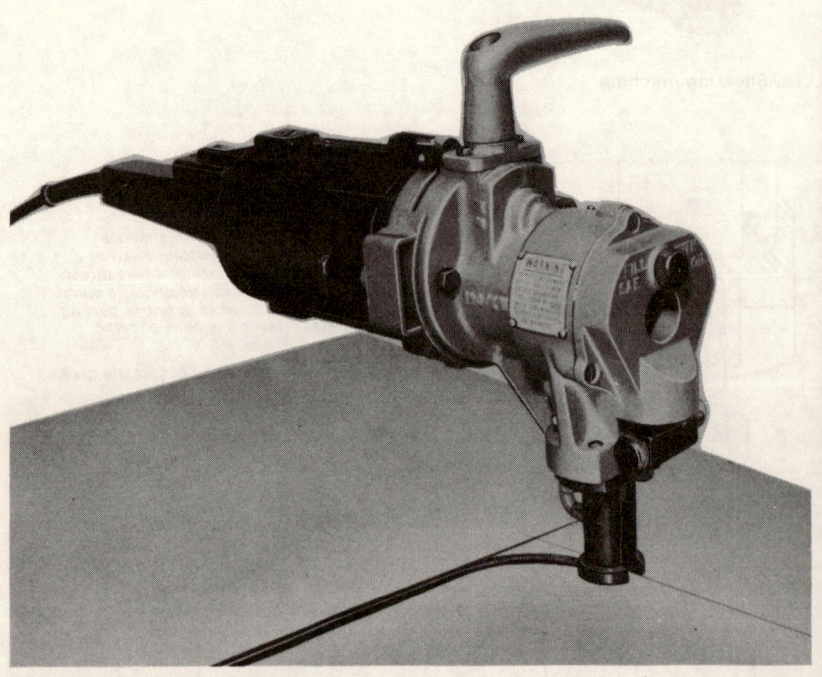

(a) **Nibbling machine**

Fig. 10.31   Portable electric nibbling machine

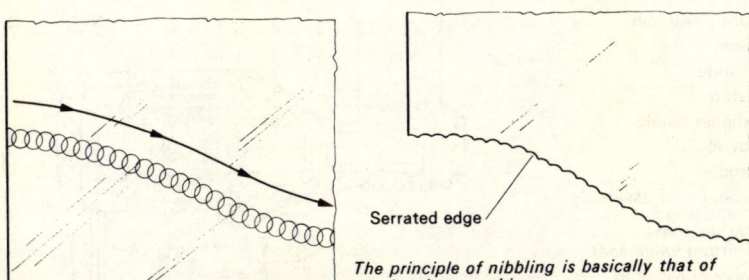

Serrated edge

The principle of nibbling is basically that of overlapping punching

The width of the cut produced by nibbling machines is determined by the diameter of the punch in relationship with the thickness of the material to be cut.   For example:

Capacity of machine —2mm Width of cut —8mm Approximate cutting speed -1·8mm/min
Capacity of machine —3·2mm Width of cut 9·5mm Approximate cutting speed -1·5mm/min

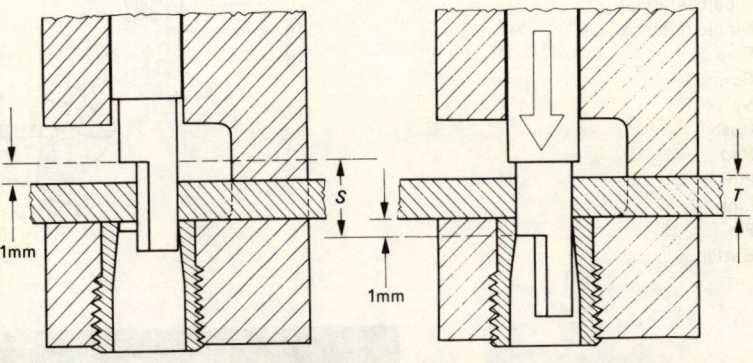

Setting of nibbling punch S = Length of stroke T = Metal thickness

The stroke is adjusted to give movement of approximately 1mm above the material and 1mm through the material, as shown in the diagram above

(b) **Principle of operation**

Fig. 10.31 (*continued*)

1 Spring retainer
2 Bush
3 Cylinder
4 Piston
5 Exhaust shield
6 Dowel
7 Handle
8 Valve chest assembly
  comprising:
    Valve chest
    Bottom valve seat
9 Valve chest plug
10 Valve
11 Grubscrew
12 Body
13 Valve
14 'O' ring
15 Nipple (standard)
16 Nipple (alternative)
17 Plug
18 Spring
19 Valve
20 Bush
21 Rod
22 Pin
23 Link
24 Pin
25 Button
26 Pin

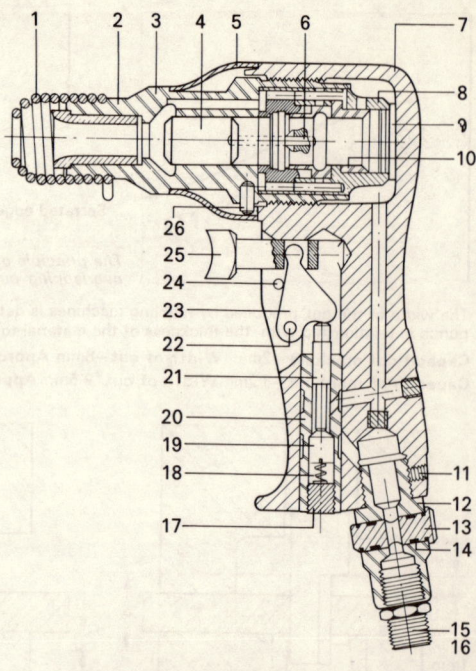

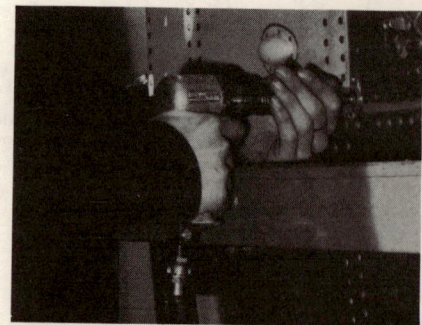

Fig. 10.32 Portable pneumatic riveting machine

Problems

Section A

1. The best way to cut clean, large diameter holes in thin sheet metal is to use a: (*a*) twist drill;   (*b*) punch and die;   (*c*) trepanning tool; (*d*) hole saw.

2. A parallel hand reamer has a: (*a*) bevel lead only;   (*b*) taper lead only;   (*c*) bevel lead and a taper lead;   (*d*) no lead.

3. The most suitable cold chisel for cutting oil grooves in a flat surface is a: (*a*) half round chisel;   (*b*) diamond point chisel;   (*c*) flat chisel; (*d*) cross cut chisel.

4. When cutting sheet metal the teeth of a hacksaw blade should have a: (*a*) fine pitch and staggered set;   (*b*) coarse pitch and staggered set; (*c*) fine pitch and wave set;   (*d*) coarse pitch and wave set.

5. To obtain a good fit between mating surfaces they should be finished by: (*a*) scraping;   (*b*) filing;   (*c*) chipping;   (*d*) sawing.

Section B

6. Describe in detail, with the aid of sketches, how a surface is scraped flat using a surface plate as a reference.

7. With the aid of clear sketches explain how the following sheet metal operations are carried out: (*a*) forming a beaded edge;   (*b*) forming a wired edge;   (*c*) making a grooved joint;   (*d*) making a knocked up joint.

8. *Shearing* machines and *nibbling* machines are both portable, powered hand tools used for cutting sheet metal: (*a*) explain with the aid of sketches the principle by which each machine operates;   (*b*) discuss the advantages and limitations of each machine.

9. Describe how the following bench fitting processes are carried out. Sketches may be used to clarify your answers: (*a*) use of taps to thread a blind hole in mild steel;   (*b*) use of a button die to thread the end of a mild steel rod;   (*c*) use of a taper pin reamer and tapered pin to fasten a collar to a shaft;   (*d*) use of a vee block to hold a circular component in a vice;   (*e*) use of a flat chisel to cut sheet metal in a vice.

10. (*a*) Sketch the plan view of a flat file and indicate its main features and length.
(*b*) By means of sketches, show the difference between single cut and double cut files.
(*c*) Sketch the following files in profile and section and explain where they would be used: (i) flat file;   (ii) hand file;   (iii) three square file;   (iv) square file;   (v) half round file.

Chapter 11

Drills and drilling machines

11.1 The twist drill

The drill does not produce a precision hole. Its sole purpose is to remove the maximum volume of metal in a minimum period of time. The hole drilled is never smaller than the diameter of the drill, but it is often larger. Dimensional inaccuracy of the hole is brought about by the drill flexing as a result of incorrect point grinding. The hole is often out of round especially when opening up an existing hole with two-flute drill. The finish is often rough and the sides of the hole scored. Thus the drill should only be considered as a roughing out tool and, if a hole of accurate size, roundness and good finish is required, the hole should be finish machined by means of a reamer or by single point boring.

The modern twist drill is made from a cylindrical blank by machining two helical grooves into it to form the flutes. The flutes run the full length of the body and have several functions:

(a)   they provide the rake angle;
(b)   they form the cutting edges;
(c)   they provide a passage for the coolant;
(d)   they facilitate swarf removal.

The flutes are not parallel to the axis of the drill but are slightly tapered, becoming shallower towards the shank of the drill as shown in Fig. 11.1(a).

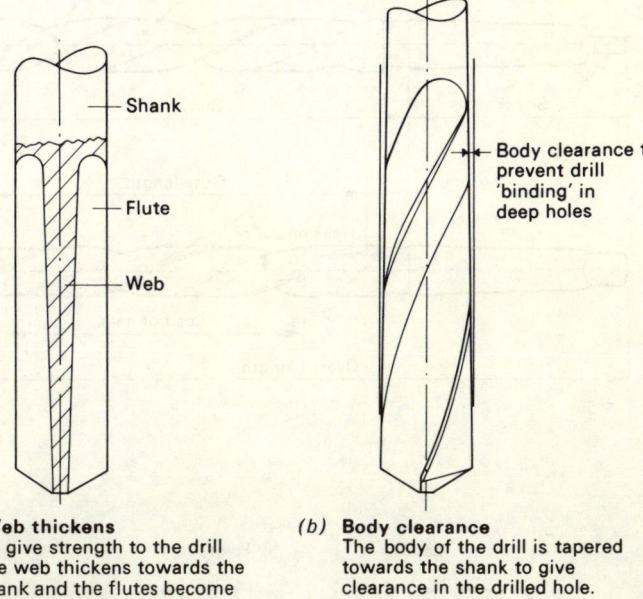

(a) **Web thickens**
To give strength to the drill the web thickens towards the shank and the flutes become shallower. Point thinning becomes necessary as the drill is ground back.

(b) **Body clearance**
The body of the drill is tapered towards the shank to give clearance in the drilled hole.

Fig. 11.1   Taper in the twist drill body

This allows the web to be thicker at the shank than at the point of the drill and provides a compromise between strength and cutting efficiency. A thick web would give maximum strength, but a thin web is required at the point to give an efficient 'chisel edge' for drilling from the solid. Thus a drill that has worn down in length from repeated re-grinding requires 'point thinning' to compensate for the thickening of the web.

The lands are also ground with a slight taper so that the overall diameter of the drill is less at the shank than at the point as shown in Fig. 11.1(*b*). This prevents the drill binding in the hole with a consequent increase in drill life and efficiency. The effects shown in Fig. 11.1 have been exaggerated for clarity.

Figure 11.2 shows a typical twist drill and names its more important features. Although a taper shank drill is shown, the same nomenclature applies to parallel shank drills.

11.2   Twist drill cutting angles

Like any other cutting tool the twist drill must be provided with the correct tool angles. Figure 9.15 showed how these compare with the corresponding angles for a single point lathe tool.

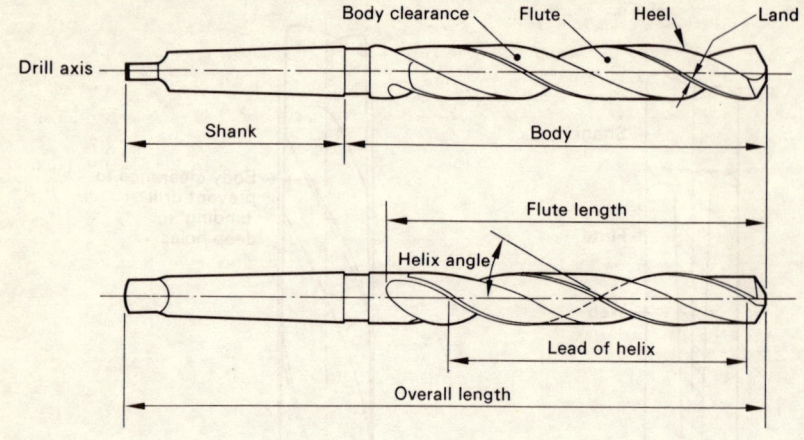

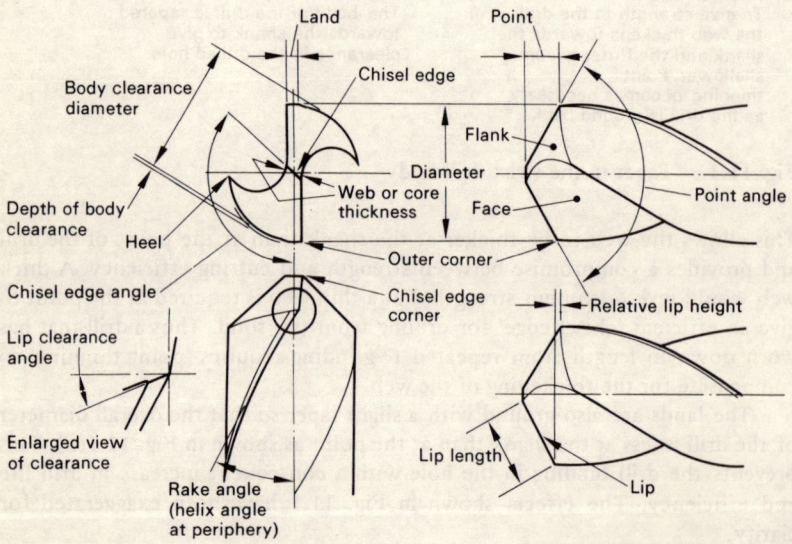

BS 328 : Part 1 : 1959

Fig. 11.2 Twist drill nomenclature

The *clearance angle* of a twist drill can be adjusted during grinding of the point. Insufficient clearance leads to rubbing and over-heating of the cutting edges. This leads to softening and early drill failure. Excessive clearance reduces the strength of the cutting edge leading to chipping and early drill failure. It will also cause the drill to dig in and chatter.

The *rake angle* of a twist drill is not so easily altered as it is formed by

be slightly modified during re-grinding.

Some control of the rake angle is possible by choosing drills with the correct helix angle for the material being cut. Figure 11.3 shows the various types available.

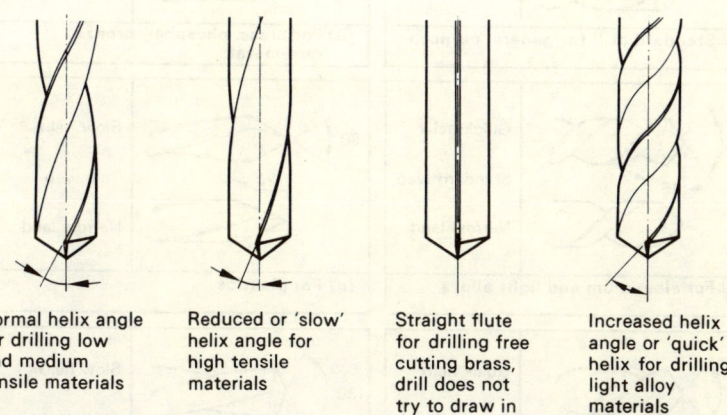

| Normal helix angle for drilling low and medium tensile materials | Reduced or 'slow' helix angle for high tensile materials | Straight flute for drilling free cutting brass, drill does not try to draw in | Increased helix angle or 'quick' helix for drilling light alloy materials |

Fig. 11.3   Helix angles

As well as varying the clearance and rake angles of the drill, its performance can also be improved by modifying the point angle from the standard 118° for certain materials. Where a large number of drills of the same size are being purchased the *web* and *land* can also be varied with advantage. Figure 11.4 shows how the point angle, web and land can be varied from different materials.

11.3   Twist drill cutting speeds and feeds

For a drill to give satisfactory performance it must operate at the correct cutting speed and feed. The conditions upon which the cutting speeds and feed rates given in this chapter are based, assume:

(*a*)   the work is rigidly clamped;
(*b*)   the machine is in good condition;
(*c*)   a coolant is used if required;
(*d*)   the drill is correctly selected and ground for the material being cut.

The rates of feed and the cutting speeds for twist drills are lower than for most other machining operations. This is because:

(*a*)   the twist drill is relatively weak compared with other cutting tools, the cutting forces being only resisted by the slender web. Further, the point of cutting is remote from the point of support (the shank) resulting in a tendency to flex and chatter;

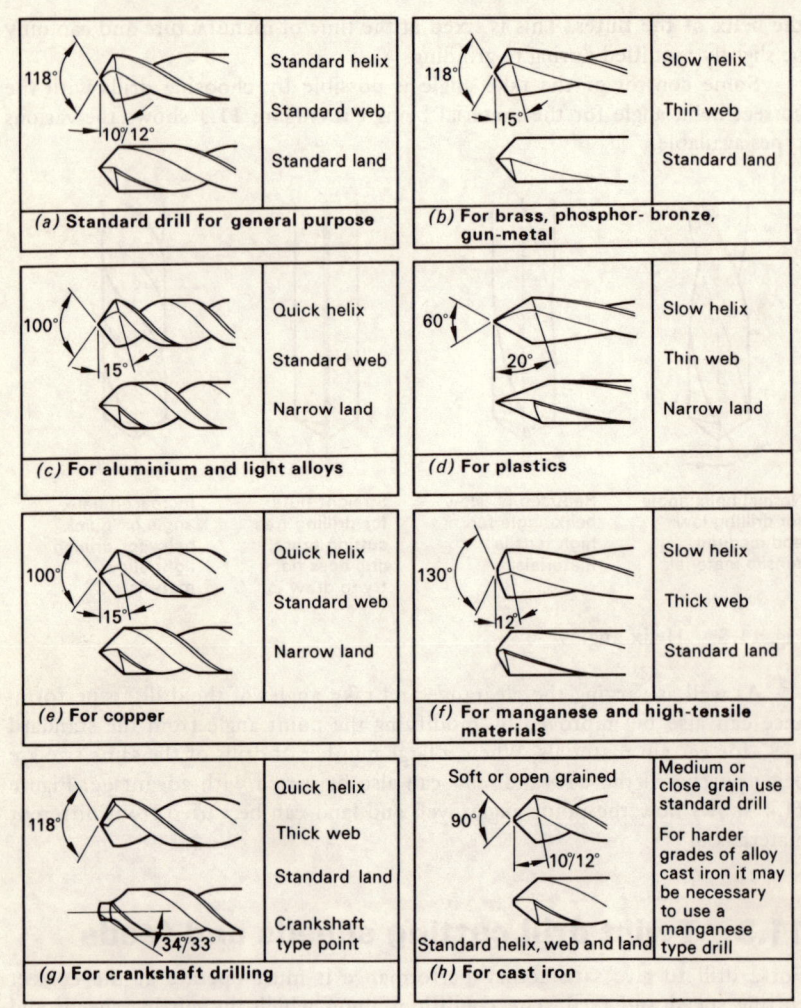

Fig. 11.4 Point angles

(b)   in deep holes it is relatively difficult for the drill to eject the chips;

(c)   the difficulty of keeping the cutting edges cool when they are enclosed in the hole. Even when a coolant is used it is difficult to apply it to the cutting edge. Not only are the flutes obstructed by the chips that are being ejected, but the flutes themselves tend to 'pump' the coolant out of the hole when the drill is rotating.

Table 11.1 gives a range of cutting speeds suitable for jobbing work using standard high-speed steel twist drills under reasonably controlled conditions. Table 11.2 gives the corresponding rates of feed. If the speed or feed recommended is not available on the drilling machine gear box, always select an alternative speed or feed that is *less* than the desired rate.

Table 11.1 Cutting speeds for HSS twist drills

Material being drilled	Cutting speed (m/min)
Aluminium	70–100
Brass	35–50
Bronze (phosphor)	20–35
Cast iron (grey)	25–40
Copper	35–45
Steel (mild)	30–40
Steel (medium carbon)	20–30
Steel (alloy-high tensile)	5–8
Thermo-setting plastic (Low speed due to abrasive properties)	20–30

Table 11.2 Feeds for HSS twist drills

Drill diameter (mm)	Rate of feed (mm/rev)
1·0–2·5	0·040–0·060
2·6–4·5	0·050–0·100
4·6–6·0	0·075–0·150
6·1–9·0	0·100–0·200
9·1–12·0	0·150–0·250
12·1–15·0	0·200–0·300
15·1–18·0	0·230–0·330
18·1–21·0	0·260–0·360
21·1–25·0	0·280–0·380

Example 11.1 *Calculate the spindle speed in rev/min for a high-speed steel drill 12 mm diameter, cutting mild steel.*

$$N = \frac{1000S}{\pi d}$$

where: N = spindle speed in rev/min
S = cutting speed in m/min
d = drill diameter (mm)
π = 3·14

From Table 11.1, a suitable cutting speed (S) for mild steel is 30 m/min, thus:

$$N = \frac{1000 \times 30}{3 \cdot 14 \times 12}$$

$$796 \cdot 2 \text{ rev/min}$$

A spindle speed between 750 and 800 rev/min would be satisfactory.

Example 11.2 *Calculate the time taken in seconds for the drill in Example 1 to penetrate a 15 mm thick steel plate.*

(*a*) From Example 1 the spindle speed has already been calculated as 796 rev/min (nearest whole number);

(*b*) from Table 11.2 it will be seen that a suitable feed for a 12 mm drill is 0·2 mm/rev.

$$t = \frac{60P}{NF}$$

$$= \frac{60 \times 15}{796 \times 0 \cdot 2}$$

$$= 5 \cdot 7 \text{ seconds}$$

(to one decimal place)

where: t = time in seconds
P = depth of penetration (mm)
N = spindle speed (rev/min)
F = feed (mm/rev)

11.4 Twist drill failures and faults

Twist drills suffer early failure or produce holes that are dimensionally inaccurate, out of round and of poor finish for the following general reasons:

(*a*) incorrect regrinding of the point;
(*b*) selection of incorrect speeds and feeds;
(*c*) abuse and mishandling.

Table 11.3 summarises the more common causes of twist drill failures and faults and suggests probable causes and remedies. Most cutting tools receive guidance from their shanks or spindles. Unfortunately, twist drills are too flexible to rely on this alone, and derive their guidance from the forces acting on the cutting edges. If the radial components of the forces acting at the two cutting edges are equal, they will push the drill point towards the axis of rotation. If the cutting edges lack symmetry (unequal length or unequal angle), or the material being drilled fails to offer uniform resistance to cutting, then the drill will 'wander' and follow a curved path.

As shown in Fig. 11.5(*a*), when drilling from the solid the drill is controlled by the chisel point. The hole will be round but may be oversize.

When opening up an existing hole, the point is floating and the drill is largely controlled by its outer corners. The diameter of the hole will be correct, but its shape may resemble the constant diameter lobed figure shown in Fig. 11.5(*b*).

This fault can be overcome by using a **core drill** for opening up an

Table 11.3   Twist drill fault-finding chart   273

Failure	Probable cause	Remedy
Damaged point	1. Do not use a hard-faced hammer when inserting the drill in the spindle 2. When removing the drill from the spindle, do not let it drop on to the hard surface of the machine table	Do not abuse the drill point
Rough hole	1. Drill point is incorrectly ground or blunt 2. Feed is too rapid 3. Coolant incorrect or insufficient	Regrind point correctly Reduce rate of feed Check coolant
Oversize hole	1. Lips of drill are of unequal length (Fig. 11.5) 2. Point angle is unequally disposed about drill axis 3. Point thinning is not central 4. Machine spindle is worn and running out of true	Regrind point correctly Recondition the machine
Unequal chips	1. Lips of drill are of unequal length 2. Point angle is unequally disposed about drill axis	Regrind point correctly
Split web (core)	1. Lip clearance angle too small 2. Point thinned too much 3. Feed too great	Regrind point correctly Reduce rate of feed

Failure	Probable cause	Remedy
Chipped lips	1. Lip clearance angle too large	Regrind point
	2. Feed too great	Reduce rate of feed
Damaged corners	1. Cutting speed too high, drill 'blues' at outer corners	Reduce spindle speed
	2. Coolant insufficient or incorrect	Check coolant
	3. Hard spot, scale, or inclusions in material being drilled	Inspect material
Broken tang	1. Drill not correctly fitted into spindle so that it slips	Ensure shank and spindle are clean and undamaged before inserting
	2. Drill jams in hole and slips	Reduce rate of feed
Broken drill	1. Drill is blunt	Regrind point
	2. Lip clearance angle too small	
	3. Drill point incorrectly ground	
	4. Rate of feed too great	Reduce rate of feed
	5. Work insecurely clamped	Re-clamp more securely
	6. Drill jams in hole due to worn corners	Regrind point
	7. Flutes choked with chips when drilling deep holes	Withdraw drill periodically and clean

existing hole. Being a multi-flute drill, there are more than two corners to give support. This gives a more truly round hole. The core drill can only open up an existing hole and cannot drill from the solid.

11.5 Blind hole drilling

A *blind* hole is one which stops part way through a component. The essential difference between drilling a *through hole* and drilling a *blind hole* is that in the latter case a means must be devised for determining when the drill has penetrated the component to a prescribed depth. Most drilling machines are

75

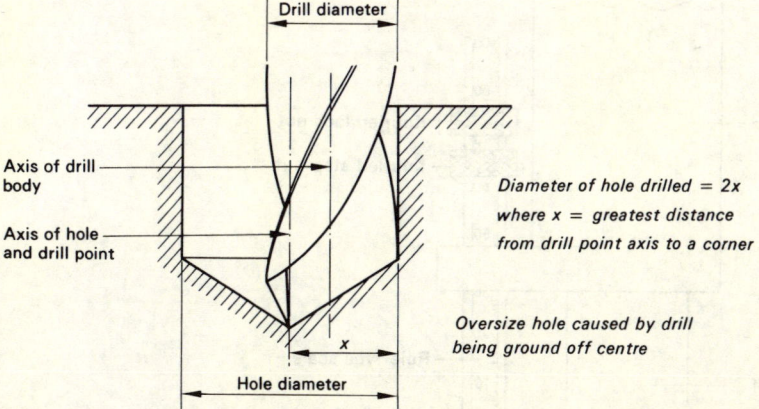

Diameter of hole drilled = 2x

where x = greatest distance
from drill point axis to a corner

Oversize hole caused by drill
being ground off centre

(a) **Oversize hole**

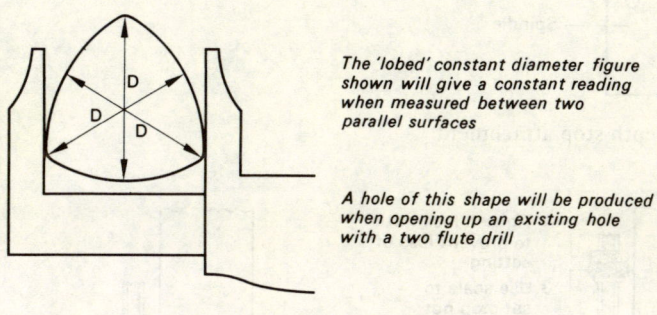

The 'lobed' constant diameter figure
shown will give a constant reading
when measured between two
parallel surfaces

A hole of this shape will be produced
when opening up an existing hole
with a two flute drill

(b) **The lobed figure**

Fig. 11.5 Hole faults

provided with an adjustable depth stop attached to the quill as shown in Fig. 11.6. It will be seen that the depth stop is engraved with a rule type scale on its front face and this scale is used, as shown in Fig. 11.7(*a*), to set the knurled stop nut and lock nut. The hole is then drilled to depth, as shown in Fig. 11.7(*b*).

Generally, blind holes are only toleranced to rule accuracy and the scale on the depth stop is an adequate indication of the hole depth. However, flat-bottomed holes are often toleranced to closer limits and an alternative technique is required to set the stop nut. The sequence of operations shown in Fig. 11.8 should be used to produce a flat-bottomed blind hole to a precise depth.

276

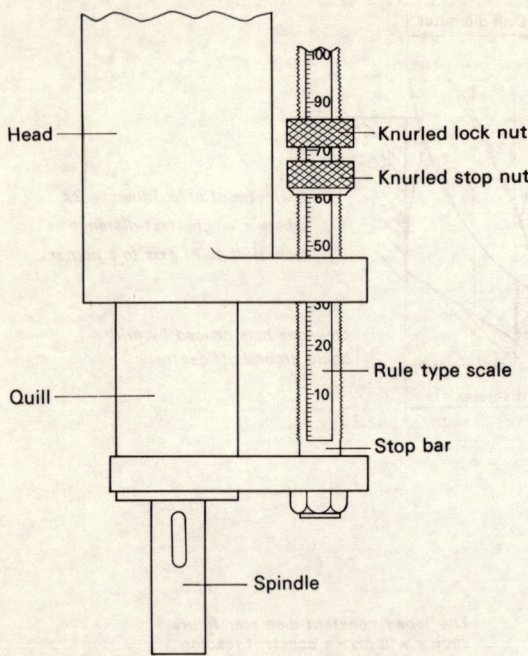

Fig. 11.6 Depth stop attachment

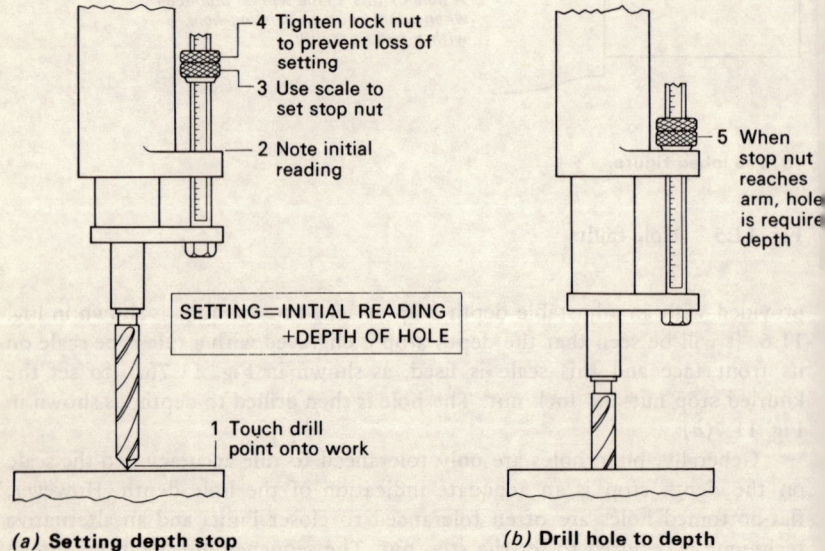

(a) Setting depth stop

(b) Drill hole to depth

Fig. 11.7 Use of depth stop to drill blind hole

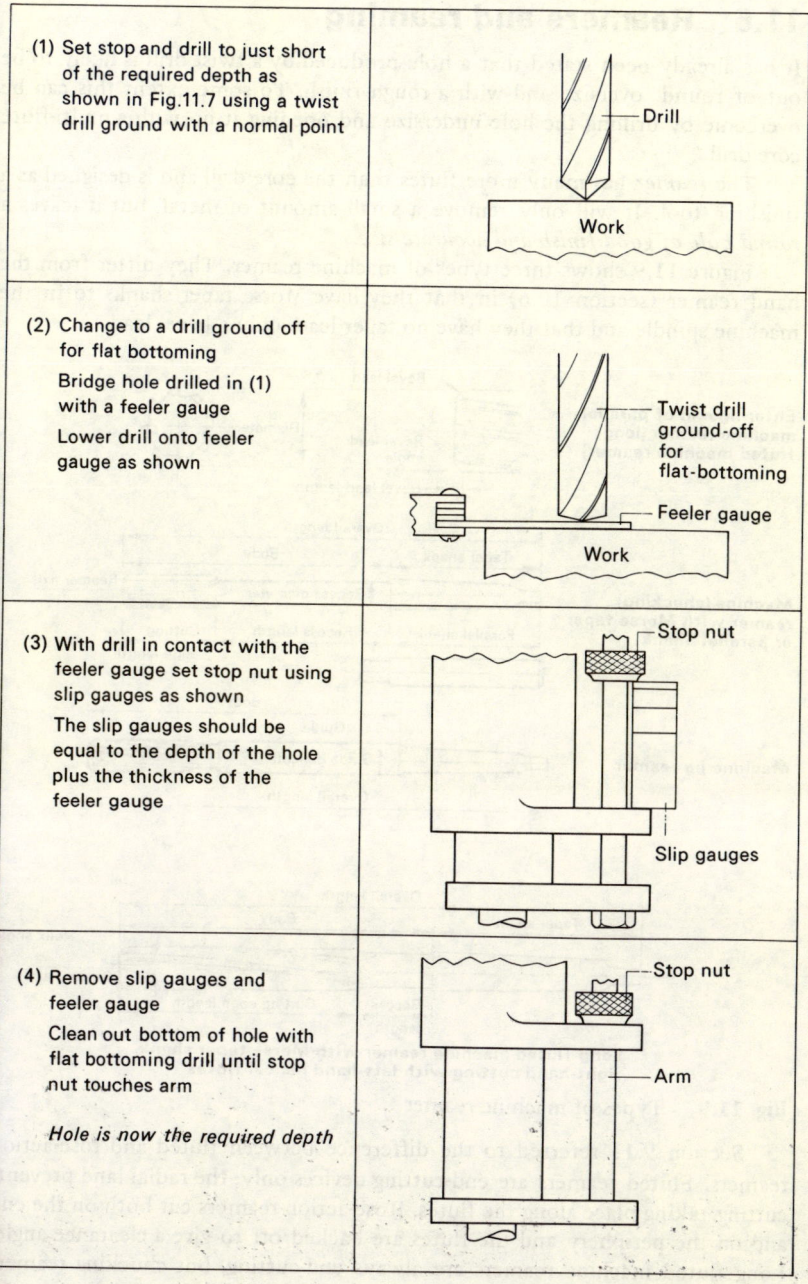

(1) Set stop and drill to just short of the required depth as shown in Fig.11.7 using a twist drill ground with a normal point

Drill

Work

(2) Change to a drill ground off for flat bottoming

Bridge hole drilled in (1) with a feeler gauge

Lower drill onto feeler gauge as shown

Twist drill ground-off for flat-bottoming

Feeler gauge

Work

(3) With drill in contact with the feeler gauge set stop nut using slip gauges as shown

The slip gauges should be equal to the depth of the hole plus the thickness of the feeler gauge

Stop nut

Slip gauges

(4) Remove slip gauges and feeler gauge

Clean out bottom of hole with flat bottoming drill until stop nut touches arm

Hole is now the required depth

Stop nut

Arm

Fig. 11.8   Precision depth setting

11.6 Reamers and reaming

It has already been stated that a hole produced by a twist drill is likely to be out of round, oversize and with a rough finish. To some extent this can be overcome by drilling the hole undersize and opening it up with a multi-flute core drill.

The *reamer* has many more flutes than the core drill and is designed as a finishing tool. It will only remove a small amount of metal, but it leaves a *round hole of good finish and accurate size.*

Figure 11.9 shows three types of machine reamer. They differ from the hand reamer (section 10.6) in that they have Morse taper shanks to fit the machine spindle and that they have no taper lead, only a bevel lead.

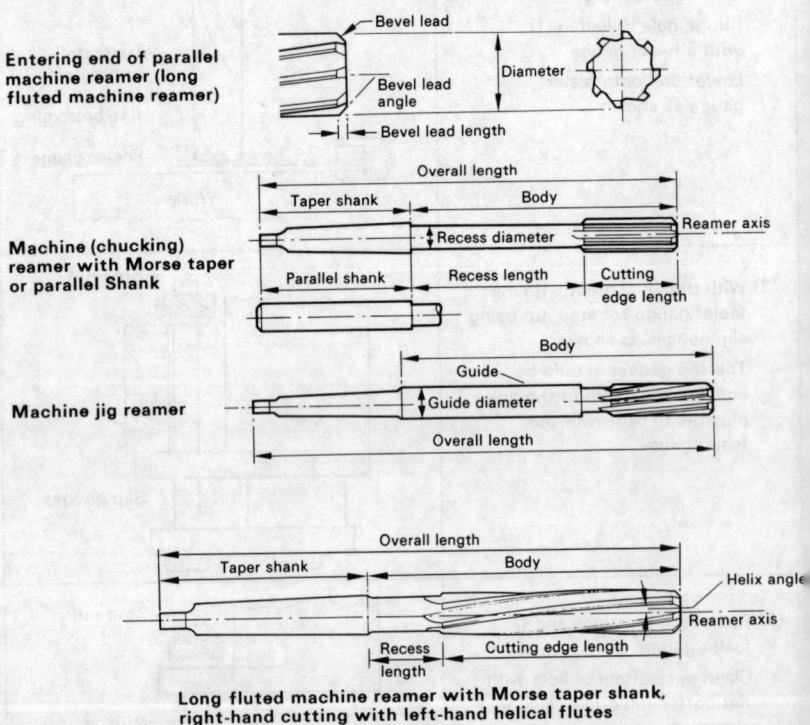

Fig. 11.9   Types of machine reamer

Section 9.11 referred to the difference between fluted and rose-action reamers. Fluted reamers are end-cutting devices only; the radial land prevent cutting taking place along the flutes. Rose-action reamers cut both on the en and on the periphery and the flutes are backed off to give a clearance angle Long fluted machine reamers are always end-cutting, but chucking reamer may have a fluted or a rose-cutting action. Fluted reamers give the best result with steels and similar ductile materials, whereas rose-action reamers ar better for cast iron and bronze materials. These materials tend to close on th

reamer and peripheral cutting prevents seizure and broken tooling. Similarly
rose-action reamers are preferable when reaming plastic materials.

Although standard reamers are made for right-hand cutting, they have flutes with a left-hand helix. This serves two purposes:

(a) to prevent the reamer being drawn into the hole by the screw action of the helix;

(b) to eject the chips ahead of the reamer and prevent them being drawn back up the hole, where they would mark the finished surface.

The reamer always follows the axis of the existing hole: it cannot correct positional errors. If the original hole is out of position or out of alignment with its datum, then these errors must be corrected by single point boring before reaming.

11.7 Miscellaneous operations

In addition to drilling holes, the following operations are also performed on the drilling machine;

(a) trepanning;
(b) countersinking;
(c) counterboring;
(d) spot facing.

(a) Trepanning

Not only is it dangerous to try to cut large diameter holes in sheet metal with a twist drill, but the resulting hole will not be satisfactory. There will be insufficient thickness of metal to guide the drill point and to resist the cutting forces. The hole will be out of round and jagged.

One way to overcome this problem is to use a *trepanning cutter*. Instead of cutting away all the metal in the hole as swarf, the trepanning cutter merely removes a thin annulus of metal. This leaves a clean cut hole in the stock and a disc of metal slightly smaller than the hole. The principle of trepanning is shown in Fig. 11.10(a).

The simplest type of trepanning cutter is the traditional *tank cutter* shown in Fig. 11.10(b). This has a number of disadvantages, and the type of cutter shown in Fig. 11.10(c) is superior where a number of holes of the same size have to be cut.

(b) Countersinking

Figure 11.11(a) shows a typical *rose-bit* used for countersinking. Since the bit is conical in form it is self-centring and does not require a pilot to ensure axial alignment with the hole being countersunk.

Countersinking is used for the following purposes.

1. To provide a recess for the head of a countersunk screw so that a flush surface is left after the screw is installed.

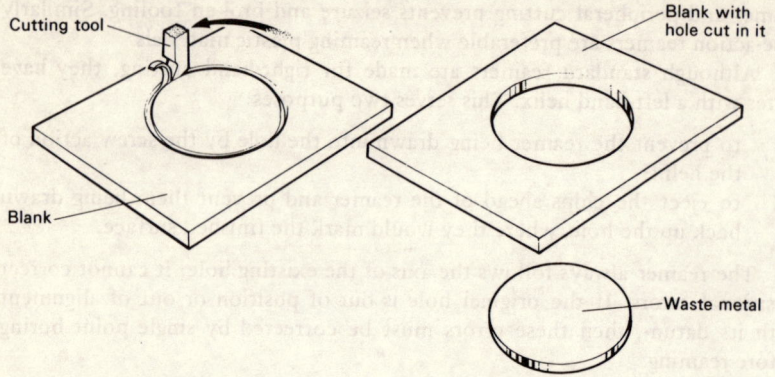

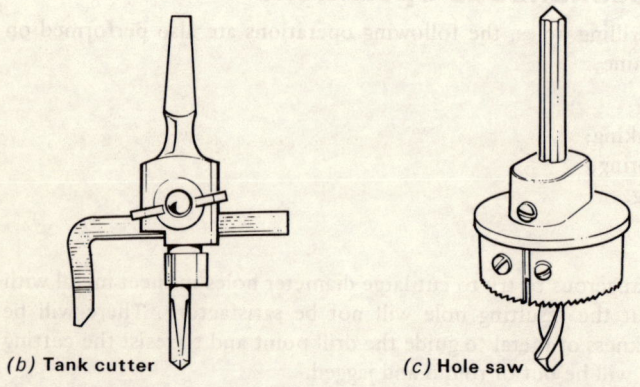

(a) **Principle of trepanning**

(b) **Tank cutter**   *(c)* **Hole saw**

Large diameter hole cutters for sheet-metal

Fig. 11.10   Trepanning large holes

2.   To deburr a hole after drilling. Burred holes are dangerous to handle, and mating components with burred holes will not seat together correctly.
3.   To chamfer sharp corners in order to make the component safer to handle and less likely to crack when hardened.

(c) Counterboring

Figure 11.11(*b*) shows a typical counterboring cutter. It is similar in design to an end-mill but is fitted with a pilot to ensure axial alignment with the hole being counterbored.

Counterboring is used to provide a recess to accept the head of a cheese-head or of a cap-head screw so that a flush surface is left after the screw is installed. Screws that are left 'proud' of the surface (not recessed) are a constant source of danger. Whenever possible screw heads and nuts should be recessed into the job by countersinking or counterboring.

(d) Spot facing

Figure 11.11(c) shows a typical spot facing cutter. It is used to provide a locally machined, flat seating for nuts or bolt-heads being pulled down onto otherwise rough castings or forgings.

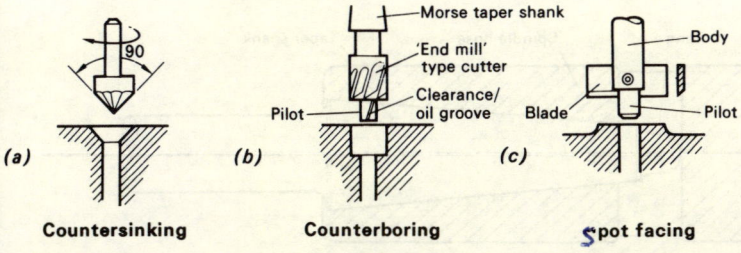

(a) Countersinking **(b)** Counterboring **(c)** Spot facing

Fig. 11.11   Cutters, (a) countersinking, (b) counterboring and (c) spot facing

11.8   Restraint and location when drilling

To successfully drill a hole in a component so that it is correctly positioned, four basic conditions must be satisfied.

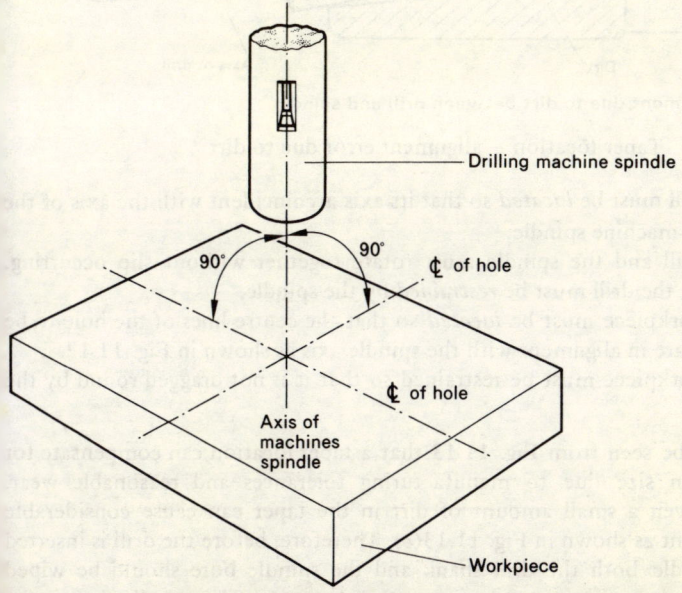

Fig. 11.12   Basic drilling alignments

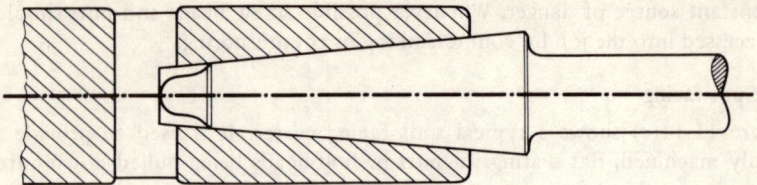

(a) **Spindle and shank maintain axial alignment under maximum metal conditions**

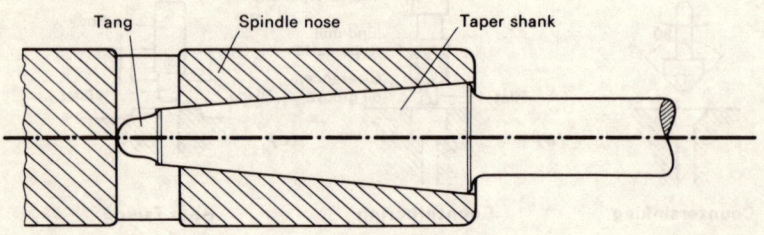

(b) **Spindle and shank maintain axial alignment under minimum metal conditions**

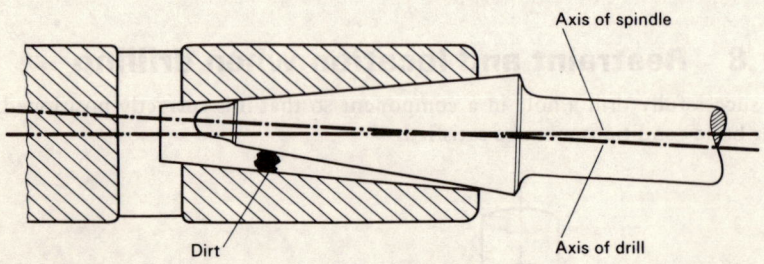

(c) **Misalignment due to dirt between drill and spindle**

Fig. 11.13 Taper location — alignment error due to dirt

1. The drill must be *located* so that its axis is coincident with the axis of the drilling machine spindle.
2. The drill and the spindle must rotate together without slip occurring. That is, the drill must be *restrained* by the spindle.
3. The workpiece must be *located* so that the centre lines of the hole to be drilled are in alignment with the spindle axis as shown in Fig. 11.12.
4. The workpiece must be restrained so that it is not dragged round by the drill.

It will be seen from Fig. 11.13 that a taper location can compensate for variations in size due to manufacturing tolerances and reasonable wear. However, even a small amount of dirt in the taper can cause considerable mis-alignment as shown in Fig. 11.13(c). Therefore, before the drill is inserted in the spindle both the drill shank and the spindle bore should be wiped clean. This also prevents undue wear and damage to the spindle due to the drill slipping. The narrow angle of taper of the drill shank causes it to wedge in the spindle of the drilling machine. This provides sufficient restraint to prevent the drill dropping out of the spindle and to prevent relative movement (slip) between the drill and the spindle as they rotate. If the drill 'digs in' to the workpiece or the drill seizes in the hole, the frictional restraint of the taper becomes inadequate and slip occurs. This results in the spindle bore and the drill shank becoming scored up so that proper location and restraint of the drill becomes impossible.

Straight shank drills and other small cutting tools are often held in self-centring chucks. The principle of the drill chuck as a device for locating and restraining small drills is shown in Fig. 11.14. It will be seen that the chuck and its arbor rely upon a system of concentric tapers to create axial alignment.

The same principles of location and restraint can be applied to the workpiece. Figure 11.15(*a*) shows the *restraints* acting on a simple component held in a machine vice. The geometric alignments necessary to *locate* the component correctly, relative to the axis of the drilling machine spindle, are shown in Fig. 11.15(*b*). Components that are too large to be held in a machine vice are usually clamped directly to the machine table.

Cylindrical components are more difficult to hold since only line contact between flat and round surfaces is possible. It is advisable to insert a vee block between a cylindical component and the fixed jaw of the vice to provide a three-point location.

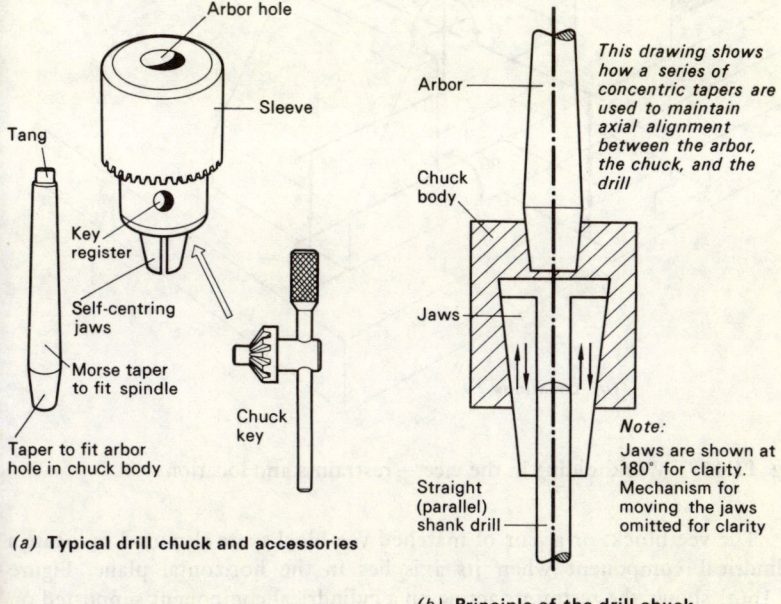

(a) **Typical drill chuck and accessories**

(b) **Principle of the drill chuck**

Fig. 11.14 The drill chuck

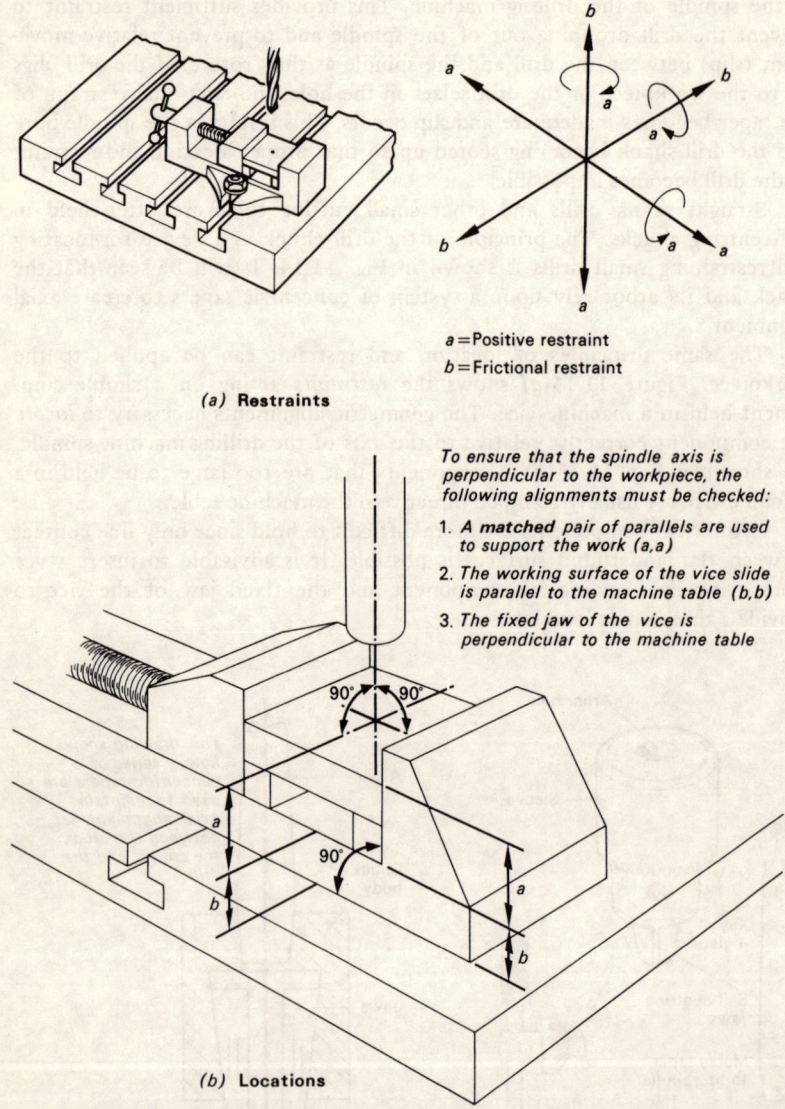

a = Positive restraint

b = Frictional restraint

(a) Restraints

To ensure that the spindle axis is
perpendicular to the workpiece, the
following alignments must be checked:

1. *A matched* pair of parallels are used
   to support the work (a,a)

2. The working surface of the vice slide
   is parallel to the machine table (b,b)

3. The fixed jaw of the vice is
   perpendicular to the machine table

(b) Locations

Fig. 11.15   Workholding in the vice — restraints and locations

The vee block, or a pair of matched vee blocks, are also used to locate a
cylindrical component when its axis lies in the horizontal plane. Figure
11.16(*a*) shows the *restraints* acting on a cylindrical component supported on
vee blocks. The geometrical alignments necessary to *locate* the component
correctly, relative to the drilling machine spindle, are shown in Fig. 11.16(*b*).

285

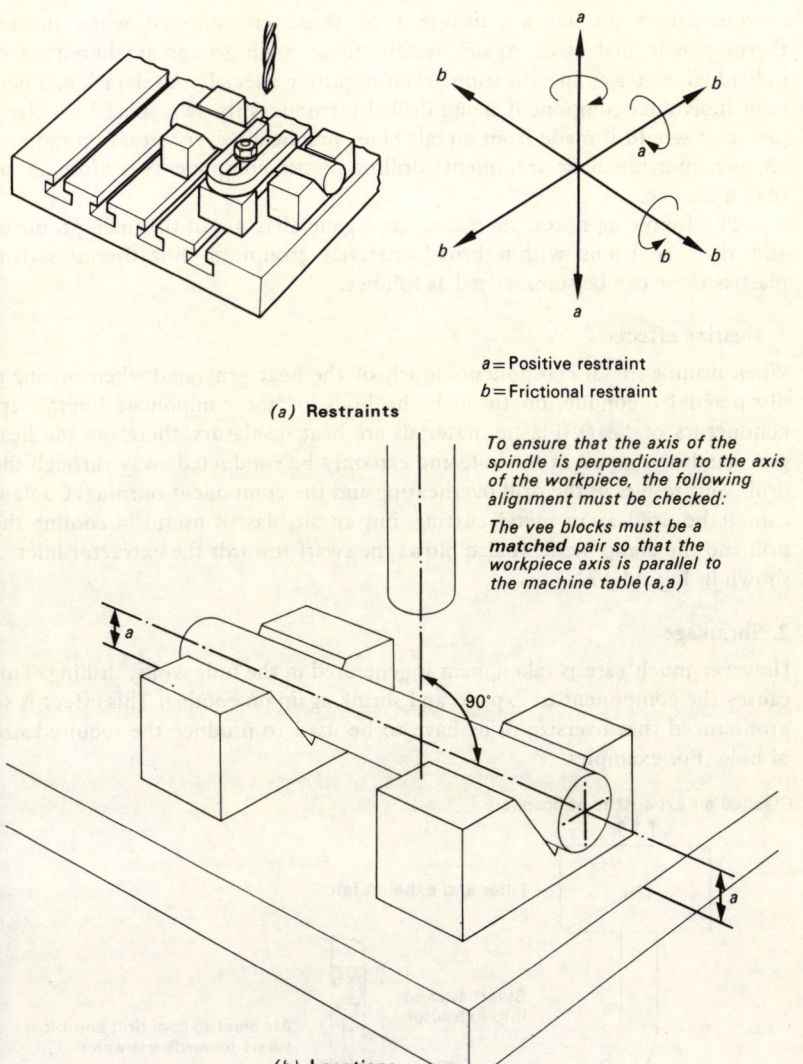

a = Positive restraint
b = Frictional restraint

(a) **Restraints**

To ensure that the axis of the spindle is perpendicular to the axis of the workpiece, the following alignment must be checked:

The vee blocks must be a **matched** pair so that the workpiece axis is parallel to the machine table (a,a)

90°

(b) **Locations**

Fig. 11.16 Workholding cylindrical work on the drilling machine table

11.9 Drilling plastic materials (thermo-setting)

The problems encountered when drilling plastic materials are completely different from those encountered when drilling metals. The situation is further complicated by the fact that the problems encountered when drilling thermo-setting plastics are different to those encountered when drilling thermoplastic materials. Again, within these main groups are hundreds of individual material specifications each requiring special consideration. Then, each individual component being drilled introduces its own special problems just as it would if made from metal. Thus, practically every situation requires its own individualistic treatment: drilling plastic components is more an art than a science.

The following notes, therefore, are a generalisation of the main problems and their solutions within broad material groupings. For thermo-setting plastics these can be summarised as follows.

1. Heating effects

When drilling metal components much of the heat generated when cutting is dissipated by conduction through the body of the component (metals are conductors of heat). Plastic materials are heat insulators, therefore the heat generated is trapped in the hole and can only be conducted away through the drill. This results in the drill overheating and the component burning. Coolant cannot be used as in metal cutting, but an air blast is useful in cooling the drill and the component. It also blows the swarf towards the extractor inlet as shown in Fig. 11.17.

2. Shrinkage

However much care is taken, heat is generated in the hole whilst drilling. This causes the component to expand and shrink again on cooling. This effect is so pronounced that oversize drills have to be used to produce the required size of hole. For example:

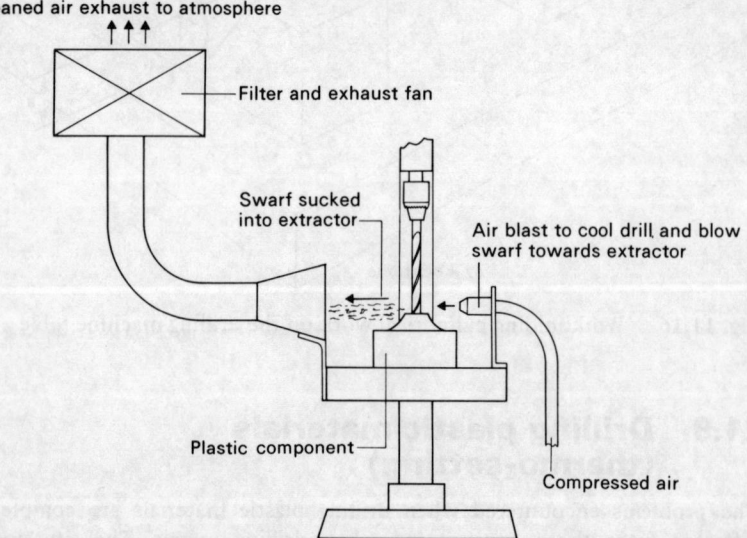

Cleaned air exhaust to atmosphere

Filter and exhaust fan

Swarf sucked into extractor

Air blast to cool drill and blow swarf towards extractor

Plastic component

Compressed air

Fig. 11.17   Air blast and extraction

2 BA tapping size drill — steel   — 3·7 mm

2 BA tapping size drill — thermo-setting plastic — 3·9 mm

3. Feed

Two problems affect the rate of feed. First, the heating effect already mentioned, and second the tendency for the swarf to clog in the flutes of the drill. To overcome these the drill is fed into the workpiece with a 'woodpecker' action. That is, instead of a continuous uniform rate of feed as usually employed when drilling metals, the drill is fed in quite rapidly and then momentarily withdrawn so that the plastic can 'breathe'. This allows the air blast to cool the drill, and eject the swarf. The hole tends to shrink in size and is cleared as the drill is given its next increment of feed. This prevents the hole closing on the body of the drill as well as preventing the drill and workpiece overheating. This method of feeding is shown diagrammatically in Fig. 11.18.

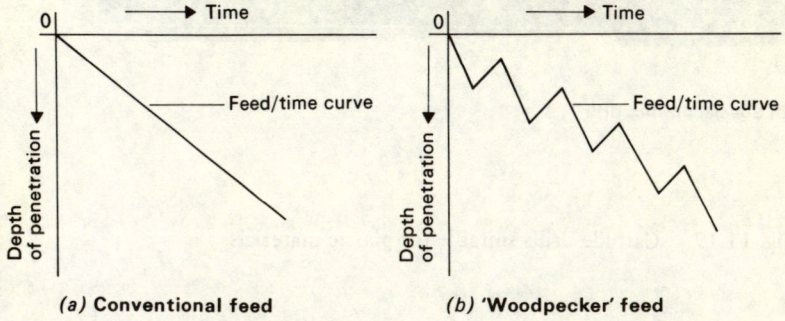

Fig. 11.18   'Woodpecker' feed action

4. Abrasion

Thermo-setting plastics, especially when mineral fillers are used such as asbestos and mica, are highly abrasive and wear out high-speed steel drills very rapidly. Therefore it is usual to use carbide-tipped drills. Usually an ordinary 'masonry' drill, as sold by ironmongers for drilling brickwork, is adequate as they tend to drill oversize and this allows for the shrinkage after drilling. Where a precision hole is required, precision ground carbide tipped drills are used. Examples of both types of drill are shown in Fig. 11.19. It will be seen that the specially designed plastic drill, shown in Fig. 11.19(b), has a slow spiral, an $80°$ point angle, wide flutes to prevent clogging and a generous clearance angle so that it cuts freely. It will also have a thin web to reduce the axial force (feed force) on the workpiece. Often the flutes are polished to help prevent clogging. The general point angle of $118°$ is suitable for opening up cored holes, but the $80°$ angle is preferable for drilling from the solid as it helps to prevent chipping the moulding as the drill breaks through as well as reducing the feed force.

(*a*) Masonry drill

(*b*) Special plastics drill

Fig. 11.19   Carbide drills suitable for plastic materials

5. Cutting speed

This tends to be kept to a fairly low value to reduce the heating effect and wear from abrasion. For holes up to 12 mm diameter the cutting speed would be in the range of 35 m/min to 45 m/min depending on the type of plastic being drilled and the filler used.

6. Workholding

Components made from thermo-setting plastics tend to be brittle and great care has to be taken in clamping to avoid cracking them by using too much force or distorting the component.

11.10   Drilling plastic materials (thermoplastic)

It is not so usual to drill thermoplastic materials as it is to drill thermo-setting plastics. This is because it is much easier to *core* the holes into components made from thermoplastic during the moulding process. There are many more

thermoplastic materials than there are thermo-setting plastic materials and the
properties vary much more widely between each sub-group of materials.
Therefore it is only possible to generalise in the space available in this
chapter. However, the following notes give some indication of the main
problems and their solution within that general group of materials referred to
as thermoplastics.

1. Heating effects

As stated in section 11.9, all plastic materials are excellent thermal insulators.
Thermoplastics are no exception and, therefore, any heat generated when
cutting cannot be dissipated by conduction. This causes the material to
overheat rapidly, soften and weld itself to the drill. Since softening occurs at
temperatures only a little above the boiling point of water, overheating is a
problem only affecting the component, the temperature being too low to
affect the temper of an ordinary high-speed steel drill. Coolant cannot be
used, only an air blast as shown in Fig. 11.17.

2. Shrinkage

This effect is even more acute than with thermo-setting plastics and allowance
has to be made when selecting drill sizes where the diameter of the finished
hole is critical. There is also the problem of the material closing on the body
of the drill. The additional friction this causes results in premature overheating and complete clogging of the drill flutes, which, in turn, results in
either a ruined component, or a broken drill, or both.

3. Feed

The importance of using a 'woodpecker' feed action to allow the hole to
'breathe' was fully discussed in section 11.9. This action needs to be even
more exaggerated when drilling thermoplastics. The drill is run much faster
and each feed increment is very coarse, resulting in a 'bodging' action. This
results in the required amount of material being removed before the temperature can build up to the softening point for the plastic concerned. It is usual
to employ the special high-speed steel drills that have been marketed for
drilling thermoplastic materials. These drills have slow helix angles, very wide
flutes, thin webs and acute points. The flutes are polished to reduce clogging.
They are ground with a generous clearance angle and are given a keen cutting
edge. The total geometry of the drill is designed to enable them to cut as
freely as possible and to allow the swarf to have rapid egress from the hole, so
as to keep the heat generated to a minimum. Figure 11.20 shows two typical
plastic drills alongside an ordinary jobber drill so that their geometries can be
compared.

4. Abrasion

With few exceptions, thermoplastic materials do not contain abrasive fillers
and high-speed steel cutting tools perform satisfactorily. This is not only

(a) Metal cutting twist drill

(b) Drill for thermoplastic materials

(c) Solid carbide drill for glass fibre reinforced plastic materials

Fig. 11.20 Twist drills for plastics and metals compared

more economical than using carbide-tipped tools, but it is easier to service high-speed steel tools and provide the keen cutting edge necessary to ensure free cutting without overheating.

The notable exception is the drilling of glass-reinforced-polyester materials (GRP). The glass fibre is incredibly abrasive and high-speed steel tooling is almost immediately destroyed by being worn undersize. Even carbide-tipped tools are not satisfactory as the steel body of the drill is rapidly worn down. The only satisfactory way to drill GRP components, on a production basis, is to use solid carbide drills. These are very expensive and very fragile and require careful and expert handling. Whereas all the previous drills shown for drilling plastic materials have slow helixes to prevent the 'screwing action' of the helix from drawing them into the workpiece, drills designed for GRP have a quick helix to help eject the swarf rapidly. The material offers sufficient resistance to cutting to prevent the drill being drawn in, and rapid removal of the swarf helps to reduce the wear of the drill body due to abrasion. Figure 11.20(c) shows a typical solid carbide drill.

Since most thermoplastic materials are not abrasive, they can be cut with a much higher speed than thermo-setting materials. For holes up to 12 mm diameter the cutting speed would be in the range of 200 to 250 m/min with a fine feed for carbide tooling and 80 to 100 m/min with a coarse feed for high-speed steel drills. The acute 'woodpecker' feed action prevents heat build-up normally associated with high cutting speed. Their use not only gives economical production rates, but the whole exercise is geared to rapid material removal before the temperature has time to rise to the softening point of the plastic.

When drilling GRP components, the cutting speed is dropped to the lower range, even though carbide tooling is being used, due to its highly abrasive nature. With the lower cutting speed care must be taken to avoid overfeeding the fragile solid carbide drill with the ensuing, expensive breakages.

6. Workholding

Components made from thermoplastics are less brittle than those made from thermo-setting plastics and are less liable to be cracked when clamped. However, they tend to be easily distorted due to their inherent flexibility. Where holes are only being opened up to size, having previously been cored, it is quite normal to 'hand-hold' the component in a simple fixture as the cutting forces are sufficiently low as not to endanger the operator.

11.11　Drilling thin plate

For a hole of reasonable form to be produced the drill must be cutting its full diameter before the point breaks through. This is because, at the moment of breakthrough, the point ceases to locate the drill and the body lands have to take over as shown in Fig. 11.21(*a*). When drilling sheet and thin plate the drill breaks through too soon and *grabs*, forming a *lobed* hole. This effect is shown in Fig. 11.21(*b*). Not only is an out-of-shape hole produced, but the resulting, ragged edges are very dangerous. Further, when the drill grabs, small drills are often broken and large drills may twist up the workpiece, pulling it clear of its clamps, and causing it to spin round in a highly dangerous manner.

To overcome these problems the drill point has to be ground to a very obtuse point as shown in Fig. 11.21(*c*). This enables the body to enter the hole before the point breaks through. The actual point angle will depend on the drill diameter and the thickness of the plate. It must be remembered that the efficiency of the drill decreases as the point angle becomes flatter.

To increase the efficiency of an obtuse angled drill and reduce the axial cutting force, the point should be thinned to reduce the length of the chisel point as shown in Fig. 11.21(*d*).

The drill can also be ground with a pilot point as shown in Fig. 11.21(*e*). This is much more difficult to produce, but it enables large holes to be produced in thin plate.

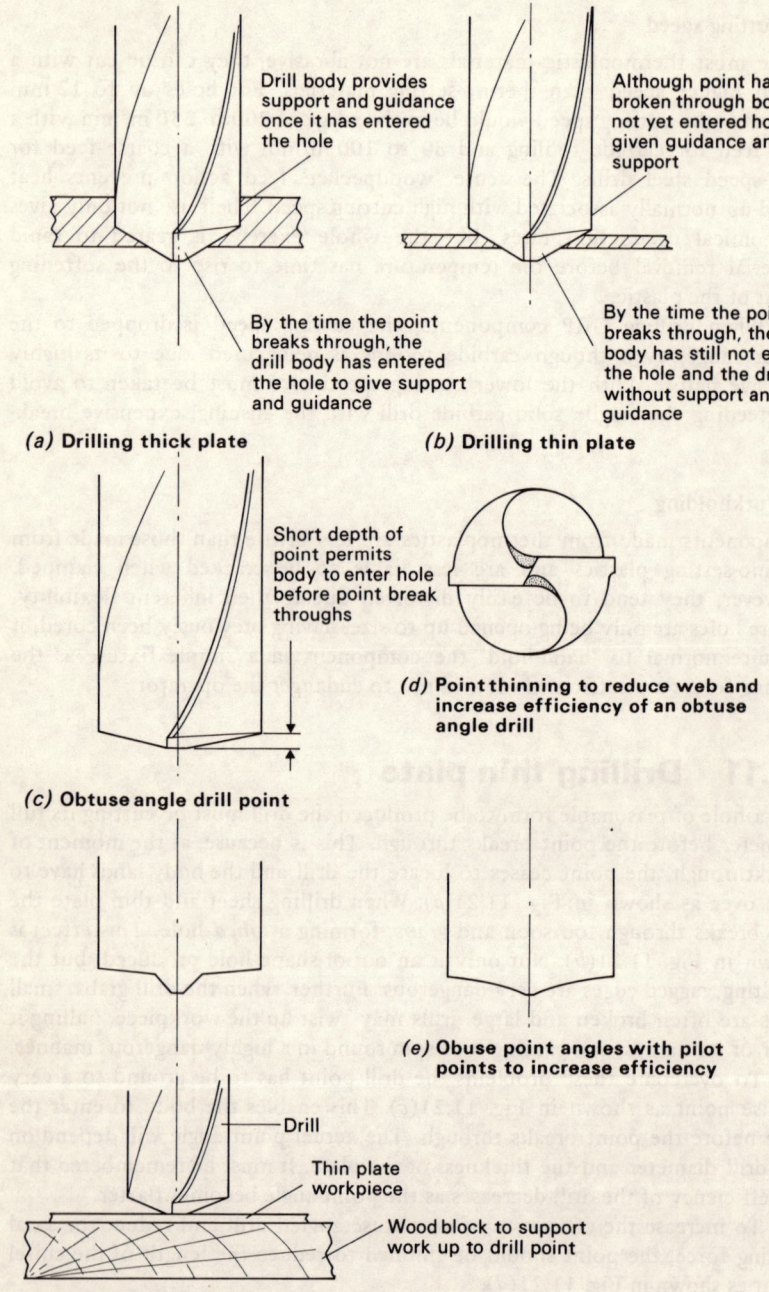

Drill body provides support and guidance once it has entered the hole

Although point has broken through body has not yet entered hole and given guidance and support

By the time the point breaks through, the drill body has entered the hole to give support and guidance

By the time the point breaks through, the drill body has still not entered the hole and the drill is without support and guidance

(a) Drilling thick plate

(b) Drilling thin plate

Short depth of point permits body to enter hole before point break throughs

(c) Obtuse angle drill point

(d) Point thinning to reduce web and increase efficiency of an obtuse angle drill

(e) Obuse point angles with pilot points to increase efficiency

Drill

Thin plate workpiece

Wood block to support work up to drill point

(f) Thin plate and sheet metal requires support right up to the drill point

Fig. 11.21   Point grinding of twist drills for thin plate

When drilling plate it is essential that it is supported right up to the cutting edge. Components should be supported on a wood block as shown in Fig. 11.21(*f*).

11.12 Basic alignments of the drilling machine

Figure 11.22 shows the basic alignments of the spindle axis and workpiece. The geometry of the drilling machine is designed to maintain these alignments.

The *spindle* locates and rotates the drill and is itself located in precision bearings in a *sleeve* that can move in the body of the drilling machine so that it travels to or from the workpiece in a path parallel to the axis of the spindle. The combined spindle and sleeve assembly is known as the 'quill'.

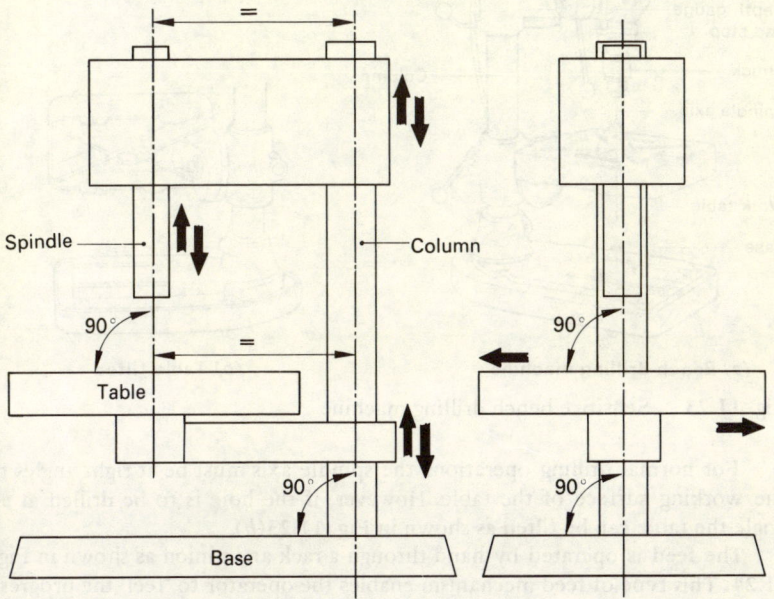

Fig. 11.22 The drilling machine — basic geometry

Reference to Fig. 11.22 shows that the spindle axis is perpendicular to the surface of the workpiece. It must, therefore, be perpendicular to the surface of the machine work table. This table is adjustable up and down a precision ground column in order that work of various thickness can be accommodated. To preserve the table alignment the axis of the spindle and the axis of the column must be mutually parallel. Further, the axis of the column is perpendicular to the machine base.

These requirements build up into the skeleton of a drilling machine as shown in Fig. 11.22, which also shows the geometrical alignments and movements described above.

11.13   The bench drilling machine

The simplest type of drilling machine is the bench drilling machine shown in Fig. 11.23(*a*). It is capable of accepting drills up to 12·5 mm diameter either in a chuck or directly mounted in the taper nose of the spindle. Variation in spindle speed is achieved by altering the belt position on the stepped pulleys.

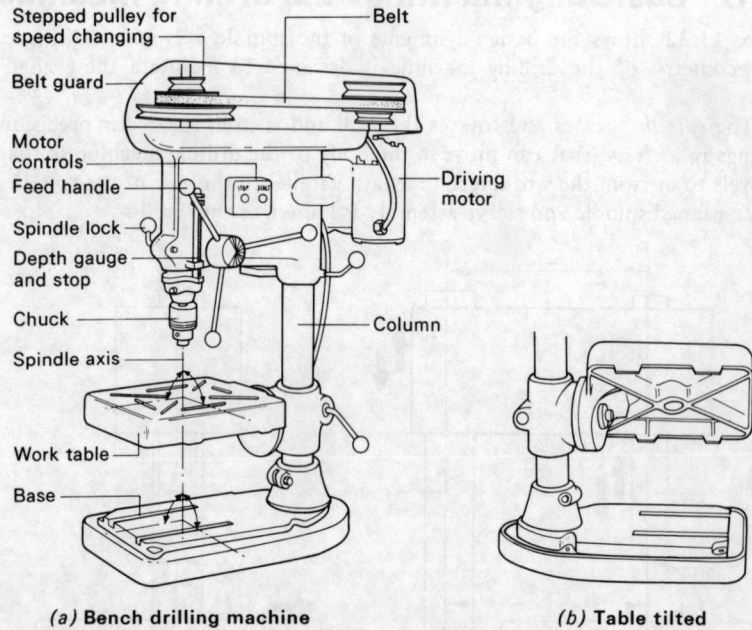

(a) **Bench drilling machine**   *(b)* **Table tilted**

Fig. 11.23   Sensitive bench drilling machine

For normal drilling operations the spindle axis must be at right angles to the working surface of the table. However, if the hole is to be drilled at an angle the table can be tilted as shown in Fig. 11.23(*b*).

The feed is operated by hand through a rack and pinion as shown in Fig. 11.24. This type of feed mechanism enables the operator to 'feel' the progress of the drill through the material being cut. Hence the name 'sensitive feed'.

11.14   The pillar drilling machine

Basically, this is an enlarged version of the 'sensitive bench' machine previously described. It is floor-mounted and very much more heavily constructed. The spindle is driven by a much more powerful electric motor and the spindle speed is varied by a gearbox. Sensitive rack and pinion feed is provided for setting up and starting the drill. Power feed is provided for the actual drilling operation, the rate of feed being controlled by an auxiliary gearbox. The spindle is bored to a Morse taper and can accept drills inserted

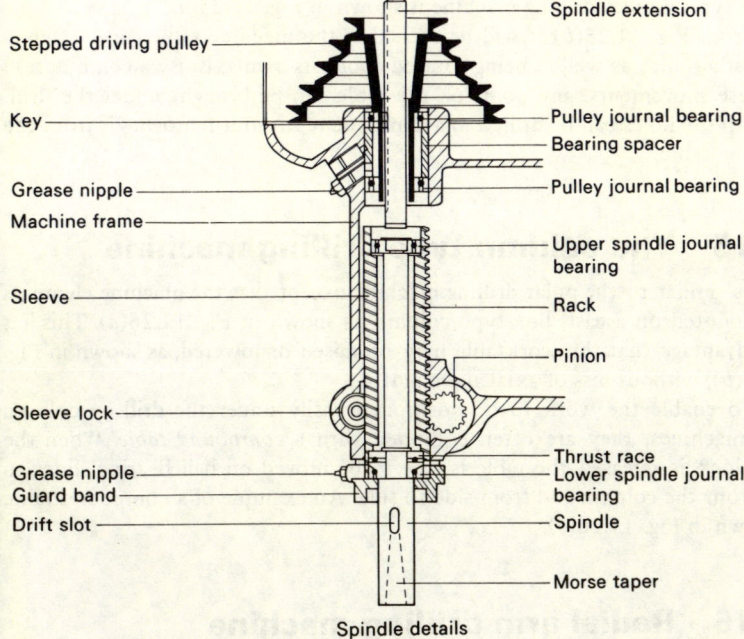

Spindle details

Fig. 11.24 Sensitive feed mechanism

directly into the spindle as well as drill chucks. Holes up to 50 mm diameter can be drilled from the solid on the larger machines.

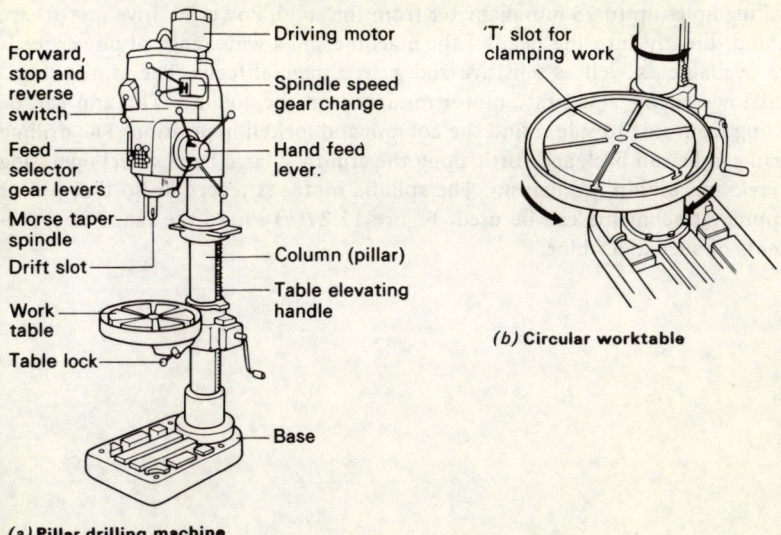

(a) Pillar drilling machine

(b) Circular worktable

Fig. 11.25 Pillar drilling machine

A typical pillar drilling machine is shown in Fig. 11.25(*a*).

From Fig. 11.25(*b*) it will be seen that the machine table can be swung about the pillar as well as being rotated about its own axis. By a combination of these movements, any point on the table can be brought under the drill. Thus all the holes can be drilled in a component without removing it from the machine table.

11.15   The column type drilling machine

This is similar to the pillar drilling machine except that the machine elements are mounted on a cast, box-type column as shown in Fig. 11.26(*a*). This has the advantage that the worktable may be raised or lowered, as shown in Fig. 11.26(*b*), without loss of axial alignment.

To enable the work to be positioned easily under the drill on column type machines, they are often equipped with a *compound table.* When the table lock is released the table is free to be moved on ball bearing slides to and from the column and from side to side. An example of a compound table is shown in Fig. 11.26(*c*).

11.16   Radial arm drilling machine

For heavy work it is often more convenient to move the drilling head about over the work than to move the work. The radial arm drilling machine provides such a facility and an example is shown in Fig. 11.27(*a*). Drilling machines such as this represent the most powerful machines available, often drilling holes up to 75 mm diameter from the solid. Powerful drive motors are geared directly into the head of the machine and a wide range of power feeds are available as well as sensitive and geared manual feeds. The arm is raised and lowered by a separate motor mounted on the column. The arm can be swung from side to side round the column and locked in position. The drilling head can be run back and forth along the arm by a large hand wheel operating a rack and pinion mechanism. The spindle motor is reversible so that power tapping attachments can be used. Figure 11.27(*b*) shows the range of movements of such a machine.

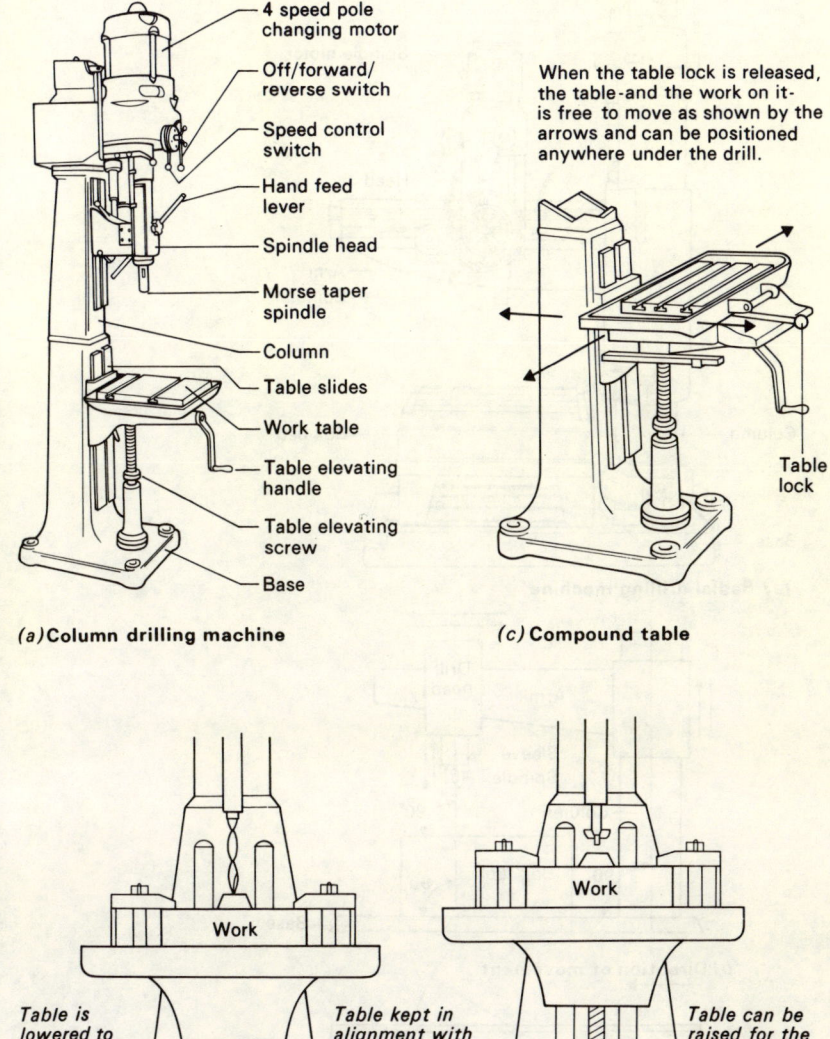

(a) Column drilling machine

- 4 speed pole changing motor
- Off/forward/reverse switch
- Speed control switch
- Hand feed lever
- Spindle head
- Morse taper spindle
- Column
- Table slides
- Work table
- Table elevating handle
- Table elevating screw
- Base

When the table lock is released, the table – and the work on it – is free to move as shown by the arrows and can be positioned anywhere under the drill.

Table lock

(c) Compound table

Table is lowered to accommodate the length of the drill.

Table kept in alignment with spindle axis by slide ways.

Work

Table can be raised for the shorter spot-facing cutter without losing alignment.

Work

(b) Table movement

Fig. 11.26   Column type drilling machine

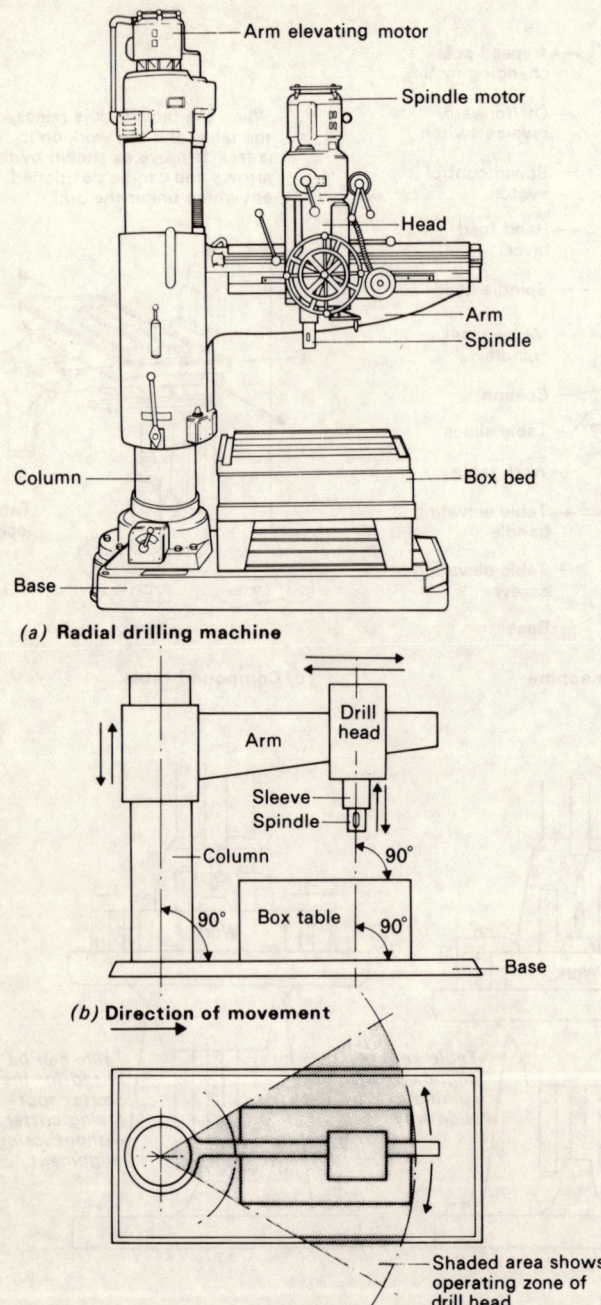

(a) Radial drilling machine

(b) Direction of movement

Shaded area shows operating zone of drill head

Fig. 11.27 Radial arm drilling machine

Problems

Section A

1. The spindle speed for a 10 mm diameter twist drill cutting at 30 m/min is: (*a*) 942 rev/min; (*b*) 955 rev/min; (*c*) 1047 rev/min; (*d*) 3000 rev/min.

2. The time taken to drill through a 10 mm thick plate at 300 rev/min and a feed of 0·2 mm/rev is: (*a*) 3 seconds; (*b*) 6 seconds; (*c*) 10 seconds; (*d*) 15 seconds.

3. A twist drill has its point thinned in order to: (*a*) reduce the hole diameter; (*b*) increase the rake angle; (*c*) locate in the centre punch mark; (*d*) reduce the axial (feed) pressure.

4. A reamer is used to correct the: (*a*) size and roundness of a drilled hole; (*b*) size and position of a drilled hole; (*c*) finish and position of a drilled hole; (*d*) finish and depth of a drilled hole.

5. Which of the following cutters should be used in a drilling machine to recess the head of a cap screw? (*a*) countersink bit; (*b*) spot-facing cutter; (*c*) counterbore; (*d*) centre-drill.

Section B

6. Describe, with the aid of sketches, how the link shown in Fig. 8.10 should be set up, drilled and reamed on a pillar drilling machine. Pay particular attention to work holding and safety. The blank will have been already marked out as described in section 8.7.

7. The casting shown in Fig. 8.17 is to have a 25 mm diameter hole drilled in the centre of boss 'B' and a 12 mm diameter hole drilled in the centre of each of the bosses marked 'A'.
    (*a*) The faces marked   have already been machined.
    (*b*) The hole centres have already been marked out on the bosses.
    Draw up an operation schedule for drilling the holes on a small radial arm drilling machine, paying particular attention to the: (i) method of work holding and clamping; (ii) method of locating the hole centres; (iii) limited feed pressure that can be brought to bear on the unsupported casting under the bosses.

8. (*a*) Make a neat freehand sketch of a machine reamer, and on it indicate the: (i) taper shank; (ii) bevel lead; (iii) flutes; (iv) body length.
    (*b*) Why do the flutes of a reamer have a **left-hand** helix?
    (*c*) What is the advantage of a reamer having; (i) diametrically opposed cutting edges; (ii) irregularly spaced cutting edges?
    (*d*) Which faults in a drilled hole is a reamer used to correct?
    (*e*) State **two** reasons why a reamer might leave a poor finish.

9. (*a*) Make a neat freehand sketch of a twist drill, and on it indicate the: (i) clearance angle; (ii) rake angle; (iii) shank; (iv) chisel point.
    (*b*) With the aid of sketches explain why the diameter of a twist drill and

the web of a twist drill both taper towards the shank, but in opposite directions.

(c) Sketch a typical application of a countersink cutter and of a counterbore cutter. Explain why the latter requires a pilot whilst the former does not.

10. (a) Sketch the feed mechanism of a bench drilling machine and explain why it is referred to as a 'sensitive' feed mechanism.

(b) Sketch a radial arm drilling machine. Name the more important parts and controls and show how its range of movements allows a number of holes to be drilled in a large casting without re-setting of the casting.

(c) With the aid of sketches explain how the table movements of a pillar drilling machine allow a component to be positioned under the drill without unclamping the component from the table.

Chapter 12

The centre lathe

12.1 Constructional features

The centre lathe is a machine tool designed to produce cyclindrical, conical and plane (flat) surfaces using a single point tool. It also produces screw threads. These surfaces and threads may be external or internal to the component. Figure 12.1 shows a typical, modern centre lathe and names the more important features. It will be seen that it is built up from a number of basic units which are accurately aligned during manufacture in order that precision turned components may be produced.

1. The bed

A typical lathe bed is a strong, bridge-like member, made of high grade cast iron and is heavily ribbed to give it rigidity. Its upper surface carries the main slideways. Since these slideways locate, directly or indirectly, most of the remaining units, they are responsible for the fundamental alignments of the machine. For this reason the bed slideways must be manufactured to high dimensional and geometrical tolerances. Further, the lathe must be installed with care in order to avoid distortion of the bed.

2. The headstock

The headstock, or 'fast-head' as it is sometimes called, is a box-like casting

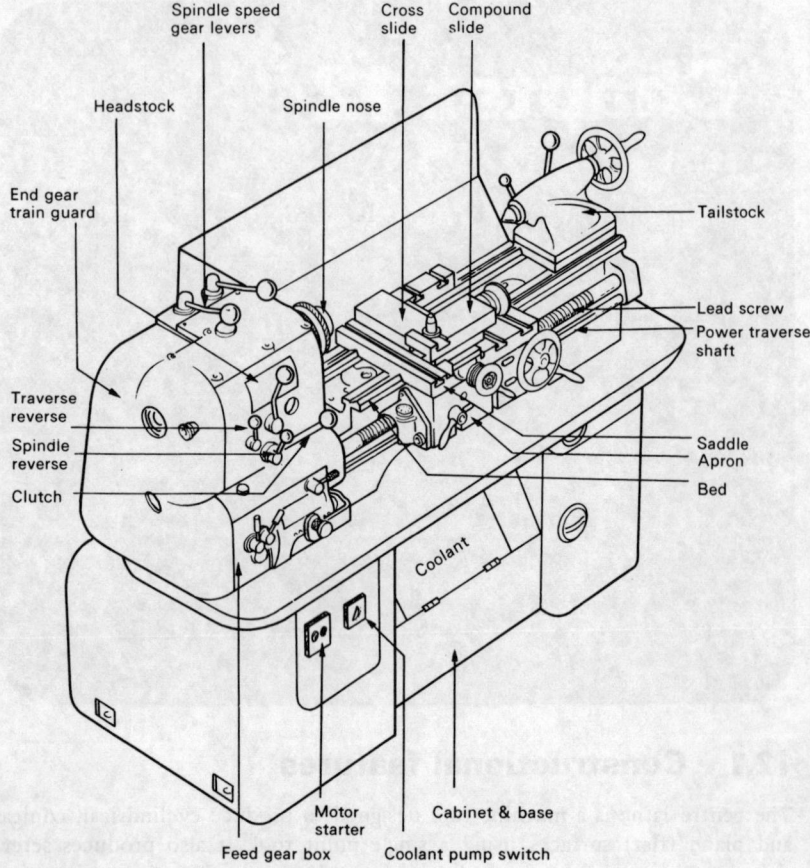

Fig. 12.1   The centre lathe

supporting the *spindle* and containing a gearbox by which the spindle may be rotated at various speeds.

The spindle is machined from a massive, hollow, alloy-steel forging and its purpose is to carry and rotate the various work-holding devices and, therefore, the work itself. The spindle is hollow to accept bar stock and its nose is bored internally to a Morse-taper to accept the adaptor sleeve that, in turn, accepts the Morse-taper shank of the live centre. The periphery of the spindle nose is machined to provide a mounting for the work-holding devices. Figure 12.2 shows three typical spindle nose mountings. Mountings based on taper locations are to be preferred as these are self-aligning and compensate for normal wear without loss of accuracy.

The spindle is located in strong and accurate bearings so that its axis, and the axis of the work-holding devices mounted upon it, is parallel to the bed slideways of the machine both in the horizontal and vertical planes. The

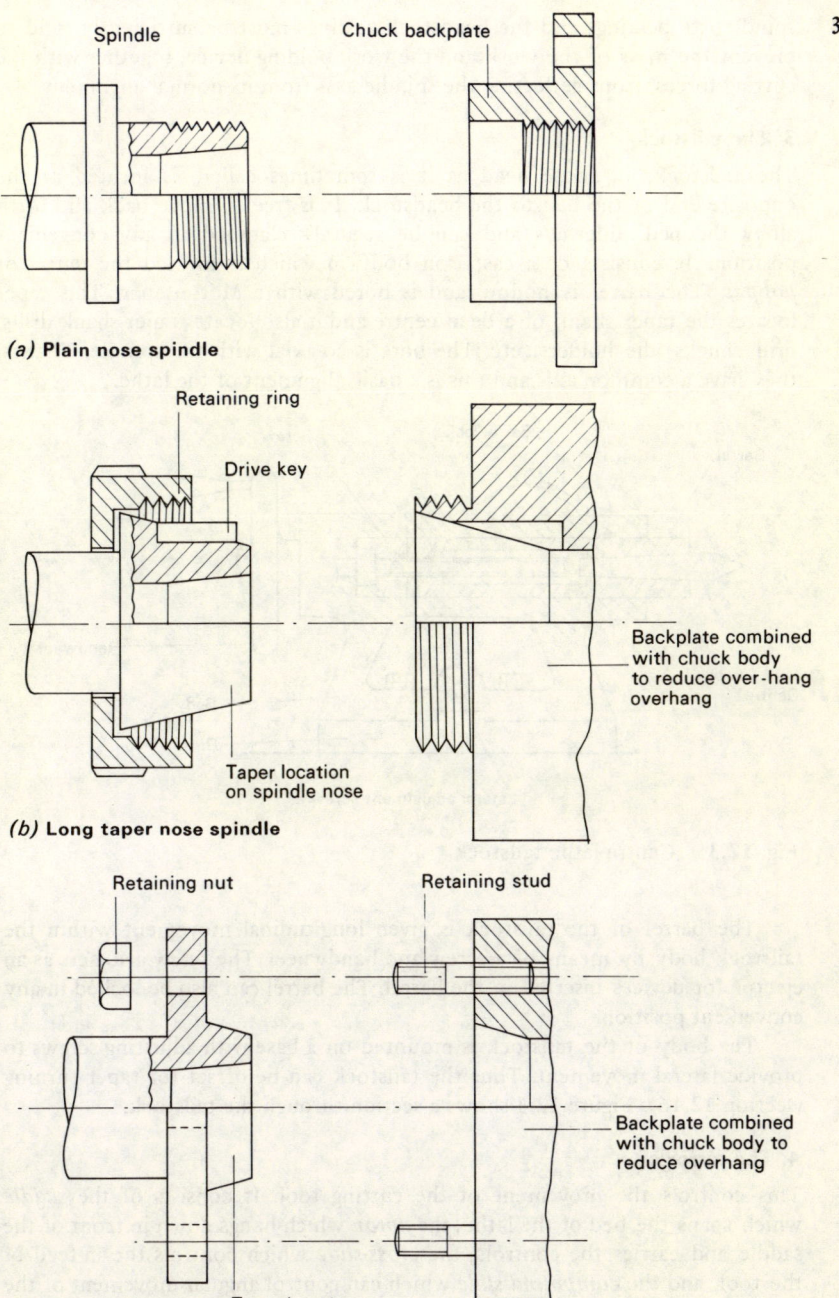

(a) **Plain nose spindle**

(b) **Long taper nose spindle**

(c) **Short taper nose spindle**

Fig. 12.2 Spindle nose mountings

spindle, its bearings and the headstock castings must be sufficiently rigid to prevent the mass of the work and the work-holding device, together with the cutting forces, from deflecting the spindle axis from its normal alignment.

3. The tailstock

The tailstock, or loose head as it is sometimes called, is located at the opposite end of the bed to the headstock. It is free to move back and forth along the bed slideways and can be securely clamped in any convenient position. It consists of a cast iron body in which is located the *barrel* or *poppet.* The barrel is hollow and is bored with a Morse-taper. This taper locates the taper shank of a dead centre and it also locates taper shank drills, drill chucks, die holders, etc. The bore is co-axial with the spindle. That is, they have a common axis and this is a basic alignment of the lathe.

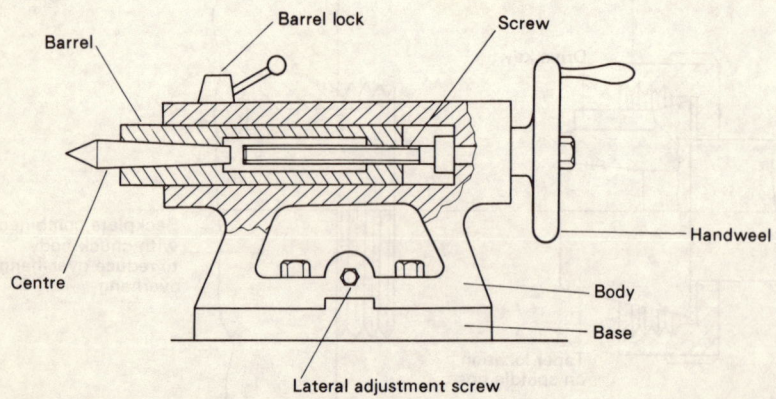

Fig. 12.3 Centre lathe tailstock

The barrel of the tailstock is given longitudinal movement within the tailstock body by means of a screw and handwheel. The screw also acts as an ejector for devices inserted in the barrel. The barrel can also be locked in any convenient position.

The body of the tailstock is mounted on a base with adjusting screws to provide lateral movement. Thus the tailstock can be offset for taper turning (section 12.16). Figure 12.3 shows a section through the tailstock.

4. The carriage

This controls the movement of the cutting tool. It consists of the *saddle* which spans the bed of the lathe; the *apron* which hangs down in front of the saddle and carries the controls; the *cross-slide* which controls the in-feed of the tool, and the *compound slide* which can control angular movement of the tool.

The carriage may be moved back and forth along the bed slideways using either the traverse handwheel (manual operation) or the power traverse.

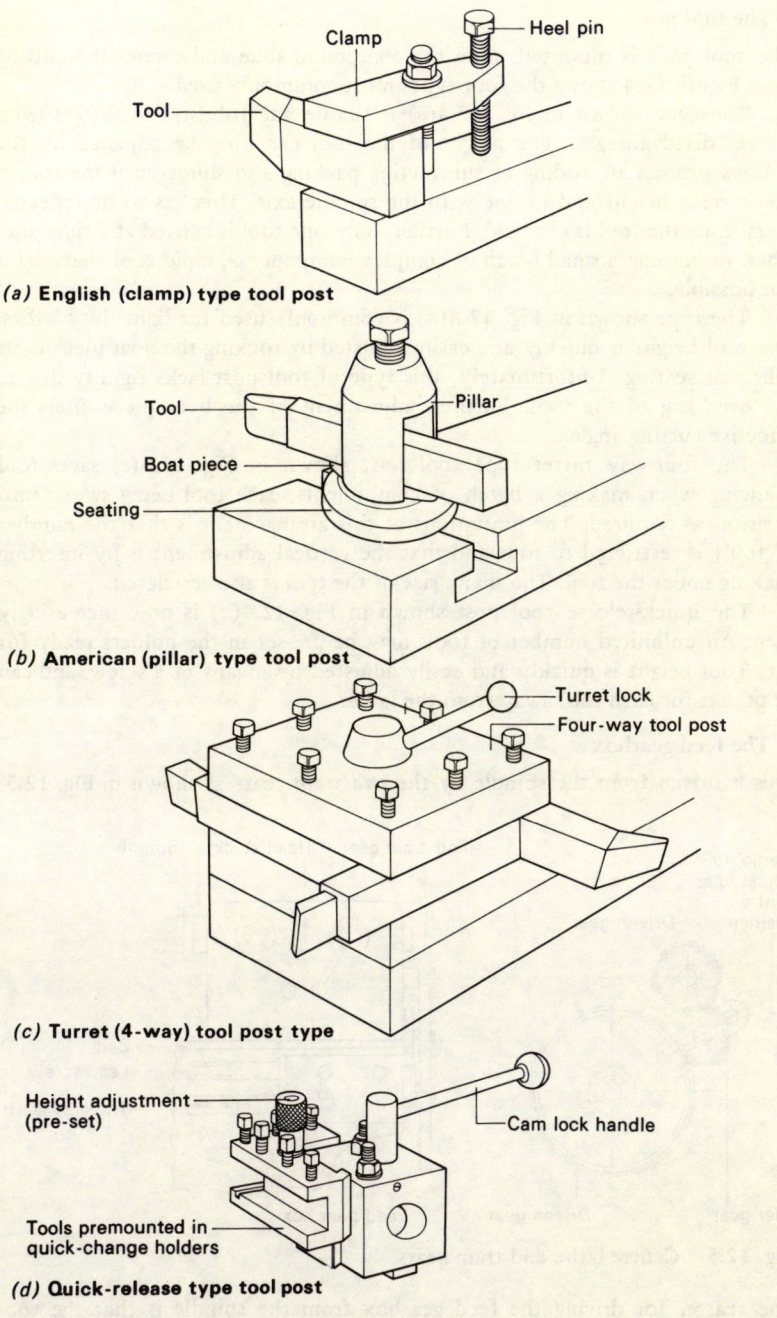

(a) **English (clamp) type tool post**

Clamp — Heel pin

Tool

(b) **American (pillar) type tool post**

Tool — Pillar

Boat piece

Seating

(c) **Turret (4-way) tool post type**

Turret lock

Four-way tool post

(d) **Quick-release type tool post**

Height adjustment (pre-set)

Cam lock handle

Tools premounted in quick-change holders

Fig. 12.4 Centre lathe tool posts

5. The tool post

The tool post is mounted upon the compound slide and carries the cutting tool. Figure 12.4 shows the four types most commonly used.

The type shown in Fig. 12.4(a) is simple and robust, but suffers from several disadvantages. The height of the tool can only be adjusted by the tedious process of adding or subtracting packing and shims until the tool is the correct height and in line with the spindle axis. This has to be repeated every time the tool is changed. Further, only one tool is carried at a time and, when machining a small batch of complex components, rapid tool changing is not possible.

The type shown in Fig. 12.4(b) is commonly used for light duty lathes. The tool height is quickly and easily adjusted by rocking the boat piece in its spherical seating. Unfortunately, this type of tool post lacks rigidity due to the overhang of the tool. Further, adjustment of the boat piece alters the effective cutting angles.

The four-way turret type tool post shown in Fig. 12.4(c) saves tool changing when making a batch of components, each tool being swung into position as required. The limitations of this arrangement is that the number of tools is restricted to four and that the vertical adjustment is by inserting packing under the tool. The shank size of the tool is also restricted.

The quick-release tool post shown in Fig. 12.4(d) is now increasingly used. An unlimited number of tools may be pre-set in the holders ready for use. Tool height is quickly and easily adjusted by means of a screw, and can be pre-set for each tool away from the lathe.

6. The feed gearbox

This is driven from the spindle by the *end train* gears as shown in Fig. 12.5.

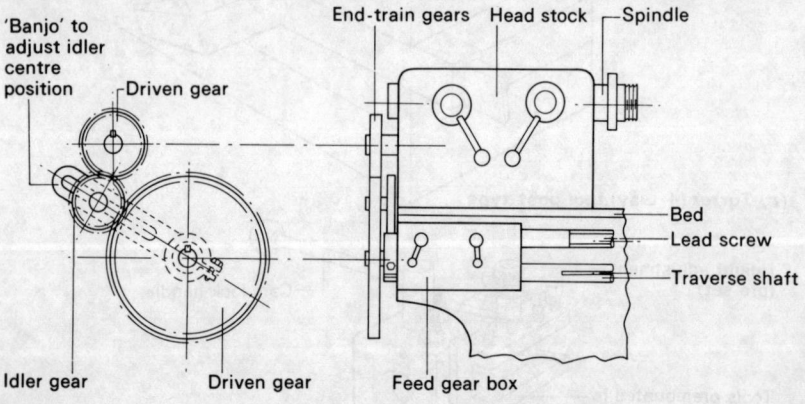

Fig. 12.5 Centre lathe end train gears

The reason for driving the feed gearbox from the spindle is that the tool movement per revolution of the spindle must remain constant even if the spindle speed is changed. The feed gearbox has three functions.

1. To control the speed at which the saddle is driven along the bed when power traversing.
2. To control the speed at which the cross slide moves across the saddle when power cross-traversing.
3. To control the speed of the lead screw when cutting screw threads, and thus controlling the lead of the screw being cut.

Note: Power traverse is provided by the traverse shaft and the lead screw should only be used when screw cutting to preserve its accuracy.

Many lathes are provided with a slipping clutch, or a shear pin, in the traverse shaft drive to prevent damage to the gearbox if too heavy a cut is taken.

The lead screw is usually provded with a dog-clutch so that it can be disconnected when not in use to prevent wear. Figure 12.6 shows a typical, Norton-type feed gearbox.

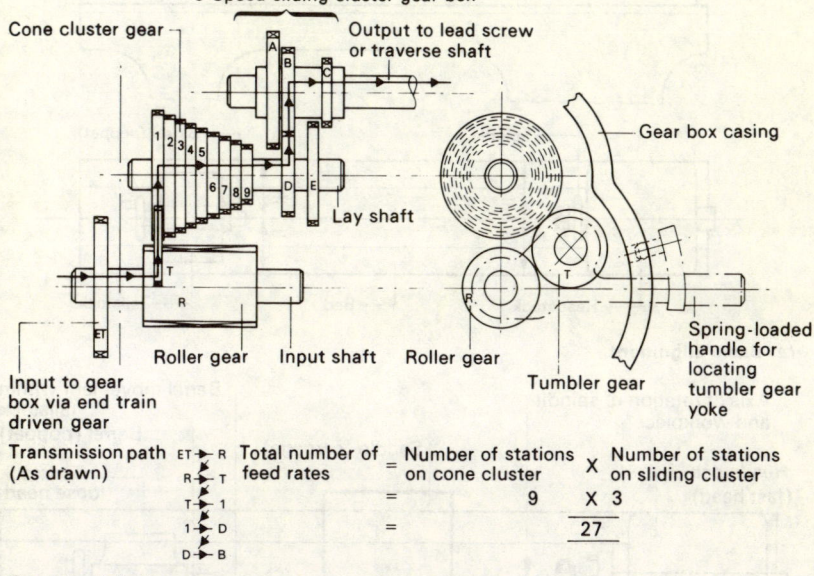

Transmission path (As drawn)	ET → R R → T T → 1 1 → D D → B	Total number of feed rates	=	Number of stations on cone cluster	X	Number of stations on sliding cluster
			=	9	X	3
			=			27

Principle of the Norton–type gear box
The Roller Gear (R), which is driven by the input shaft, can be engaged with any station (1–9) on the cluster cone gear by means of the tumbler gear (T). The tumbler gear is carried on a movable yoke which is located in the gearbox casing by a peg in the spring-loaded handle. The peg locates in holes in the gear box casing.

Fig. 12.6   Centre lathe − feed gear box

12.2   Basic alignments and movement

Figure 12.7(*a*) shows the basic alignment of the headstock, tailstock, spindle and bed slideways. It will be seen that the common spindle and tailstock axis is parallel to the bed slideways in both the vertical and horizontal planes. This is the basic alignment of the lathe and all other alignments are referred to it.

The movement of the carriage alone is shown in Fig. 12.7(*b*). It will be seen that it moves the cutting tool in a path parallel to the spindle axis and this produces cylindrical surfaces.

The cross-slide on top of the carriage is aligned at 90° to the spindle axis as shown in Fig. 12.7(*c*). Since this slide moves the tool in a path at right angles to the spindle axis, it is used for producing plane surfaces. This operation is called 'facing' and is used to machine the flat ends of components. The cross-slide is also used to control the depth of cut, or in-feed, of the cutting tool and for this purpose the handwheel is fitted with a micrometer dial.

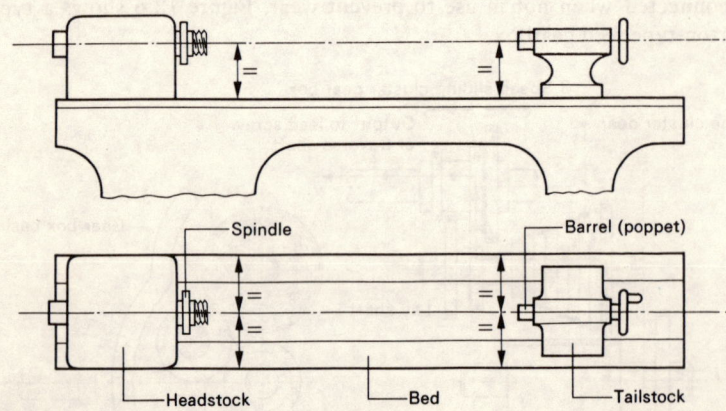

Spindle — Barrel (poppet)

Headstock — Bed — Tailstock

(a) **Basic alignment**

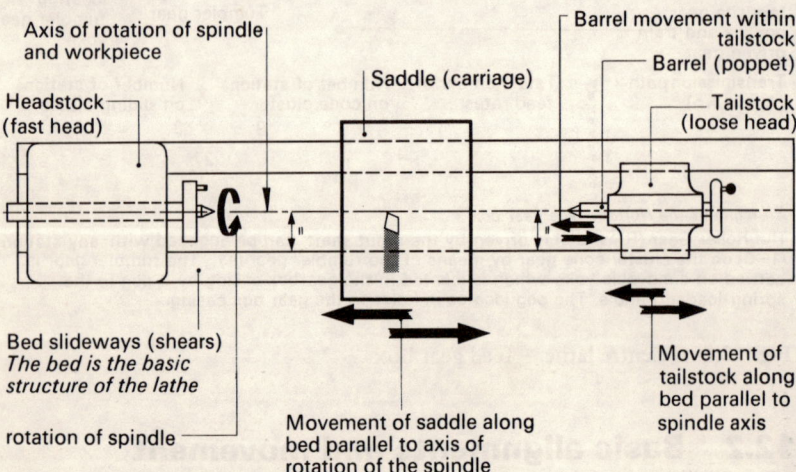

Axis of rotation of spindle and workpiece

Saddle (carriage)

Barrel movement within tailstock

Barrel (poppet)

Headstock (fast head)

Tailstock (loose head)

Bed slideways (shears)
The bed is the basic structure of the lathe

rotation of spindle

Movement of saddle along bed parallel to axis of rotation of the spindle

Movement of tailstock along bed parallel to spindle axis

(b) **The carriage or saddle provides the basic movement of the cutting tool parallel to the work axis**

Fig. 12.7 Centre lathe — basic geometry

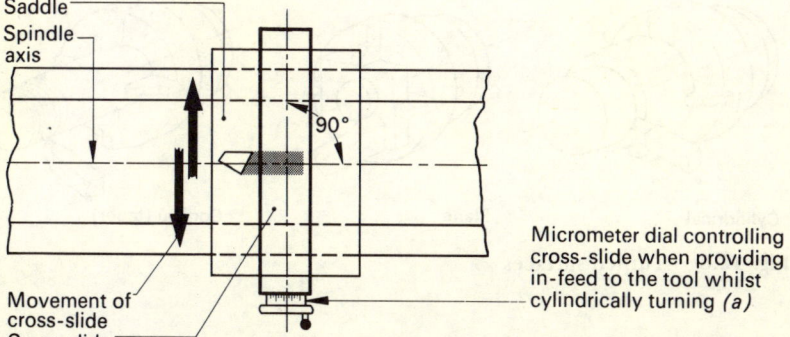

Saddle

Spindle axis

90°

Movement of cross-slide
Cross-slide

Micrometer dial controlling cross-slide when providing in-feed to the tool whilst cylindrically turning *(a)*

(c) The cross-slide

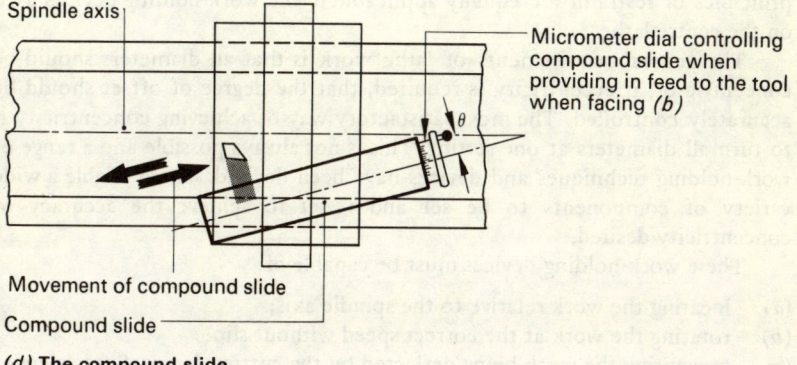

Spindle axis

Micrometer dial controlling compound slide when providing in feed to the tool when facing *(b)*

θ

Movement of compound slide

Compound slide

(d) The compound slide

Fig. 12.7 *(continued)*

The top or compound-slide is located on top of the cross-slide and can be set at an angle to the spindle axis as shown in Fig. 12.7(*d*). Since it moves the tool in a path that is at an angle to spindle axis it is used to produce conical or tapered components. Taper turning is considered more fully in section 12.18. When set parallel to the spindle axis the compound-slide can be used for controlling the in-feed of the cutting tool when facing the component from the cross-slide. For this purpose the handwheel of the compound-slide is fitted with a micrometer dial.

The surfaces produced by these basic alignments and movements are summarised in Fig. 12.8. Since these surfaces depend upon the movements and geometrical alignments of the slideways for their production, and are independent of the shape of the cutting tool, they are said to be *generated*. The generation and forming of surfaces will be considered in detail, for a range of machine tools, in the unit for *Manufacturing Technology I* (level 2).

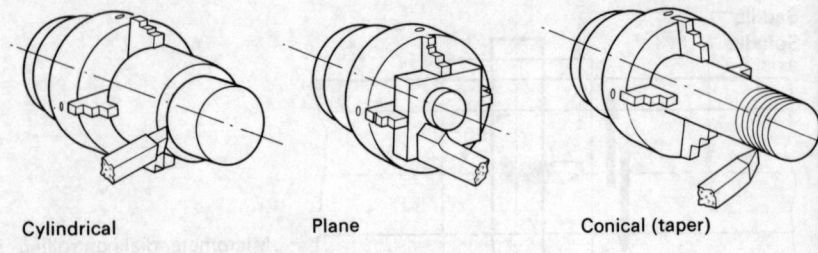

| Cylindrical | Plane | Conical (taper) |

Fig. 12.8   Turned surfaces

12.3   Work-holding on the centre lathe

The use of locations and clamps to restrain the workpiece during a metal cutting process has already been introduced in Chapter 10 and 11. The same principles of restraint are equally applicable to the work-holding devices used on the centre lathe.

The general requirements of lathe work is that all diameters should be concentric or, if eccentricity is required, that the degree of offset should be accurately controlled. The most satisfactory way of achieving concentricity is to turn all diameters at one setting. This is not always possible and a range of work-holding techniques and devices have been devised. These enable a wide variety of components to be set and re-set to achieve the accuracy of concentricity desired.

These work-holding devices must be capable of:

(a)   locating the work relative to the spindle axis;
(b)   rotating the work at the correct speed without slip;
(c)   preventing the work being deflected by the cutting forces. Some slender work requires additional support;
(d)   holding the work sufficiently rigidly so that it will not spin out of the machine, be ejected by the cutting forces, yet not be crushed or distorted by the work-holding device.

The following are the normal methods of work-holding on the centre lathe. They exclude special turning fixtures.

12.4   Between centres

This is the traditional method work-holding from which the centre lathe gets its name, and is shown in Fig. 12.9(a). It will be seen that the component is located between the centres and is driven by the catchplate and carrier. Since the driving mechanism can 'float' it has no influence on the accuracy of location of the workpiece. The centres, themselves, have Morse-taper shanks and are located in taper sockets in the spindle and tailstock barrel. The use of taper locations ensures axial alignment irrespective of variations due to manufacturing tolerance and wear. Figure 12.9(b) shows the restraints acting

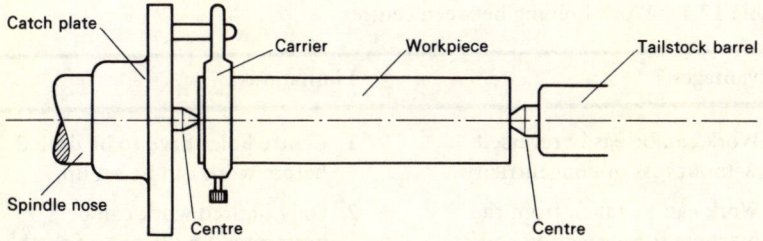

(a) **Workholding between centres**

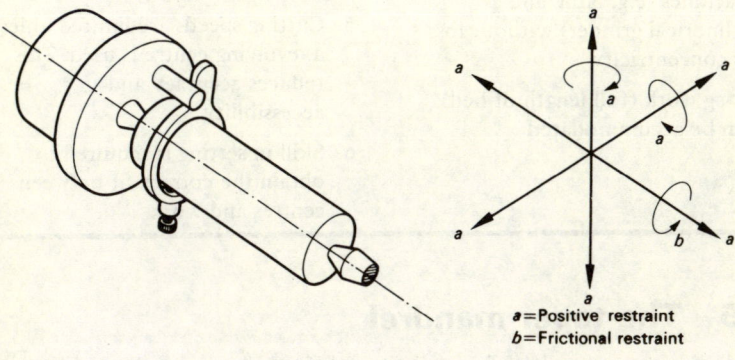

a = Positive restraint
b = Frictional restraint

(b) **Restraints**

Fig. 12.9 Work-holding between centres

upon a component held between centres. The advantages and limitations of this method of work-holding are listed in Table 12.1.

For parallel turning, the axis of the spindle centre and the axis of the tailstock centre must be coincident with each other and parallel to the main bed slideways. The tailstock is provided with lateral (sideways) adjustment to achieve this end.

When turning between centres a trial cut should be taken along the component. It is then checked at each end with a micrometer caliper. If the readings are the same, then the component is a true cylinder and the roughing and finishing cuts may be taken. If the readings are different, then the tailstock has to be adjusted as shown in Fig. 12.10(*a*). Further trial cuts may be taken until the diameter is constant along the length of the component.

A more convenient method of bringing the axis into alignment by lateral adjustment is shown in Fig. 12.10(*b*). This can only be done if a parallel test bar is available. Note that a mandrel cannot be used as it has a built-in taper and would give false readings. (See section 12.5.)

Table 12.1   Work-holding between centres

Advantages	Limitations
1. Work can be easily reversed without loss of concentricity	1. Centre holes have to be drilled before work can be set up
2. Work can be taken from the machine for inspection and easily re-set without loss of concentricity	2. Only limited work can be performed on the end of the bar
3. Work can be transferred between machines (e.g. lathe and cylindrical grinder) without loss of concentricity	3. Boring operations cannot be performed
	4. There is lack of rigidity
4. Long work (full length of bed) can be accommodated	5. Cutting speeds are limited unless a revolving centre is used. This reduces accuracy and accessibility
	6. Skill in setting is required to obtain the correct fit between centres and work

12.5   The taper mandrel

If the diameter of a hole is small, it is difficult to make the boring tool sufficiently rigid to true up the run out of the previously drilled hole. Under these conditions it is better to rough out the external diameters; drill and ream the bore to provide a round hole of good finish; mount the component on a mandrel as shown in Fig. 12.11(a) and finish turn the external diameters which will then be true with the mandrel axis. A mandrel press for mounting and dismounting the component is shown in Fig. 12.11(b).

The mandrel is tapered so that the further the component is forced on, the more firmly it is held in place. Therefore, the direction of cutting should always be towards the 'plus end' of the mandrel.

Figure 12.12 shows the restraints acting on a component mounted on a mandrel. The restraints shown are those relative to the mandrel. The restraints acting on the mandrel itself are the same as for any component mounted between centres. Table 12.2 lists the advantages and limitations of this method of work-holding.

12.6   The self-centring chuck

Figure 12.13(a) shows the construction of a typical three-jaw chuck. It will be seen that the scroll not only clamps the component in place, it also locates the component as well. This is fundamentally bad practice, since any wear in

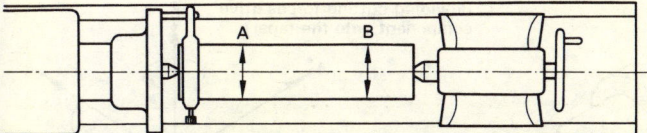

For parallel cylindrical turning the axis of the headstock spindle must be in
alignment with the tailstock barrel. If this is so, then the diameter of the
component at 'A' will be the same as the diameter at 'B'

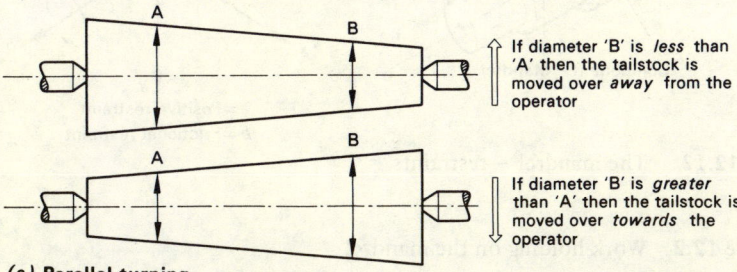

If diameter 'B' is *less* than
'A' then the tailstock is
moved over *away* from the
operator

If diameter 'B' is *greater*
than 'A' then the tailstock is
moved over *towards* the
operator

(a) **Parallel turning**

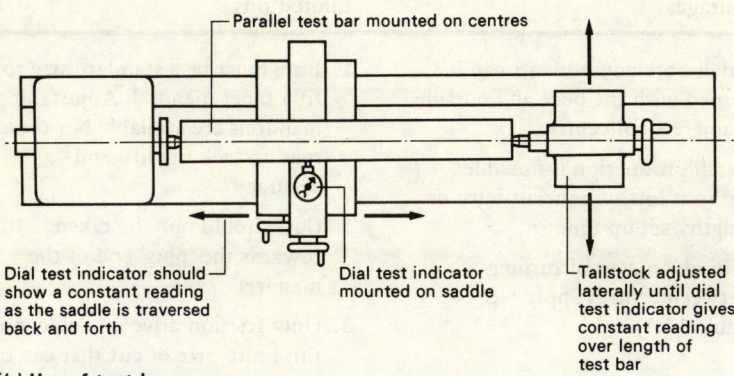

Parallel test bar mounted on centres

Dial test indicator should
show a constant reading
as the saddle is traversed
back and forth

Dial test indicator
mounted on saddle

Tailstock adjusted
laterally until dial
test indicator gives
constant reading
over length of
test bar

(b) **Use of test bar**

Fig. 12.10 Parallel turning – setting the tailstock

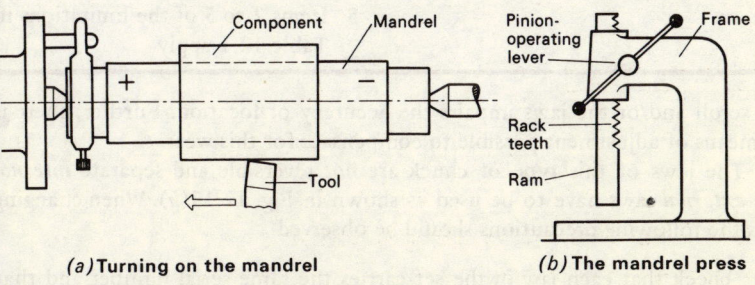

Component   Mandrel

Tool

Pinion-
operating
lever

Frame

Rack
teeth

Ram

(a) **Turning on the mandrel**

(b) **The mandrel press**

Fig. 12.11 Mandrel work

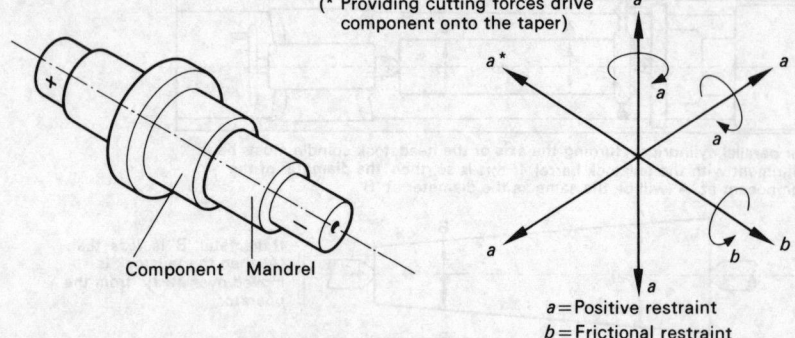

(* Providing cutting forces drive component onto the taper)

a = Positive restraint
b = Frictional restraint

Fig. 12.12   The mandrel — restraints

Table 12.2   Work-holding on the mandrel

Advantages	Limitations
1. Small-bore components can be turned with the bore and outside diameters concentric	1. Bore must be a standard size to fit a taper mandrel. Adjustable mandrels are available but these tend to lack rigidity and accuracy
2. Batch production is possible without loss of concentricity or lengthy set-up time	2. Cuts should only be taken towards the 'plus' end of the mandrel
3. The advantages of turning between centres apply (see Table 12.1)	3. Only friction drive available, and this limits size of cut that can be taken
	4. Special mandrels can be made but this is not economical for one-off jobs
	5. Items 2 to 5 of the limitations in Table 12.1 apply

the scroll and/or the jaws impairs the accuracy of location. Further, there is no means of adjustment possible to compensate for this wear.

The jaws of this type of chuck are *not* reversible and separate *internal* and *external* jaws have to be used as shown in Fig. 12.13(b). When changing jaws the following precautions should be observed:

(a)   check that each jaw in the set carries the same serial number and that this serial number is the same as that on the chuck body;

(b)   insert the jaws sequentially; number 1 jaw in number 1 slot, etc.

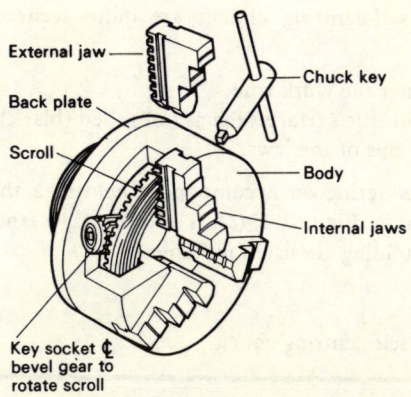

(a) **Construction**

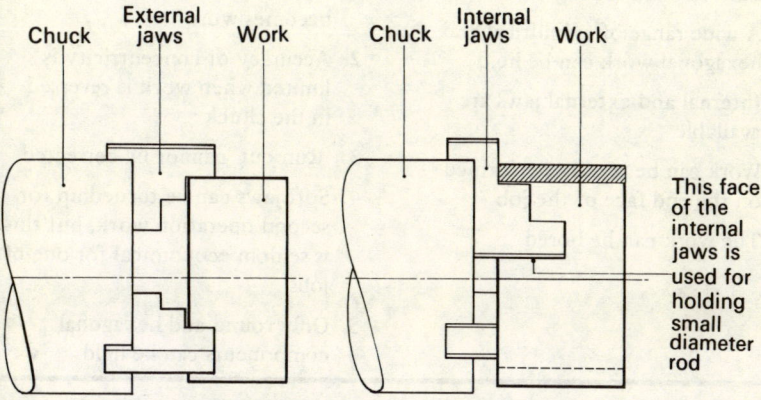

(b) **Internal and external work holding**

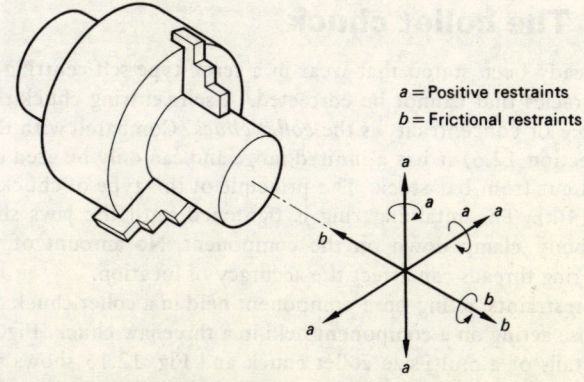

(c) **The self-centring chuck - restraints**

Fig. 12.13 The three-jaw, self centring chuck

When new, self-centring chucks are quite accurate. To preserve this accuracy, *never*:

(*a*)   try to hammer the work true;
(*b*)   hold on an untrue surface such as hot rolled (black) bar;
(*c*)   hold on the tips of the jaws.

The restraints acting on a component held in a three-jaw self-centring chuck are shown in Fig. 12.13(*c*). The advantages and limitations of this method of work-holding are listed in Table 12.3.

Table 12.3   The self-centring chuck

Advantages	Limitations
1. Ease of work setting	1. Accuracy decreases as chuck becomes worn
2. A wide range of cylindrical and hexagonal work can be held	2. Accuracy of concentricity is limited when work is reversed in the chuck
3. Internal and external jaws are available	3. 'Run out' cannot be corrected
4. Work can be readily performed on the end face of the job	4. Soft jaws can be turned up for second operation work, but this is seldom economical for one-off jobs
5. The work can be bored	5. Only round and hexagonal components can be held

12.7   The collet chuck

It has already been stated that wear in a scroll-type self-centring chuck leads to inaccuracies that cannot be corrected. A self-centring chuck that retains a high degree of concentricity is the *collet chuck*. Compared with the three-jaw chuck (section 12.6) it has a limited range and can only be used on bar stock or blanks cut from bar stock. The principle of this type of chuck is shown in Fig. 12.14(*a*). The retaining ring is tightened until the jaws sliding in the tapered body clamp down on the component. No amount of wear in the retaining ring threads can affect the accuracy of location.

The restraints acting on a component held in a collet chuck are the same as for those acting on a component held in a three-jaw chuck. Figure 12.14(*b*) shows details of a multi-size collet chuck and Fig. 12.15 shows a split collet chuck. The advantages and limitations of these two types of chuck are considered in Tables 12.4 and 12.5 respectively.

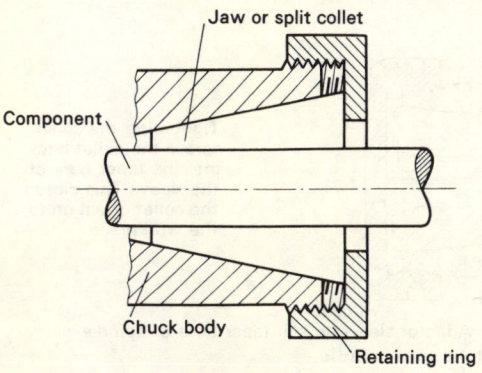

Jaw or split collet

Component

Chuck body

Retaining ring

(a) **Collet chuck–Principle**

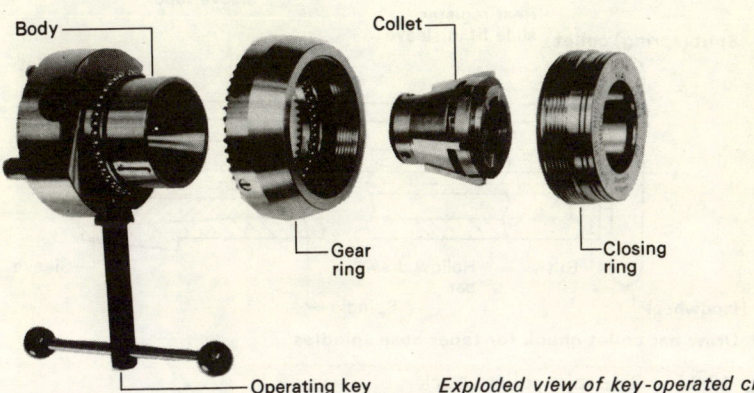

Body

Collet

Gear ring

Closing ring

Operating key

Exploded view of key-operated chuck

E type collet with standard plain gripping blades

(b) **The multi-size collet chuck**

Fig. 12.14 The collet chuck

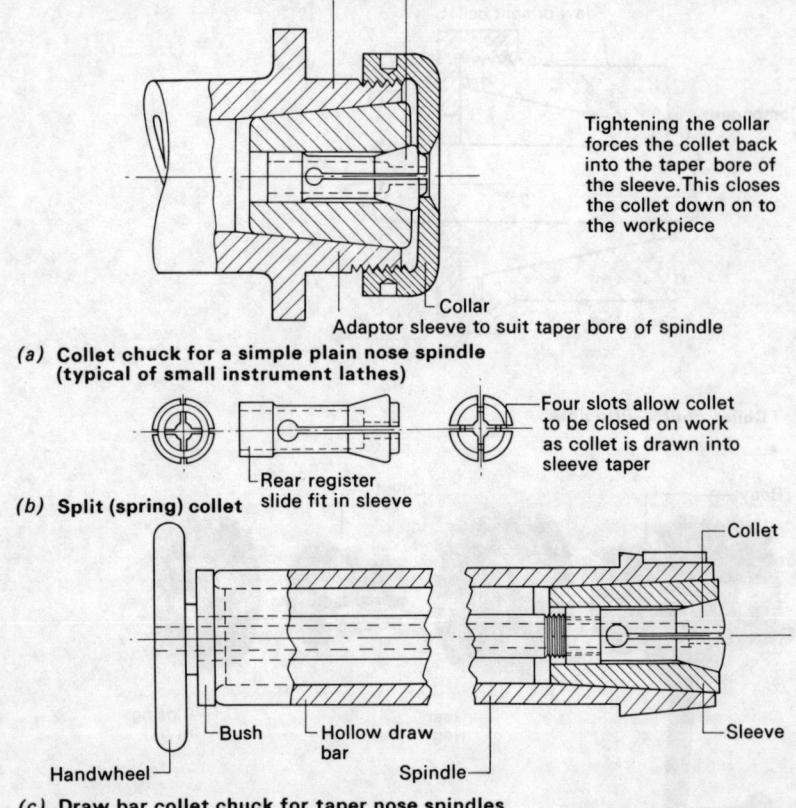

(a) **Collet chuck for a simple plain nose spindle**
(typical of small instrument lathes)

Spindle nose — Split (spring) collet

Tightening the collar forces the collet back into the taper bore of the sleeve. This closes the collet down on to the workpiece

Collar
Adaptor sleeve to suit taper bore of spindle

Four slots allow collet to be closed on work as collet is drawn into sleeve taper

Rear register slide fit in sleeve

(b) **Split (spring) collet**

Collet

Bush — Hollow draw bar
Handwheel — Spindle — Sleeve

(c) **Draw bar collet chuck for taper nose spindles**

Fig. 12.15 The split collet chuck

Table 12.4 The split collet chuck

Advantages	Limitations
1. Very high accuracy of concentricity	1. Only accurately turned, ground or drawn rod can be held in a collet
2. Accuracy maintained over long periods of use	2. Separate collets have to be used for each size of rod. Range of adjustment very small
3. Simple, compact and reliable	
4. Very quickly loaded	3. Although simple, initial cost is high due to the large number of collets that have to be bought
5. Considerable gripping power	
6. Unlikely to mark or damage work	

Table 12.4 (*continued*)   319

Advantages	Limitations
7. Work can be removed and replaced without loss of accuracy	4. Range of sizes that can be held limited by bore of spindle
8. Work can be turned externally, internally (bored) and end faced	5. Work can only be held on external surfaces
9. No overhang from spindle nose reduces chatter and geometrical inaccuracy. Very useful where work has to be 'parted off'	6. Only collets with circular or hexagonal jaws are available from stock. Other sections have to be made to special order (costly)
	7. Special adaptor sleeve required to suit bore of spindle nose

Table 12.5   The multi-jaw collet chuck

Advantages	Limitations
1. High accuracy of concentricity	1. Only accurately turned, ground or drawn rod can be held in a collet
2. Accuracy maintained over long periods of use	
3. Quickly loaded	2. Only collets for round or hexagonal rods are available
4. Considerable gripping power	
5. Unlikely to mark or damage work	3. Work can only be held on external diameters
6. Work can be removed and replaced without loss of accuracy	4. Chuck body and collets, complex and expensive
7. Work can be turned externally, internally (bored) and end faced	5. Chuck body has the overhang of a three-jaw or four-jaw chuck. Hence no improvement in rigidity of work-holding
8. Each collet has a wide range of adjustment	
9. A wide range of sizes can be accommodated by few collets	6. Range of sizes that can be held are greater than for a split collet, but smaller than for a conventional chuck
10. Collet jaws have a parallel movement which improves their gripping power	
11. Chuck body will fit any standard spindle nose	

12.8   The four-jaw independent chuck

This type of chuck, which is shown in Fig. 12.16, is much more heavily constructed than the self-centring chuck and has much greater holding power. Each jaw is moved independently by a square thread screw and is reversible. These chucks are used for holding:

(a)   irregularly shaped work;
(b)   work that must be trued up to run concentrically;
(c)   work that must be deliberately offset to run eccentrically.

Eccentrically mounted work must be balanced when using either the four-jaw chuck or the face plate (see section 12.9).

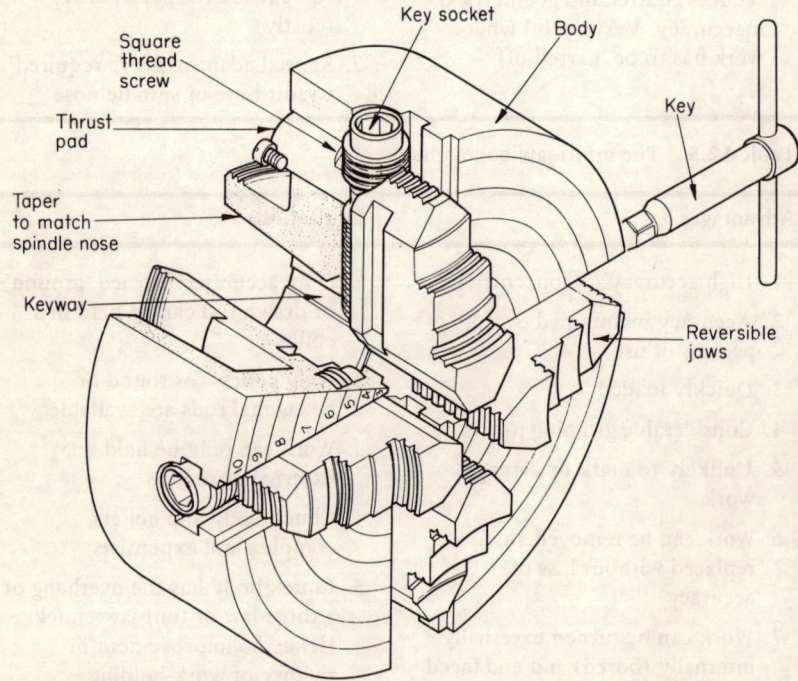

Fig. 12.16   The four-jaw chuck

The jaws of the four-jaw chuck *can be reversed* for holding externally on the workpiece and, therefore, separate internal and external jaws are not required (see section 12.6). Since the jaws are moved independently in this type of chuck, the component has to be set to run concentrically with the spindle axis. If a smooth or previously machined surface is available a dial test indicator (DTI) may be used as shown in Fig. 12.17(a). Alternatively, if a centre point is to be picked up for drilling and boring, a floating centre and dial test indicator may be used as shown in Fig. 12.17(b). Rough work may be set as shown in Fig. 12.17(c).

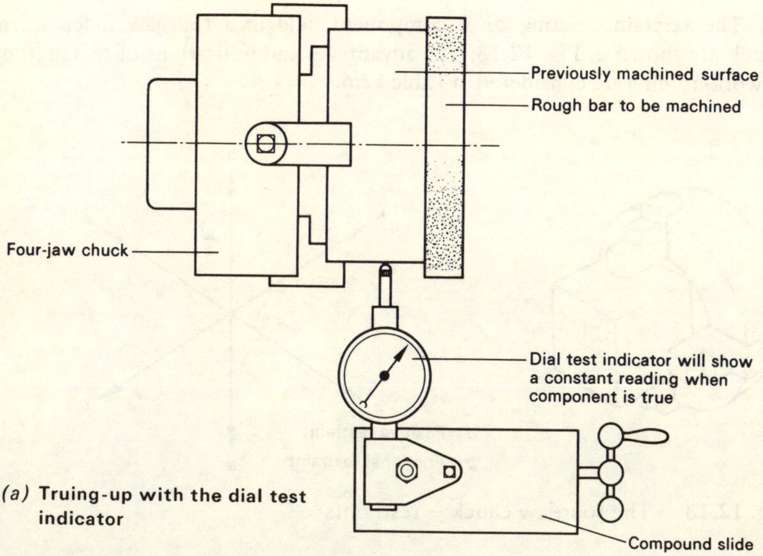

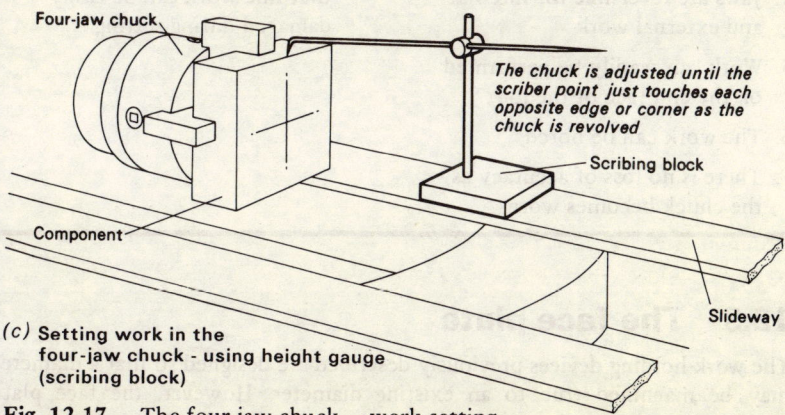

Previously machined surface

Rough bar to be machined

Four-jaw chuck

Dial test indicator will show a constant reading when component is true

(a) Truing-up with the dial test indicator

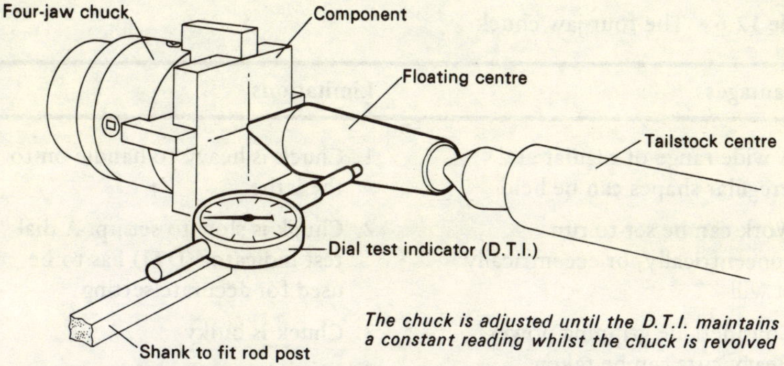

Compound slide

Four-jaw chuck

Component

Floating centre

Tailstock centre

Dial test indicator (D.T.I.)

The chuck is adjusted until the D.T.I. maintains a constant reading whilst the chuck is revolved

Shank to fit rod post

(b) Setting work in the four-jaw chuck – using D.T.I. and centre

Four-jaw chuck

The chuck is adjusted until the scriber point just touches each opposite edge or corner as the chuck is revolved

Scribing block

Component

Slideway

(c) Setting work in the four-jaw chuck – using height gauge (scribing block)

Fig. 12.17 The four-jaw chuck – work setting

The restraints acting on a component held in a four-jaw independent chuck are shown in Fig. 12.18. The advantages and limitations of this method of work-holding are considered in Table 12.6.

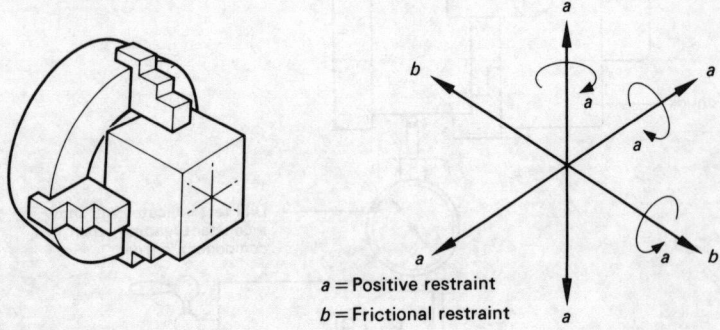

a = Positive restraint
b = Frictional restraint

Fig. 12.18   The four-jaw chuck — restraints

Table 12.6   The four-jaw chuck

Advantages	Limitations
1. A wide range of regular and irregular shapes can be held	1. Chuck is heavy to handle on to the lathe
2. Work can be set to run concentrically, or eccentrically at will	2. Chuck is slow to set up. A dial test indicator (DTI) has to be used for accurate setting
3. Considerable gripping power. Heavy cuts can be taken	3. Chuck is bulky
4. Jaws are reversible for internal and external work	4. The gripping power is so great that fine work can be easily damaged during setting
5. Work can readily be performed on the end face of the job	
6. The work can be bored	
7. There is no loss of accuracy as the chuck becomes worn	

12.9   The face plate

The work-holding devices previously described are designed so that a diameter may be machined true to an existing diameter. However, the face plate enables the component to be located so that a diameter can be turned parallel or perpendicular to a previously machined flat surface. This flat surface is the datum from which the diameter is set as shown in Fig. 12.19.

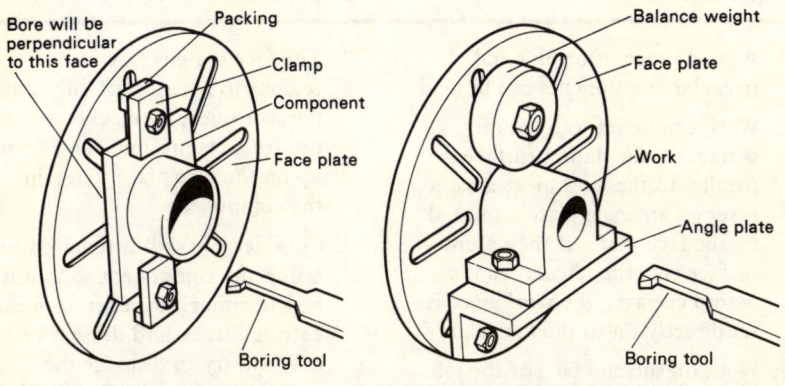

(a) Balanced work   **(b) Unbalanced work**

Fig. 12.19   The face plate

It will be seen that in the example shown in Fig. 12.19(a) the axis of the bore will be perpendicular to the datum surface. In the example shown in Fig. 12.19(b) the axis of the bore will be parallel to the datum which, in this case, is the base of the workpiece.

It will be seen that a balance weight is used to prevent out-of-balance forces damaging the spindle bearings and causing chatter when turning. It also prevents the work swinging round, after the machine has been stopped, and trapping the operator whilst he takes measurements or sets the tool.

The restraints acting on a component held on a face plate are shown in Fig. 12.20. The advantages and limitations of this method of work-holding are considered in Table 12.7.

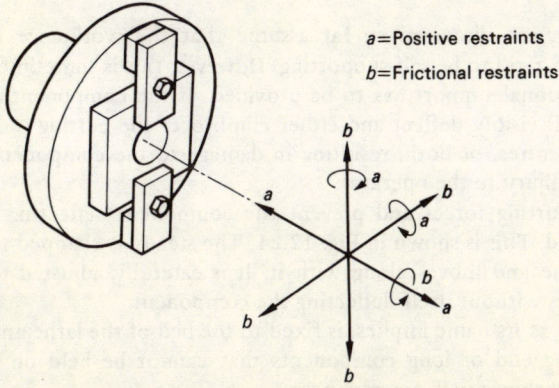

a=Positive restraints
b=Frictional restraints

Fig. 12.20   The face plate — restraints

Table 12.7 The face-plate

Advantages	Limitations
1. A wide range of regular and irregular components can be held	1. The face plate is slow and tedious to set up. Not only must the workpiece be clocked up to run true, clamps must also be set up on the face plate to retain the component
2. Work can be set to a datum surface. If the datum surface is parallel to the workpiece axis, it is set on an angle plate mounted on the face plate. If the datum surface is perpendicular to the workpiece axis, the workpiece is set directly on to the face plate	2. Considerable skill is required to clamp the component so that it is rigid enough to resist both the cutting forces, and those forces that will try tp dislodge the work as it spins rapidly round
3. Work on the end face of the job is possible	3. Considerable skill is required to avoid distorting the workpiece by the clamps
4. The work can be bored	
5. The work can be set to run concentrically or eccentrically at will	4. Irregular jobs have to be carefully balanced to prevent vibration, and the job rolling back on the operator
6. There are no moving parts to lose their accuracy with wear	
7. The work can be rigidly clamped to resist heavy cuts	5. The clamps can limit the work that can be performed on the end face

12.10 Use of steadies

The work-holding devices discussed so far assume that the workpiece is sufficiently robust and rigid to be self-supporting. However, this is sometimes not the case and additional support has to be provided. If the component is long and slender it will visibly deflect and either climb over the cutting tool or spring out of the centres, or both, resulting in damage to the component and, possibly, serious injury to the operator.

To balance the cutting forces and prevent the component deflecting a *travelling steady* is used. This is shown in Fig. 12.21. The steady is clamped to the saddle of the lathe and moves along with it. It is carefully adjusted to resist the cutting forces without itself deflecting the component.

The fixed steady, as its name implies, is fixed to the bed of the lathe and is used to support the end of long components that cannot be held on a centre. Figure 12.22(*a*) shows such a component.

The component would be set up as follows. The 80 mm diameter end

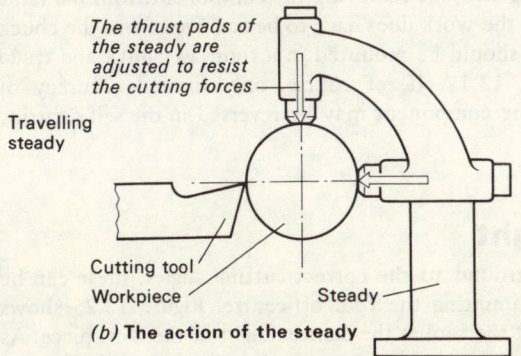

(Courtesy Colchester Lathe Co.)

(a) The travelling steady mounted on the Saddle

The thrust pads of the steady are adjusted to resist the cutting forces

Travelling steady

Cutting tool

Workpiece

Steady

(b) The action of the steady

Fig. 12.21 The travelling steady

would be filed flat. It would then be centred by marking out with a combination square, centre-finder as shown in Fig. 12.22(*b*). The centre hole would be carefully drilled with a portable electric drill.

The flange of the component would then be checked, and the opposite end held on the tailstock centre. A 'track' for the fixed steady would then be turned, as shown in Fig. 12.22(*c*) (unless a suitable diameter has to be turned anyway).

The steady would then be fixed in position on the lathe bed and closed around the component. It would then be adjusted to support the component and the tailstock would be removed. Figure 12.22(*d*) shows the steady in position ready for the end of the bar to be machined.

12.11   Concentricity

When a component is being turned it is usual to try to keep the various diameters concentric, that is, having the same axis. The meaning of concentricity and eccentricity is shown in Fig. 12.23.

In Fig. 12.23(*a*) the two diameters A and B are *concentric*; that is, they have the same centre of rotation and lie on the same axis. For example, if diameter B was rotated in a vee block and a dial test indicator was in contact with diameter A, then the dial test indicator would show a constant reading.

In Fig. 12.23(*b*) the two diameters A and B are *eccentric*; that is, they have different centres of rotation and do not lie on a common axis. The distance E between the centres of rotation is the amount of 'offset' or eccentricity. In this example, if the diameter B was rotated in a vee block and dial test indicator was in contact with diameter A, then it would show a variation in reading equal to 2E. This variation between the maximum and minimum readings of the dial test indicator is called the *throw*.

Throw = 2 × Eccentricity = 2E

The easiest way to ensure concentricity is to turn as many diameters as possible at the same setting without removing the component from the lathe as shown in Fig. 12.24. If the work does have to be removed from the chuck and turned round, then it should be mounted in a four-jaw chuck and trued up as was shown in Fig. 12.17. If, of course, only limited accuracy of concentricity is required the component may be reversed in the self-centring chuck.

12.12   Tool height

Although a tool may be ground to the correct cutting angles, these can be effectively destroyed by mounting the tool off-centre. Figure 12.25 shows why it is essential to mount the tool at the centre height of the workpiece. As well as altering the effective cutting angles, mounting the tool off-centre can also distort the profile of the component when taper turning, screw cutting and forming.

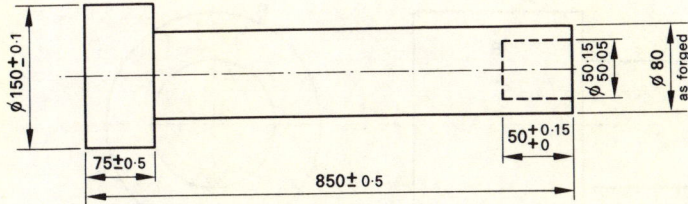

(a) Component requiring fixed steady

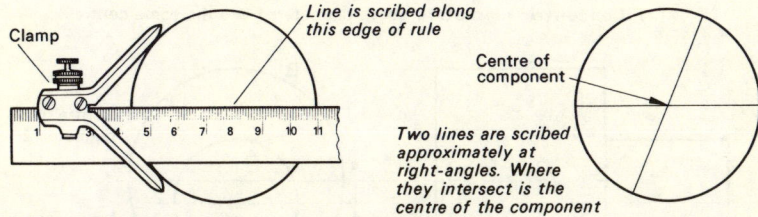

Line is scribed along this edge of rule

Clamp

Centre of component

Two lines are scribed approximately at right-angles. Where they intersect is the centre of the component

(b) Use of a centre-finder

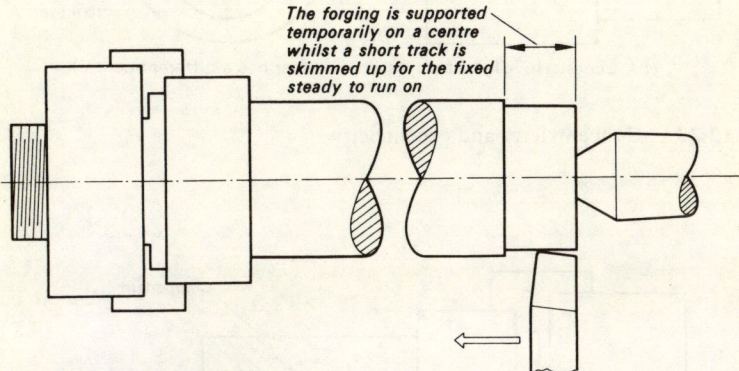

The forging is supported temporarily on a centre whilst a short track is skimmed up for the fixed steady to run on

(c) Turning track to support steady

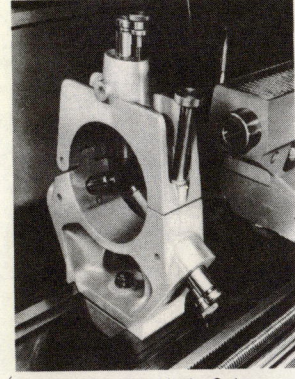

(Courtesy Colchester Lathe Co.)

(d) Component supported

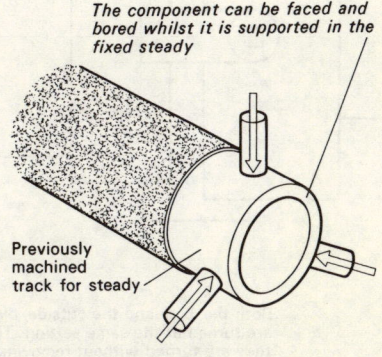

The component can be faced and bored whilst it is supported in the fixed steady

Previously machined track for steady

Fig. 12.22   The fixed steady

328

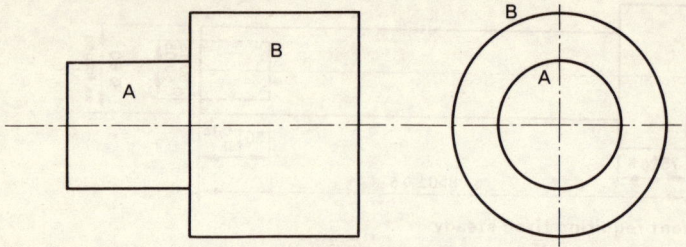

(a) **Concentric diameters** (both diameters have the same centres)

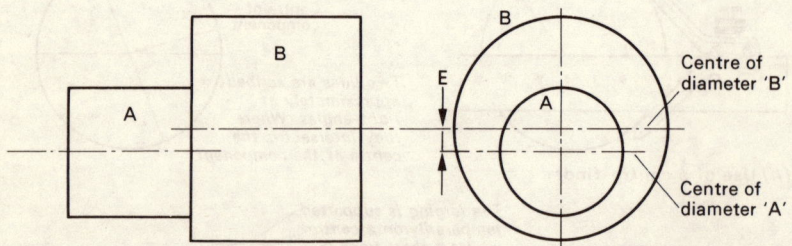

(b) **Eccentric diameters** (each diameter has a different centre)

Fig. 12.23 Concentricity and eccentricity

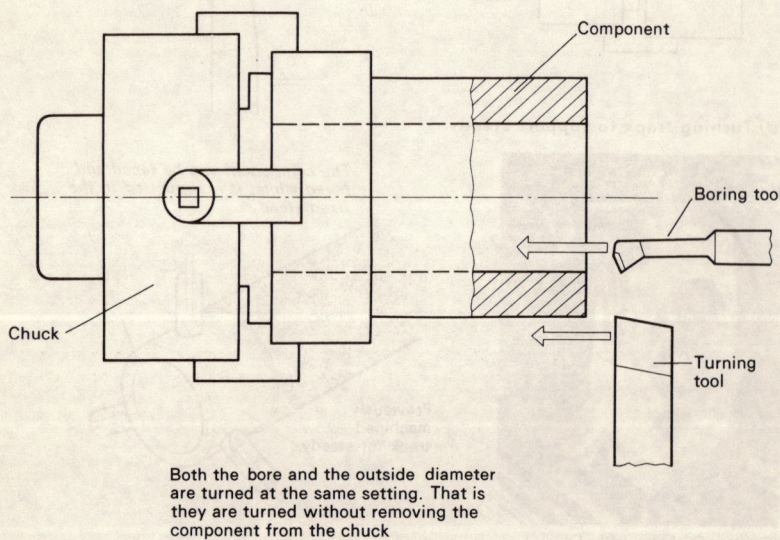

Both the bore and the outside diameter are turned at the same setting. That is they are turned without removing the component from the chuck

Fig. 12.24 Maintaining concentricity

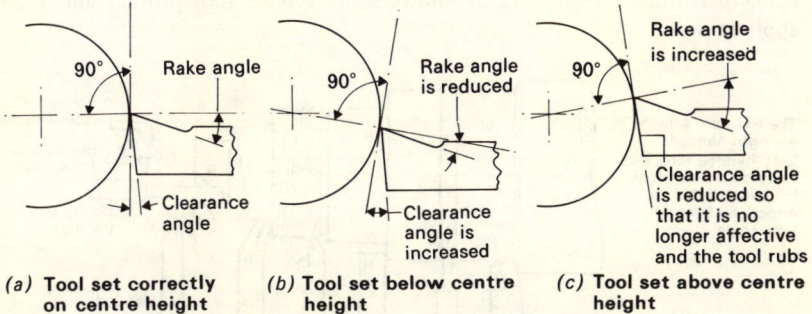

Fig. 12.25   Effect of tool height on turning tool angles

Figure 12.25(*a*) shows the tool correctly set. It will be seen that the effective cutting angles are the same as those ground on the tool.

When the tool point is set below the centre height of the job, as shown in Fig. 12.25(*b*), the effective rake angle is decreased and the effective clearance angle is increased.

When the tool point is set above the centre height of the job, as shown in Fig. 12.25(*c*), the effective rake angle is increased and the effective clearance angle is decreased.

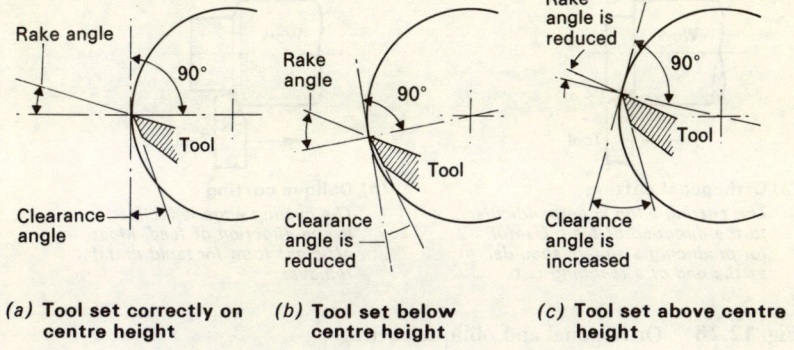

Fig. 12.26   Effect of tool height on boring tool angles

Figure 12.26 shows the effect of setting a boring tool above and below centre. It will be seen that the changes in the effective cutting angles are the reverse of those shown in Fig. 12.25.

12.13   Types of turning tool

The lathe uses a single-point cutting tool based on the principles discussed in Chapter 9. As well as being ground to the correct cutting angles, the lathe tool must also have the correct plan view (profile) according to the operation being performed. Figure 12.27 shows some typical tool profiles and their applications.

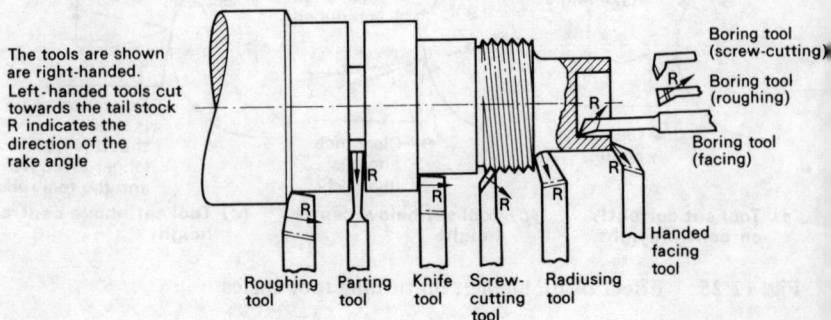

The tools are shown are right-handed.
Left-handed tools cut towards the tail stock
R indicates the direction of the rake angle

Boring tool (screw-cutting)
Boring tool (roughing)
Boring tool (facing)

Handed facing tool

Roughing tool Parting tool Knife tool Screw-cutting tool Radiusing tool

Fig. 12.27   Lathe tool profiles

Some of the tools shown are cutting *orthogonally* and some *obliquely*. These terms are explained in Fig. 12.28. It is easier to re-grind an orthogonal tool to the correct cutting angles, but generally an oblique tool cuts more easily since it reduces the chip thickness — and thus the load on the tool — for any given value of metal removed per revolution.

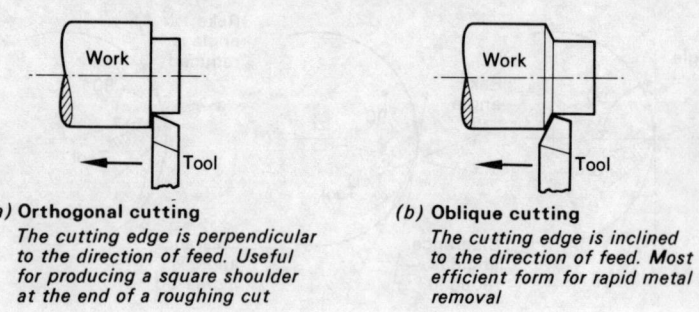

Work

Tool

(a) Orthogonal cutting
The cutting edge is perpendicular to the direction of feed. Useful for producing a square shoulder at the end of a roughing cut

Work

Tool

(b) Oblique cutting
The cutting edge is inclined to the direction of feed. Most efficient form for rapid metal removal

Fig. 12.28   Orthogonal and oblique cutting

12.14   Boring on the lathe

Hollow as well as solid components have to be produced on the lathe. Holes, concentric with the spindle axis, may be produced by:

(a)   drilling;
(b)   reaming;
(c)   boring;
(d)   a combination of the above.

The technique or combination of techniques chosen will depend upon

the size of the hole and its depth/diameter ratio, the accuracy required, and
whether a 'through' or 'blind' hole is required.

1. Use of twist drill

Unless the bore is only a shallow recess it is usual to rough out the hole with a
twist drill held in the tail stock. A start is made for the twist drill by means of
a centre drill in exactly the same way as when preparing a component for
turning between centres. Since the feed force available from the tailstock
barrel screw is limited, it is usual — when drilling large holes — to drill a pilot
bore first and then open it up with successively larger drills. The limitations
of a drilled hole are:

(a)   poor finish compared with boring or reaming;
(b)   lack of dimensional accuracy;
(c)   lack of 'roundness' or geometrical accuracy;
(d)   lack of positional accuracy as the drill tends to 'wander', especially
      when drilling deep holes in soft materials such as brass.

2. Use of reamer

The quality of the hole is greatly improved if it is drilled slightly undersize
and finished with a reamer (Fig. 11.7). The reamer should be held in a
'floating' holder so that it can follow the drilled hole without flexing. This
prevents ovality and bell-mouthing. Figure 12.29 shows a typical reamer

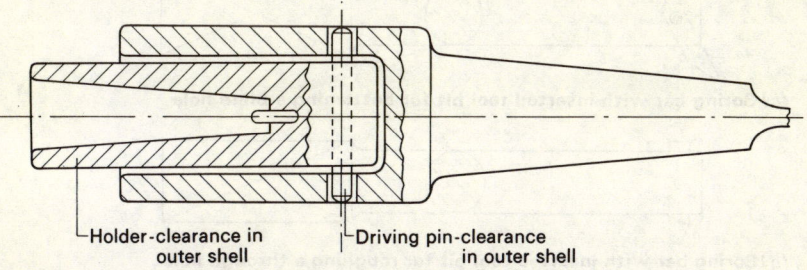

    Holder-clearance in   Driving pin-clearance
        outer shell   in outer shell

Fig. 12.29   Centre lathe — floating reamer holder

holder. The reamed hole has a good finish and a high degree of roundness.
However, the limitations of a reamed hole are:

(a)   lack of positional accuracy since the reamer follows the axis of the
      original drilled hole and reproduces any 'wander' that is present;
(b)   unless the quantity of components being produced warrants special
      tooling, only holes whose diameter is the same as a standard reamer size
      can be produced.

     Where a hole is too small to bore accurately, it is usual to drill and ream
the hole to size and then turn the outside diameter true with the bore whilst
the workpiece is mounted on a mandrel (see section 12.5). In this case the
initial 'wander' of the drilled hole is unimportant.

3. Use of boring tool

Figure 12.30 shows a solid boring tool used for small holes and a boring bar with an inserted tool-bit for larger holes. It also shows the necessity for *secondary clearance* on a boring tool to prevent the heel of the tool fouling on small diameter holes. Because of the overhang of the tool point from the tool post and the slender shank of the tool, boring tools are prone to chatter and the cut is liable to run off due to deflection of the shank. This makes boring a highly skilful operation compared with external turning and great care is required in grinding the tool and selecting appropriate feeds and speeds.

If a standard size hole is required, it is often preferable to drill, then bore

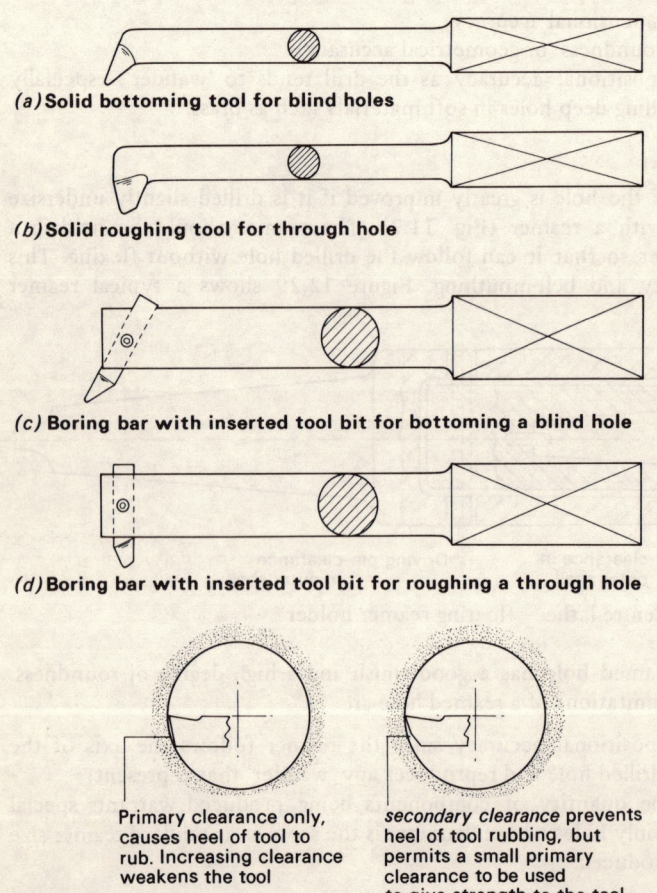

(a) Solid bottoming tool for blind holes

(b) Solid roughing tool for through hole

(c) Boring bar with inserted tool bit for bottoming a blind hole

(d) Boring bar with inserted tool bit for roughing a through hole

Primary clearance only, causes heel of tool to rub. Increasing clearance weakens the tool

secondary clearance prevents heel of tool rubbing, but permits a small primary clearance to be used to give strength to the tool

(e) Need for secondary clearance

Fig. 12.30 Centre lathe – boring tools

slightly under size to remove any 'wander' in the hole axis, and ream to size.
The reamer will give a better finish and a truer hole than a small boring tool.

Whilst boring is the only possible means of removing 'wander' and giving a high degree of positional accuracy, it suffers from the following limitations:

(*a*)   chatter and poor surface finish, especially on small diameter holes where only a slender tool can be used;

(*b*)   the hole tends to be oval and bell mouthed due to deflection of the boring tool shank. These defects become less severe as the hole diameter increases, thus allowing an increasingly more rigid boring tool.

12.15   Speeds and feeds

To keep the cost of production as low as possible during a machining process, the waste material must be removed as rapidly as possible. Unfortunately high rates of material removal usually result in a poor finish and relatively low dimensional accuracy. For example, a hole is quickly roughed out with a twist drill, but if a high degree of accuracy and finish is required the hole must be finished with a reamer which only removes a small volume of material.

The factors controlling the rate of material removal are:

(*a*)   the finish required;
(*b*)   the depth of cut;
(*c*)   the tool geometry;
(*d*)   the properties and rigidity of the cutting tool and its mounting;
(*e*)   the properties of the workpiece material;
(*f*)   the rigidity of the workpiece;
(*g*)   the power and rigidity of the machine tool.

In a machining operation the same rate of material removal may be achieved by using:

(*a*) A high rate of feed and a shallow cut as shown in Fig. 12.31(*a*). Unfortunately this not only leads to a rough finish, but imposes a greater load on the cutting tool as shown in Fig. 12.31(*b*).

(*b*) A low rate of feed and a deep cut as shown in Fig. 12.32(*a*). This gives a better finish and, if the size is correct, avoids having to take a finishing cut. It also reduces the load on the cutting tool as shown in Fig. 12.32(*b*). Unfortunately, a deep cut is liable to cause excessive chatter.

Therefore a compromise between depth of cut and rate of feed has to be arrived at. However, it is usually advisable, for the reasons given above, to use the deepest cut possible for any given set of conditions.

Table 12.8 gives suitable speeds and feeds for turning operations on a centre lathe using high-speed steel tools. These are only a guide and the actual rates used may be increased or decreased, as experience dictates, for any particular set up.

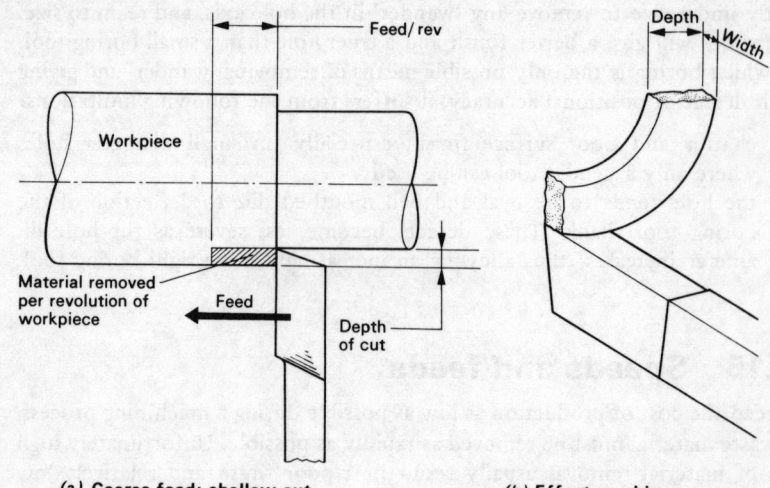

(a) **Coarse feed: shallow cut**

(b) **Effect on chip**

With coarse feed and shallow cut, the chip is bent across its deepest section. The bending force increases as the cube of the depth of the chip, i. e. doubling the depth of the chip increases the bending force eight times

Fig. 12.31 Effect of high feed rates

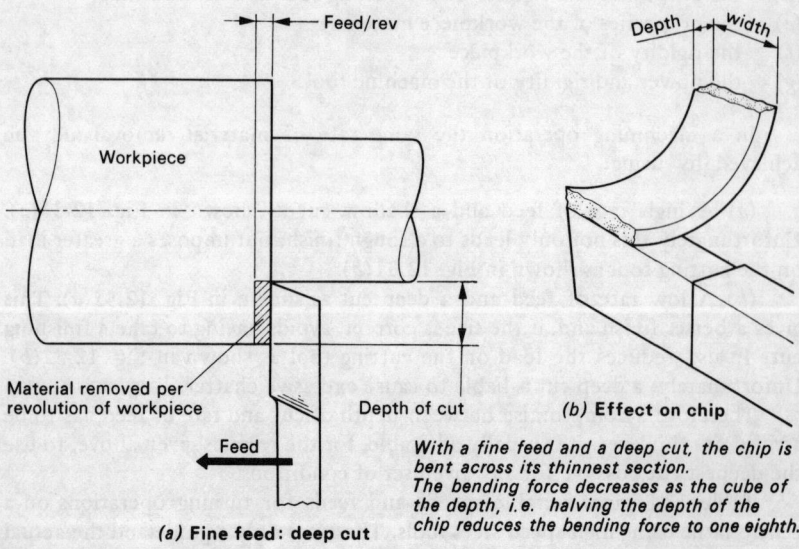

(a) **Fine feed: deep cut**

(b) **Effect on chip**

With a fine feed and a deep cut, the chip is bent across its thinnest section. The bending force decreases as the cube of the depth, i.e. halving the depth of the chip reduces the bending force to one eighth.

Fig. 12.32 Effect of deep cuts

Table 12.8 Cutting speeds and feeds for HSS turning tools

Material being turned	Feed (mm/rev)	Cutting speed (m/min)
Aluminium	0·2—1·00	70—100
Brass (alpha) (ductile)	0·2—1·00	50—80
Brass (free-cutting)	0·2—1·5	70—100
Bronze (phosphor)	0·2—1·0	35—70
Cast iron (grey)	0·15—0·7	25—40
Copper	0·2—1·00	35—70
Steel (mild)	0·2—1·00	35—50
Steel (medium carbon)	0·15—0·7	30—35
Steel (alloy-high tensile)	0·08—0·3	5—10
Thermo-setting plastic (low speed due to abrasive properties)	0·2—1·0	35—50

Notes:
1. The above feeds and speeds are for ordinary HSS tools. For *super* HSS tools the feeds would remain the same, but the cutting speeds could be increased by 15% to 20%.
2. The *lower* speed range is suitable for heavy, roughing cuts.
   The *higher* speed range is suitable for light, finishing cuts.
3. The feed is selected to suit the finish required, and rate of metal removal.

Example 12.1 *Calculate the spindle speed, to the nearest rev/min, for turning a 25 mm diameter bar at a cutting speed of 30 m/min (take π as 3.14).*

$$N = \frac{1000S}{\pi D}$$

$$= \frac{1000 \times 30}{3·14 \times 25}$$

$$= \underline{382 \text{ rev/min}}$$

(to nearest rev/min)

where: N = spindle speed
S = 30 m/min
π = 3·14
D = 25 mm

Example 12.2 *Calculate the time taken to turn a brass component 49 mm diameter by 70 mm long if the cutting speed is 44 m/min and the feed is 0·5 mm/revolution. Only one cut is taken (take π as $^{22}/_7$).*

$$N = \frac{1000S}{\pi D}$$

$$= \frac{1000 \times 44 \times 7}{22 \times 49}$$

$$= \underline{286 \text{ rev/min}}$$

(to nearest rev/min)

where: N = spindle speed
S = 44 m/min
$\pi = \frac{22}{7}$
D = 49 mm

Rate of feed	$= 0\cdot5$ mm/rev
	$= 0\cdot5 \times 286$ mm/min
	$= \underline{143}$ mm/min
Time taken to traverse 70 mm	$= \dfrac{70}{143}$ min
	$= \dfrac{70 \times 60}{143}$ sec
	$= \underline{29\cdot37}$ sec

12.16 Taper turning

Earlier in this chapter (section 12.4) great stress has been placed on the importance of maintaining the axial alignment of the headstock and tailstock, and upon the cutting tool moving parallel to this axis if a truly cylindrical component is to be produced. Consequently, a conical component is produced if these basic alignments are disturbed. Taper turning involves the controlled disturbances of these alignments so that the tool moves in a path that is inclined at a given angle to the workpiece axis. This inclination is relative; it does not matter whether the workpiece axis is offset or whether the tool path is offset. Three methods of taper turning will now be described.

1. Off-set tailstock

Using the lateral adjusting screws, the body of the tailstock and, therefore, the tailstock centre can be offset. This inclines the axis of a workpiece held between centres relatively to the path of the cutting tool as shown in Fig. 12.33(a). The advantages and limitations of this technique are summarised in Table 12.9.

2. Taper turning attachment

Another way in which tapers may be produced is by the use of a taper-turning attachment. These are usually an 'optional extra' and have to be purchased separately to the lathe. They vary in detail from make to make but a typical example is shown in Fig. 12.33(b).

The advantages and limitations of this technique are summarised in Table 12.9.

3. Compound slide

Setting over the compound slide is the simplest method of producing a taper although it has some limitations. Figure 12.33(c) shows a typical application. The advantages and limitations of this technique of taper turning is summarised in Table 12.9.

12.17 Screw cutting

Figure 12.34 shows – diagrammatically – a helix. The helix is defined as:

> The locus (path) of a point travelling around an imaginary cylinder such that its axial and circumferential velocities maintain a constant ratio.

When screw-cutting, the spindle of the lathe provides the rotational movement and the lead screw provides the axial movement necessary to

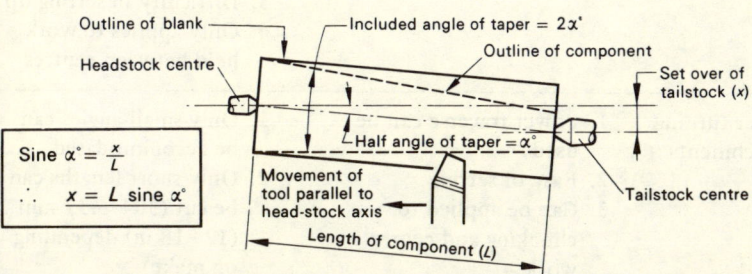

$$\text{Sine } \alpha° = \frac{x}{L}$$
$$\therefore \quad x = L \text{ sine } \alpha°$$

(a) **set over of centres**

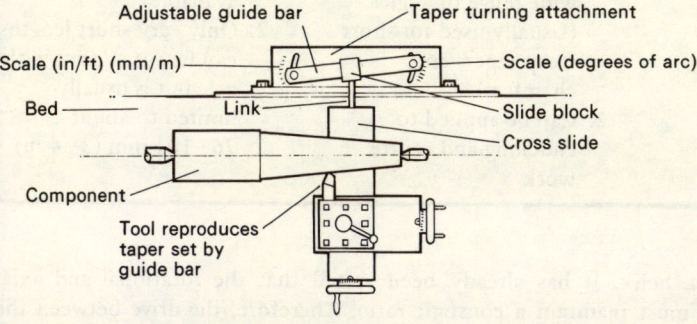

(b) **the taper turning attachment**

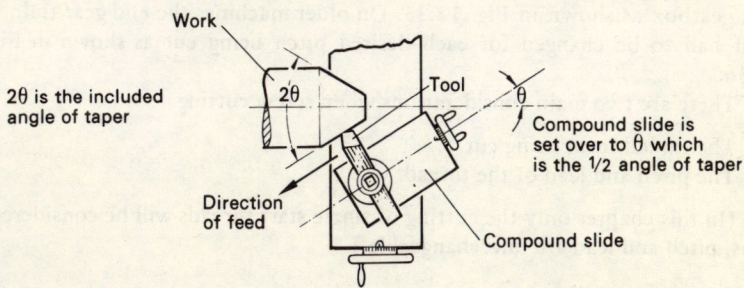

(c) **compound slide**

Fig. 12.33 Taper turning

Table 12.9 Comparison of taper turning techniques

Method	Advantages	Limitations
Set over of tailstock	1. Power traverse can be used 2. The full length of the bed used	1. Only small angles can be accommodated 2. Damage to the centre holes can occur 3. Difficulty in setting up 4. Only applies to work held between centres
Taper turning attachments	1. Power traverse can be used 2. Ease of setting 3. Can be applied to chucking and centre work	1. Only small angles can be accommodated 2. Only short lengths can be cut (304–457 mm (12–18 in) depending on make)
Compound slide	1. Very easy setting over a wide range of angles. (Usually used for short steep tapers and chamfers) 2. Can be applied to chucking and centre work	1. Only hand traverse available 2. Only very short lengths can be cut. Varies with m/c but is usually limited to about 76–101 mm (3–4 in)

generate a helix. It has already been stated that the rotational and axial velocities must maintain a constant ratio. Therefore, the drive between the spindle and the lead screw must be positive (without slip). Usually a gear train is used. On modern machines the ratio of velocities can be changed by means of a gearbox as shown in Fig. 12.35. On older machines the end gear train in itself had to be changed for each desired pitch being cut as shown in Fig. 12.36.

There are two main considerations when screw cutting.

1. The thread form being cut.
2. The pitch and lead of the thread.

(In this chapter only the cutting of single start threads will be considered. Thus, pitch and lead are interchangeable.)

1. Thread

The screw-cutting tool is ground to the correct profile using a *centre gauge*. This gauge is also used to set the tool sqare to the workpiece (Fig. 12.37).

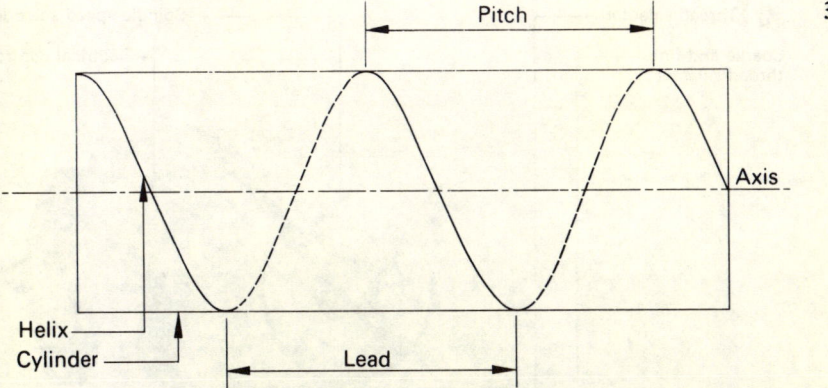

PITCH is the distance between adjacent, corresponding points on the helix measured parallel to the axis

LEAD is the axial distance a point moves along the helix in one revolution (i.e. the distance a nut would move along a bolt in one revolution)

(a) **Single start helix**

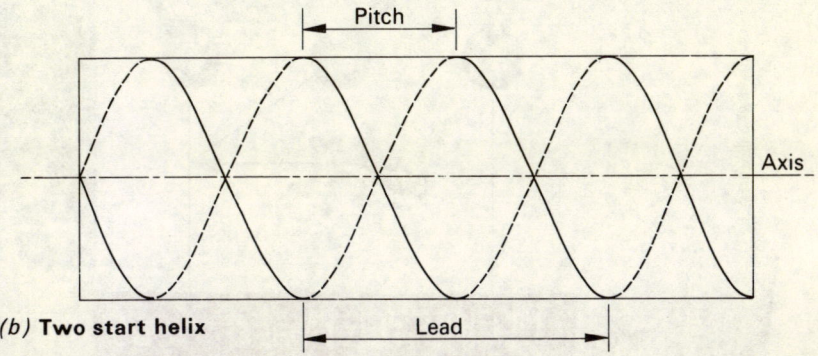

(b) **Two start helix**

Lead = pitch × number of starts

Fig. 12.34 The helix

2. Lead

The means by which the velocity of the lead screws may be set relative to the velocity of the spindle has already been discussed. Where change gears are used they may be calculated quite simply by the formula:

$$\frac{\text{Driver}}{\text{Driven}} = \frac{\text{t.p.i. lead screw}}{\text{t.p.i. to be cut}} = \frac{\text{lead to be cut}}{\text{lead of lead screw}}$$

(*Note:* t.p.i. = threads per inch)

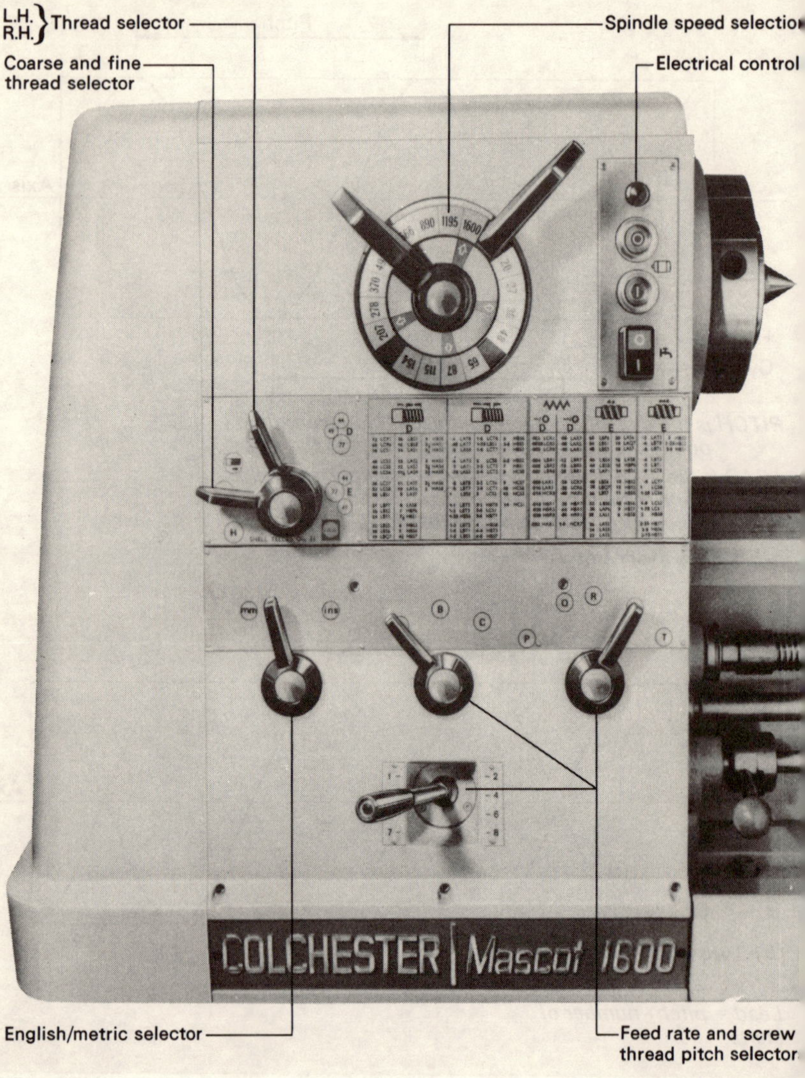

Fig. 12.35 Screw-cutting gear box

This formula can be used:

(a) for cutting English pitches on a lathe with an English lead screw;
(b) metric pitches on a lathe with a metric lead screw.

The standard gears supplied are:

20-tooth (2); 25-tooth to 120-tooth, in steps of 5 teeth.

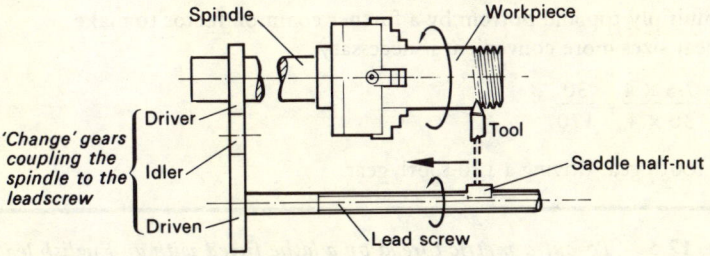

Fig. 12.36   The 'change' gear train

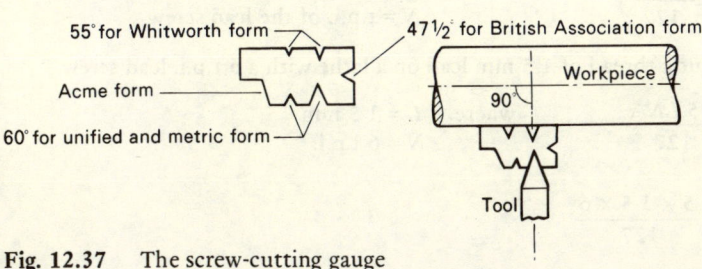

Fig. 12.37   The screw-cutting gauge

Example 12.3   *Calculate the gears to cut a 16 t.p.i. thread on a lathe with a 6 t.p.i. lead screw.*

$$\frac{\text{Driver}}{\text{Driven}} = \frac{\text{t.p.i. lead screw}}{\text{t.p.i. to be cut}}$$

$$= \frac{6}{16}$$

multiply top and bottom by 5 to fit standard gears

$$= \frac{6 \times 5}{16 \times 5} = \frac{30}{80}$$

i.e. a 30-tooth gear driving an 80-tooth gear.

Example 12.4   *Calculate the gears to cut a 1·5 mm pitch thread on a lathe with a 6 mm lead screw.*

$$\frac{\text{Driver}}{\text{Driven}} = \frac{\text{pitch to be cut}}{\text{pitch of lead screw}}$$

$$= \frac{1·5}{6·0}$$

multiply top and bottom by 5 to fit standard gears

$$= \frac{1·5 \times 5}{6·0 \times 5} = \frac{7·5}{30}$$

multiply top and bottom by a further common factor to make gear sizes more convenient if necessary

$$= \frac{7 \cdot 5 \times 4}{30 \times 4} = \frac{30}{120}$$

i.e. a 30-tooth gear driving a 120-tooth gear.

Example 12.5   *To cut a metric thread on a lathe fitted with an English lead screw, use the following formula:*

$$\frac{Driver}{Driven} = \frac{5\,LN}{127} \qquad \text{where:} \quad \begin{array}{l} L = \text{the lead to be cut in mm} \\ N = \text{t.p.i. of the lead screw} \end{array}$$

To cut a thread of 1·5 mm lead on a lathe with a 6 t.p.i. lead screw.

$$\frac{Driver}{Driven} = \frac{5\,LN}{127} \qquad \text{where:} \quad \begin{array}{l} L = 1\cdot5 \text{ mm} \\ N = 6 \text{ t.p.i.} \end{array}$$

$$= \frac{5 \times 1\cdot5 \times 6}{127}$$

$$= \frac{45}{127}$$

i.e. a 45-tooth gear driving a 127-tooth gear.

Therefore, when cutting metric threads on a lathe fitted with an English lead screw, a 127-tooth conversion gear is required in addition to the standard range of gears.

Note: In the above three examples no reference has been made to the 'hand' of the thread, since lead screws are normally right-handed.

(a)   To cut a right-hand thread, job and lead screw rotate in the same direction (one idler gear);

(b)   to cut a left-hand thread, job and lead screw rotate in opposite directions (two idler gears).

In a simple gear train the idler gears only control the direction of rotation of the first and last gear; they do not affect the speed of rotation.

There are various techniques used for actually cutting the thread. One of the most successful is the *'half angle technique'* shown in Fig. 12.38. It gets its name from the fact that the compound slide is set over to half the included angle of the thread being cut.

The lathe is set in motion and the micrometer dial on the compound slide is set to zero. The cross-slide is fed in until the tool just touches the workpiece and then the cross-slide micrometer dial is set to zero. After this the cross-slide is only used for quick withdrawal of the tool at the end of each cut, the tool being repositioned by bringing the cross-slide micrometer dial

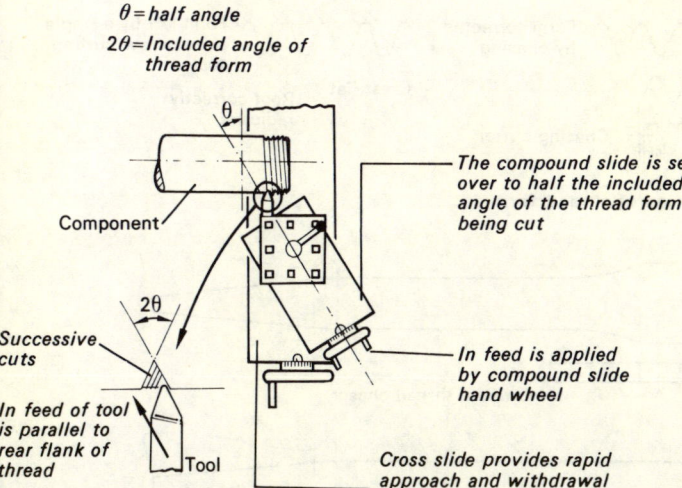

θ = half angle
2θ = Included angle of thread form

The compound slide is set over to half the included angle of the thread form being cut

Component

Successive cuts

In feed of tool is parallel to rear flank of thread

Tool

In feed is applied by compound slide hand wheel

Cross slide provides rapid approach and withdrawal of the cutting tool

Fig. 12.38 Half-angle screw cutting technique

back to zero again. The depth of cut is controlled by the compound slide. This method of screw cutting enables the tool to be ground with true cutting angles since it cuts on one lip only. The advantages of this technique over the plunge cut techniques are:

(*a*)   better finish (less chatter);
(*b*)   more rapid stock removal;
(*c*)   no need to remember the previous setting after each tool withdrawal since the cut is being applied progressively by the compound slide.

It is not possible, except for fine instrument threads, to cut the thread to its full depth in one pass of the tool. Therefore, a number of cuts must be taken, each one being successively deeper. The tool must follow the identical path of each previous cut otherwise a series of grooves will be cut side by side. The device by which the lathe operator judges the correct moment to engage the half nut with the lead screw is called the *chasing dial*. These vary with different machines and the maker's handbook should always be consulted before using the chasing dial on a strange machine.

12.18   Thread chasing

It is not possible to cut a full thread form with a single point tool. The reason for this is shown in Fig. 12.39(*a*). In order to radius the crest of the thread a multi-tooth form tool has to be used. Such a tool is referred to as a *chasing tool*. Examples of internal and external form tools are shown in Fig. 12.39(*b*). These chasers are used to trim the thread to size as shown in Fig. 12.39(*c*). Note that the chasing rest must be kept close up to the workpiece

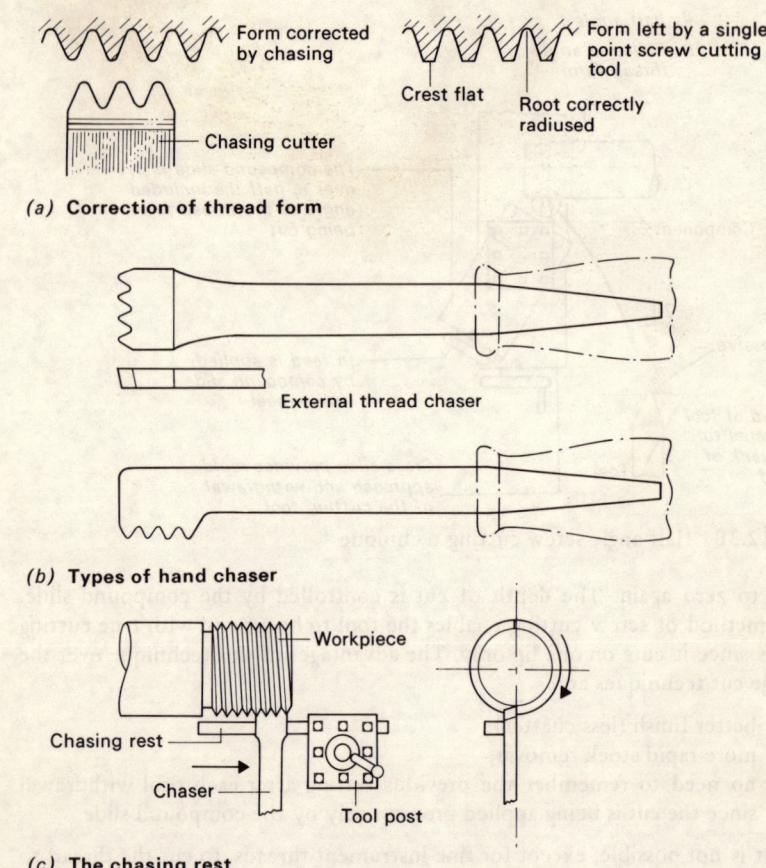

(a) **Correction of thread form**

Form corrected by chasing

Form left by a single point screw cutting tool

Crest flat

Root correctly radiused

Chasing cutter

External thread chaser

(b) **Types of hand chaser**

Workpiece

Chasing rest

Chaser

Tool post

(c) **The chasing rest**

Fig. 12.39 Thread chasing

so that the chaser cannot be dragged down between the rest and the workpiece. When chasing, the thread is cut slightly oversize, and it is then trimmed down to size with the chaser, constantly checking the thread with a ring gauge.

Alternatively, a full thread may be cut from the solid using a circular form tool as shown in Fig. 12.40(a). The form on the tool is slightly modified, to allow for the distortion caused by the rake angle, so that a true form is cut. Such form tools are very expensive compared with a single point tool and care must be taken not to chip them by incorrect usage. A button die in a tailstock die-holder may also be used to finish threads to size and even cut them from the solid when machining low strength materials. See Fig 12.40(b).

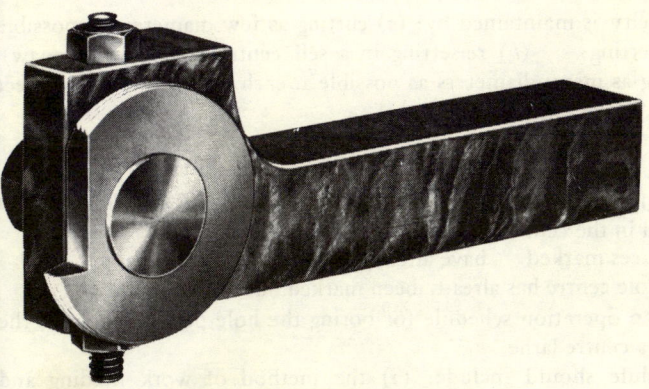

(a) **Circular form tool (screw cutting)**

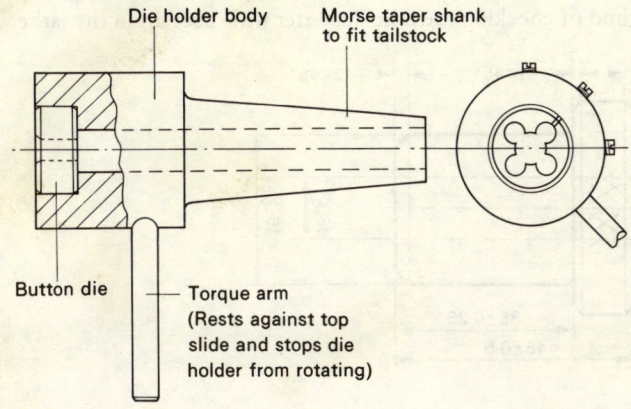

Die holder body

Morse taper shank
to fit tailstock

Button die

Torque arm
(Rests against top
slide and stops die
holder from rotating)

(b) **Button die and holder**

Fig. 12.40   Form tool and die

Problems

Section A

1. The spindle speed for turning a 40 mm diameter bar at 30 m/min is: (*a*) 239 rev/min; (*b*) 377 rev/min; (*c*) 418 rev/min; (*d*) 424 rev/min.

2. The time taken for a cut 80 mm long at 320 rev/min if the feed rate is 0·7 mm/rev is: (*a*) 10·5 s; (*b*) 21·4 s; (*c*) 105 s; (*d*) 168 s.

3. The gears to cut a 1·75 mm pitch thread on a lathe with a 6 mm pitch lead screw are given by the expression, drivers/driven equal: (*a*) 175/60; (*b*) 60/175; (*c*) 120/35; (*d*) 35/120.

4. A straight-nose roughing tool is used for cutting: (*a*) orthogonally; (*b*) obliquely; (*c*) profiles; (*d*) square shoulders.

5. Concentricity is maintained by: (*a*) cutting as few diameters as possible at each setting; (*b*) re-setting in a self centring chuck (three-jaw); (*c*) cutting as many diameters as possible at each setting; (*d*) correct alignment of the tailstock.

Section B

6. The casting shown in Fig. 8.17 is to have a 25·05 + 0·05 mm diameter hole bored in the centre of boss 'B'.
    1. The faces marked   have already been machined.
    2. The hole centre has already been marked out on the boss.

    Draw up an operation schedule for boring the hole perpendicular to the face × on a centre lathe.

    The schedule should include: (*a*) the method of work holding and clamping; (*b*) the method of locating the hole centre; (*c*) balancing the face plate if necessary; (*d*) the process of boring the hole; (*e*) a method of checking the hole diameter whilst set up in the lathe.

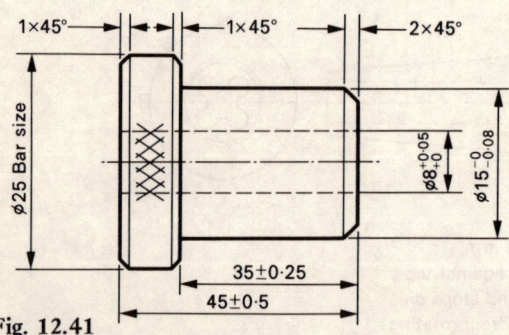

Fig. 12.41

7. Draw up an operation schedule for turning the bush shown in Fig. 12.41. To maintain concentricity the bush should be finished on a mandrel. Calculate a suitable cutting speed for turning the 15 mm diameter.

8. Name *three* basic techniques of taper turning used in the centre lathe. Tabulate the advantages and limitations of each of the techniques chosen.

9. Describe, with the aid of sketches, the cutting of a 2·5 mm pitch screw thread on a 20 mm diameter in a centre lathe fitted with a 6 mm lead screw. The thread being cut has a single start and the half angle technique should be adopted.

    Your answer should include: (*a*) calculation of suitable change gears; (*b*) setting of the cutting tool; (*c*) the technique of cutting the thread; (*d*) the technique of finishing to a correct thread form.

10. (*a*) Describe, with the aid of sketches, two methods of setting work accurately in the four-jaw (independent) chuck.

    (*b*) Sketch a travelling steady and explain how it is used to support long slender work.

Chapter 13

Shaping, milling and grinding processes

13.1 The shaping machine

Figure 13.1 shows a modern shaping machine and identifies its more important features and controls. Unlike the lathe, which produces cylindrical surfaces, the shaping machine uses a single point tool to produce plane surfaces.

To understand how the shaping machine produces a plane surface, it is easier to consider a surface parallel to the machine table. The ram of the shaping machine moves the cutting tool backwards and forwards in a straight line across the workpiece as shown in Fig. 13.2(a). Each time the tool moves forward it cuts a sliver of metal from the workpiece. Each time the tool moves backwards it lifts clear of the workpiece so that the workpiece is not damaged and the cutting edge of the tool is not blunted. During this return stroke the work moves sideways in a path perpendicular to the movement of the tool as shown in Fig. 13.2(b).

Thus the work remains stationary during the forward (cutting) stroke of the tool, and only moves across by one traverse increment during the return (non-cutting) stroke. The appearance of the machined surface is that of a succession of closely spaced straight line cuts. For this reason the surface produced by a shaping machine is often referred to as a *ruled surface*. The construction and mechanism of the shaping machine is considered in detail in *Manufacturing Technology: Level II*.

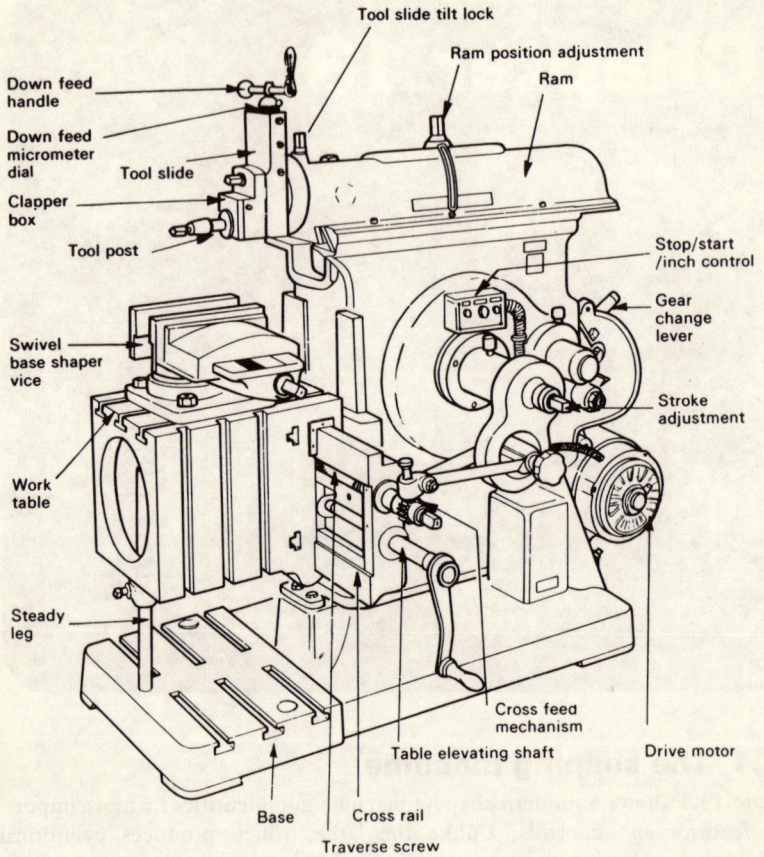

Down feed
handle

Down feed
micrometer
dial

Tool slide

Clapper
box

Tool post

Tool slide tilt lock

Ram position adjustment

Ram

Stop/start
/inch control

Gear
change
lever

Stroke
adjustment

Swivel
base shaper
vice

Work
table

Steady
leg

Cross feed
mechanism

Table elevating shaft

Drive motor

Base

Cross rail

Traverse screw

Fig. 13.1   The shaping machine

13.2 Cutting tools

One of the greatest advantages of the shaping machine for jobbing shop work
is the fact that it uses single point tools similar to lathe tools. Figure 13.3(a)
shows the geometry of a shaping machine tool. Its similarity to a lathe tool
is immediately apparent. Not only are single point tools cheap, they are
easily re-sharpened or ground to simple forms using an off-hand tool grinding
machine. They are often used for roughing out castings and removing the
scale from tool and die steels before finish machining such components on
a milling machine with its more costly cutters. Figure 13.3(b) shows some
typical shaping machine tools for a variety of operations.

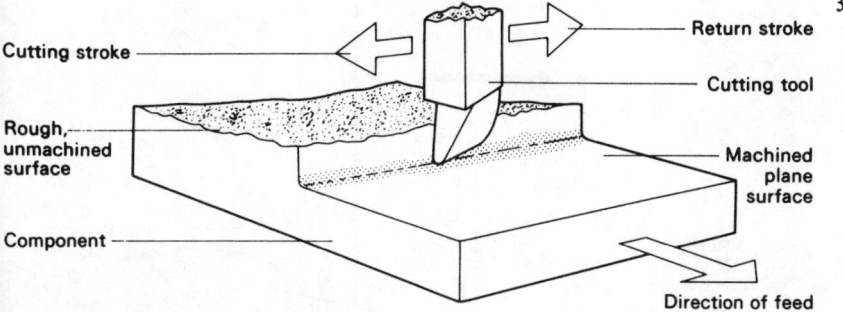

Cutting stroke

Return stroke

Cutting tool

Rough, unmachined surface

Machined plane surface

Component

Direction of feed

(a) **Cutting action of the shaping machine**

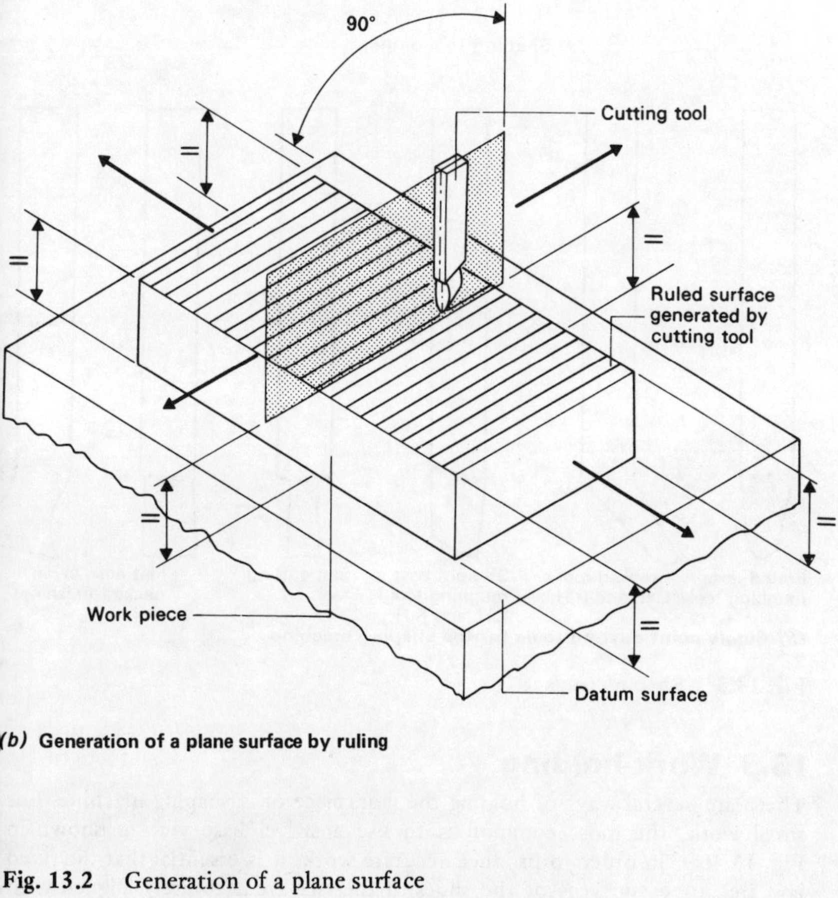

90°

Cutting tool

Ruled surface generated by cutting tool

Work piece

Datum surface

(b) **Generation of a plane surface by ruling**

Fig. 13.2   Generation of a plane surface

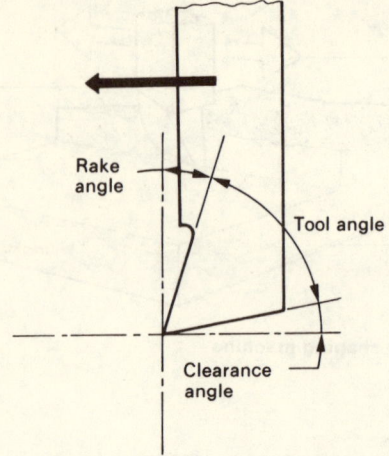

(a) **Shaping tool angles**

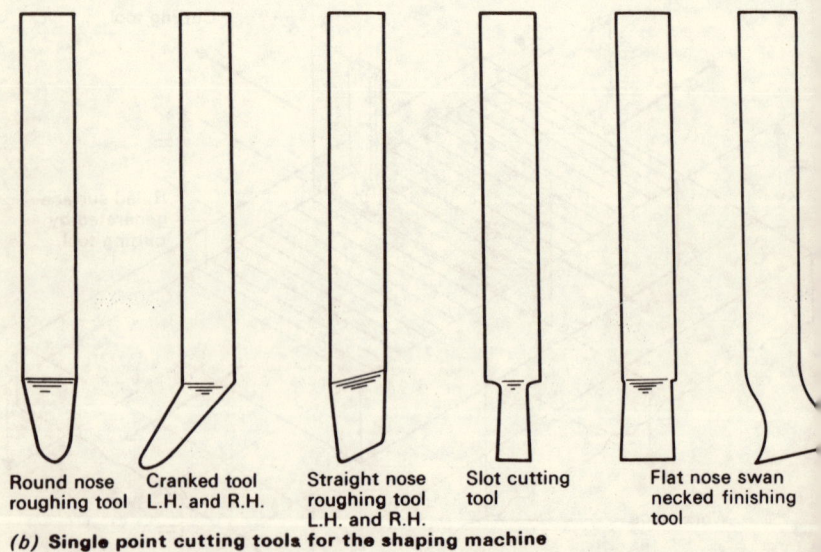

| Round nose roughing tool | Cranked tool L.H. and R.H. | Straight nose roughing tool L.H. and R.H. | Slot cutting tool | Flat nose swan necked finishing tool |

(b) **Single point cutting tools for the shaping machine**

Fig. 13.3 Shaping tools

13.3 Work-holding

There are several ways of holding the workpiece on a shaping machine. For small work, the most common is to use a swivel base vice as shown in Fig. 13.4(*a*). In order to produce accurate work, it is essential that the fixed jaw and upper surfaces of the slides of the vice are accurately aligned with the machine worktable.

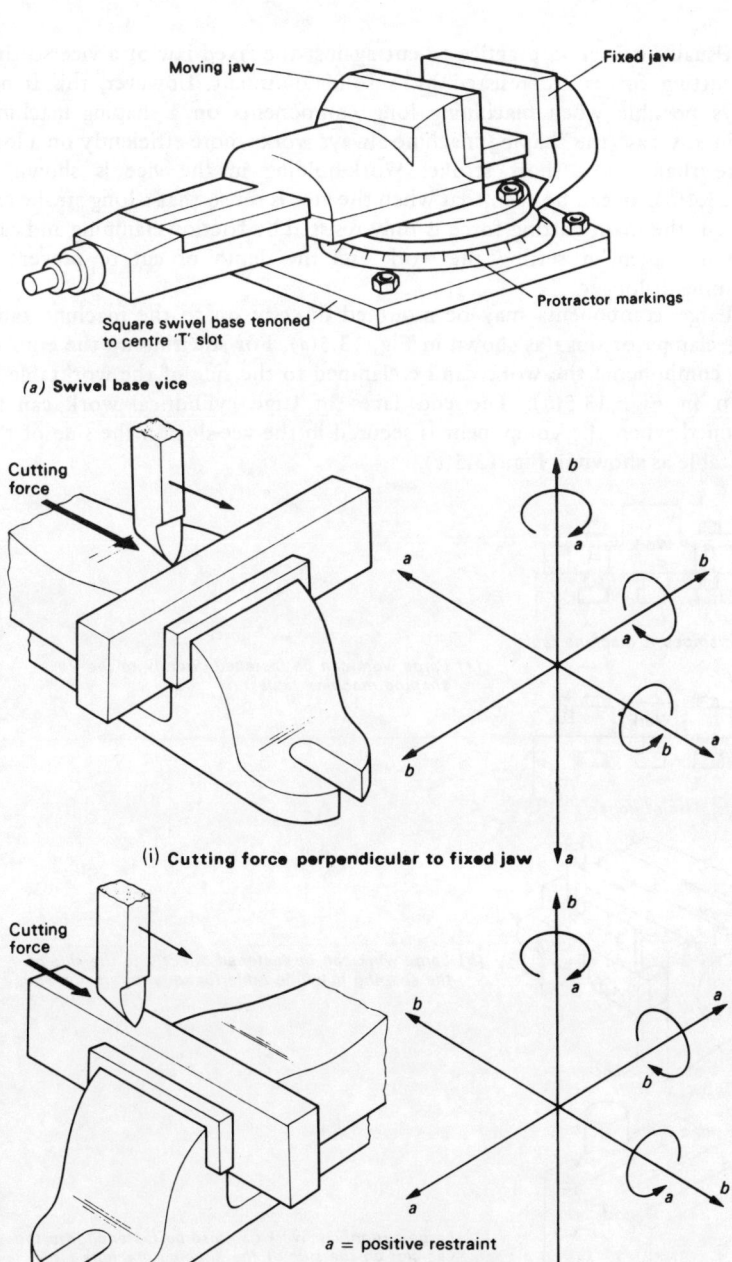

(a) Swivel base vice

(i) Cutting force perpendicular to fixed jaw

a = positive restraint
b = frictional restraint

(ii) Cutting force parallel to fixed jaw

(b) Work holding in the vice

Fig. 13.4   Machine vice – restraints

Usually it is good practice to cut against the fixed jaw of a vice so that the cutting forces are resisted by a solid abutment. However, this is not always possible when machining long components on a shaping machine, and in any case the shaping machine always works more efficiently on a long stroke than on a short stroke. Work-holding in the vice is shown in Fig. 13.4(*b*). It can be seen that when the vice is set so that a long stroke can be used, the main cutting force is only resisted by friction clamping and care must be taken in setting the work and the depth of cut to prevent it becoming dislodged.

Large components may be mounted directly on to the machine table using clamps or dogs as shown in Fig. 13.5(*a*). For squaring up the ends of large components the work can be clamped to the side of the worktable as shown in Fig. 13.5(*b*). The end faces of large cylindrical work can be machined when the component is secured in the vee-slot on the side of the worktable as shown in Fig. 13.5(*c*).

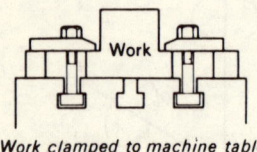

Work clamped to machine table

(a) Large work can be fastened directly to the shaping machine table

Use of dogs

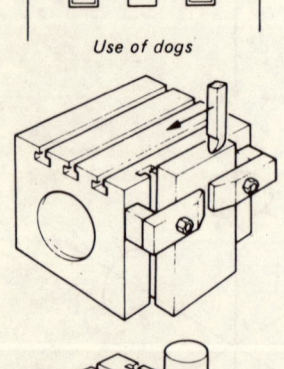

(b) Large work can be fastened directly to the side of the shaping machine table for squaring up the ends

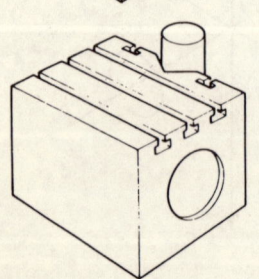

(c) Large cylindrical work can also be fastened into the vee slot on the side of the shaping machine table

Fig. 13.5   Alternative methods of work-holding

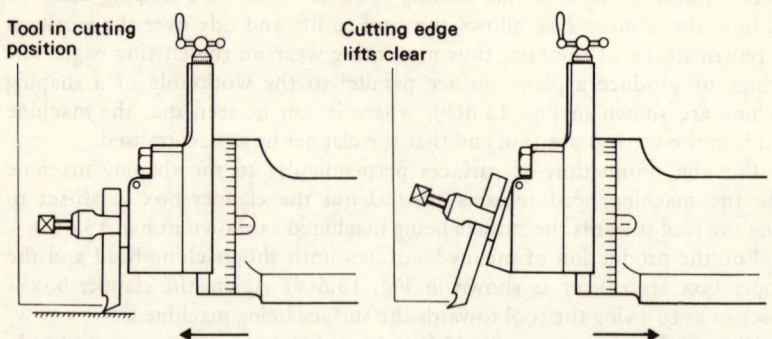

Tool in cutting
position

Cutting edge
lifts clear

*During the forward (cutting) stroke
the forces on the tool keep the
clapper box shut*

*During the return stroke the clapper box
is free to lift. This allows the tool to ride
over the work without wear or damage
to the cutting edge of the tool.*

(a) Action of the clapper box

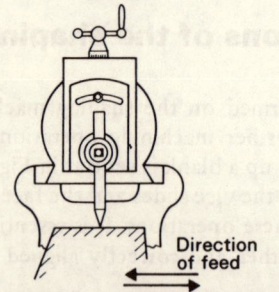

Direction
of feed

Direction
of feed

*For horizontal surfaces the clapper
box is centred on the tool slide*

(b) Setting for horizontal surfaces

*To shape a vertical surface, the clapper
box is set over so that the tool is swung
towards the surface being machined*

(c) Setting for vertical surface

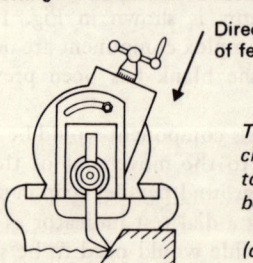

Direction
of feed

*To shape an inclined surface the
clapper box is set over so that the
tool is swung towards the surface
being machined.*

(d) Setting for inclined surface

Fig. 13.6   The clapper box action and setting the tool slide and clapper box

OP	DESCRIPTION	SET UP
1	Set vice jaws parallel to ram, using a D.T.I. mounted in the tool post. When vice is correctly set, the D.T.I. reading should be constant as it travels along the parallel strip	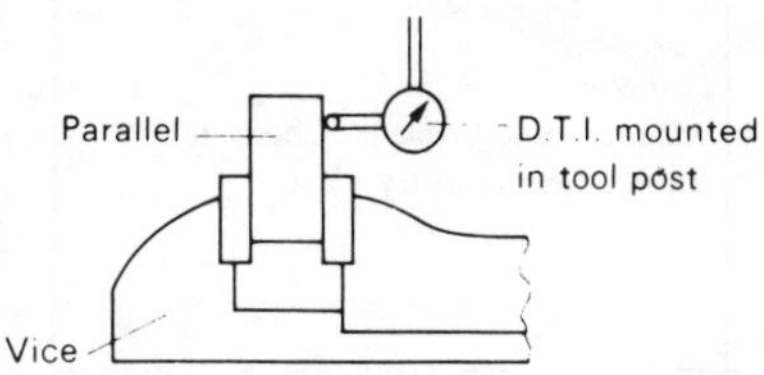
2	Set sawn blank in vice using grips. Machine upper surface 'A'	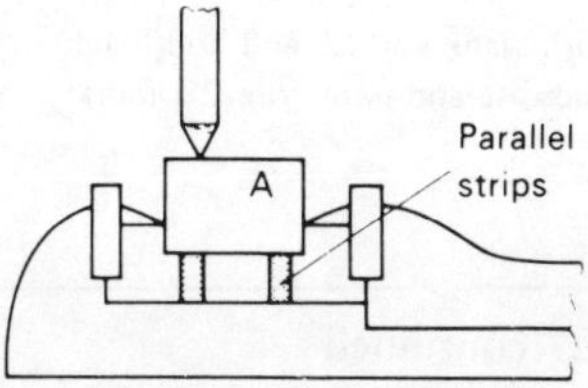
3	Turn job through 90° so that previously machined surface is against fixed jaw of vice. This ensures that surface A and B will be perpendicular to each other. Machine surface B	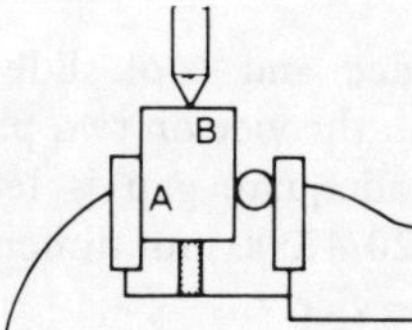
4	Turn job through 90° and machine surface C until job is 40mm thick Check thickness at each end to ensure parallelism	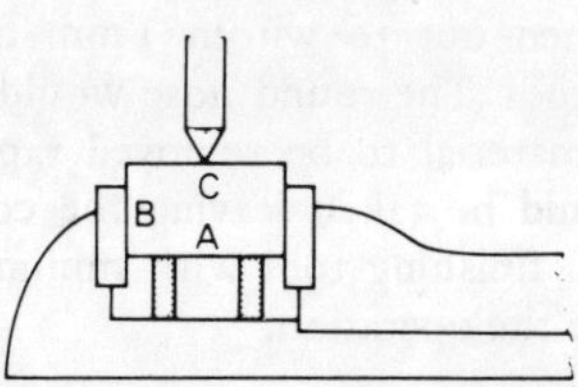
5	Turn job through 90° again and machine surface D until job is 65mm wide. Check thickness at each end to ensure parallelism	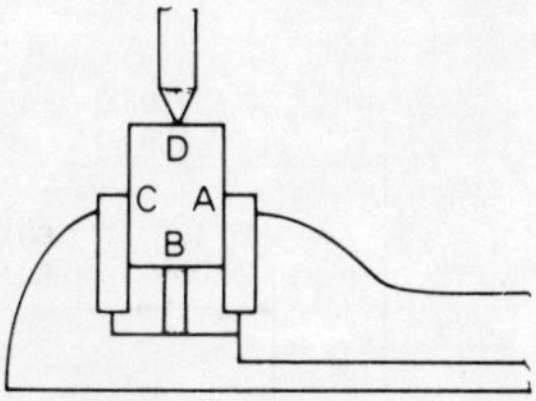
6	Turn vice through 90° and check with D.T.I. Set clapper box as shown in Fig.9.31	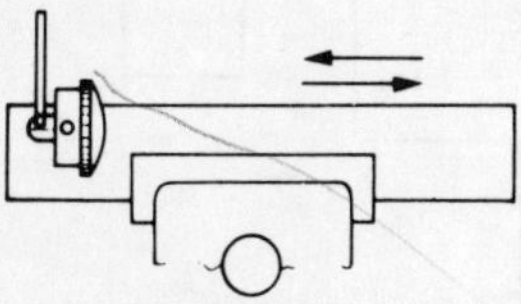

Fig. 13.7 Squaring up a blank (continued on page 356)

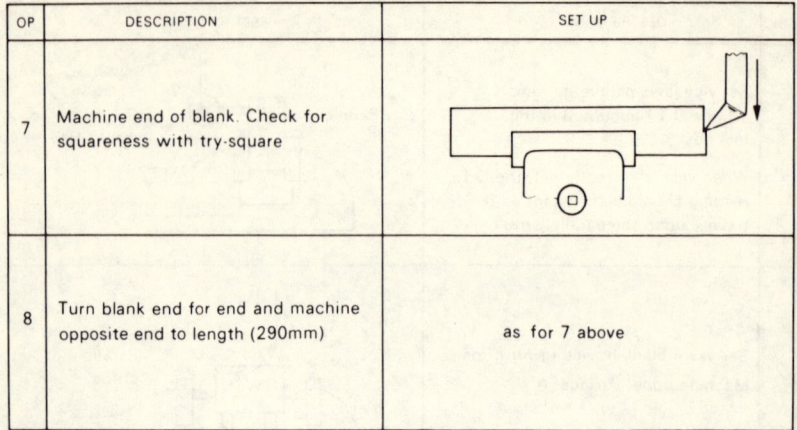

OP	DESCRIPTION	SET UP
7	Machine end of blank. Check for squareness with try-square	
8	Turn blank end for end and machine opposite end to length (290mm)	as for 7 above

Fig. 13.7 continued

Once the vice and tool slide are correctly set, the blank would be clamped firmly in the vice on two parallel strips as shown, care being taken to ensure that an adequate gap is left so that a micrometer can be used to measure the 45·20/45·00 mm dimension without disturbing the setting of the component in the vice.

The step and slot would be machined before the chamfer in order to leave a datum edge to measure from. Commencing with the step, this would be roughed out to within 1 mm of the scribed line using a round-nosed handed tool. The round nose would give strength to the tool and enable the surplus material to be removed rapidly. A series of horizontal and vertical cuts would be taken leaving the corner as sharp as the nose radius would allow. A finishing tool with minimum nose radius would then be used to complete the operation.

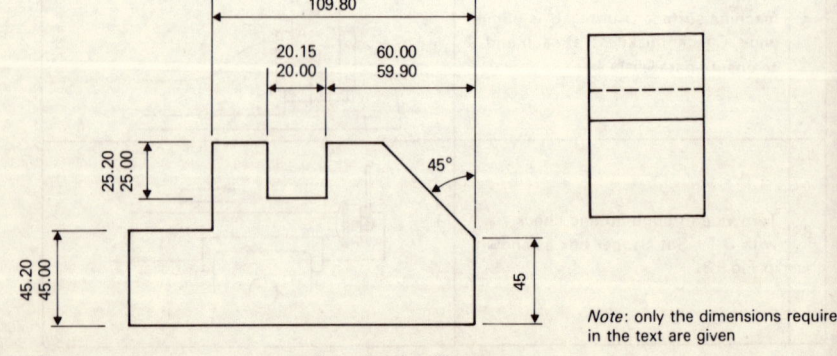

(a) Stepped component   dimensions in millimetres

Note: only the dimensions required in the text are given

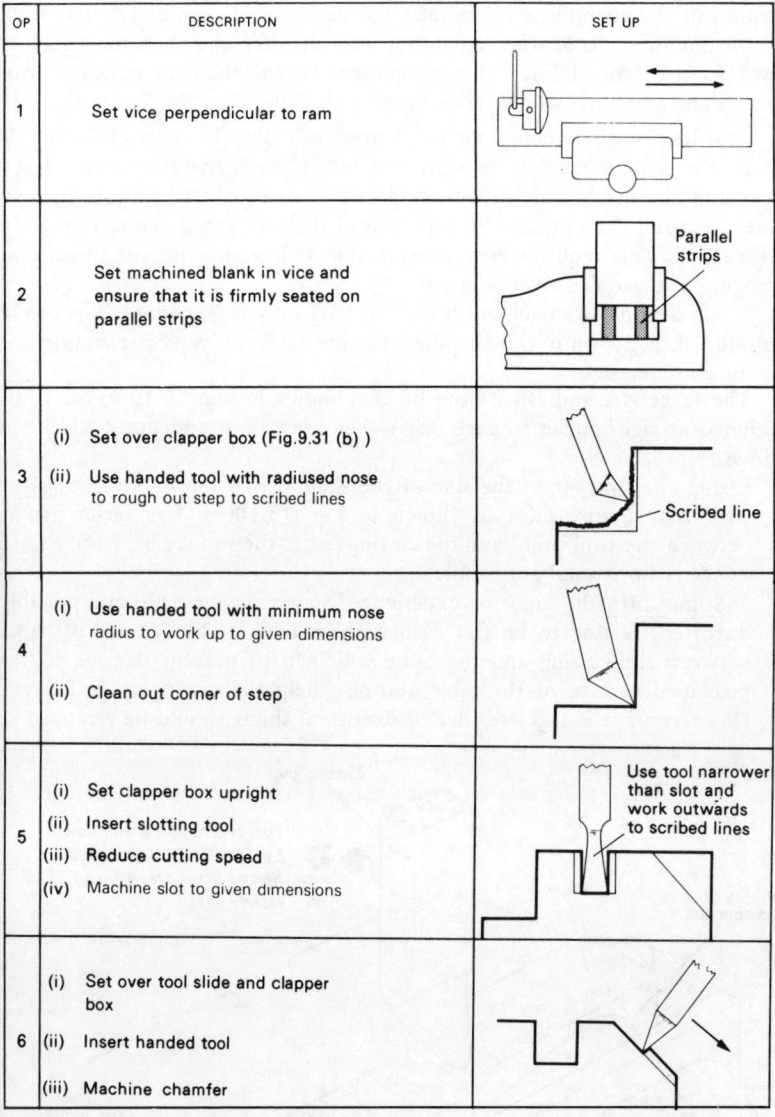

OP	DESCRIPTION	SET UP
1	Set vice perpendicular to ram	
2	Set machined blank in vice and ensure that it is firmly seated on parallel strips	Parallel strips
3	(i) Set over clapper box (Fig.9.31 (b)) (ii) Use handed tool with radiused nose to rough out step to scribed lines	Scribed line
4	(i) Use handed tool with minimum nose radius to work up to given dimensions (ii) Clean out corner of step	
5	(i) Set clapper box upright (ii) Insert slotting tool (iii) Reduce cutting speed (iv) Machine slot to given dimensions	Use tool narrower than slot and work outwards to scribed lines
6	(i) Set over tool slide and clapper box (ii) Insert handed tool (iii) Machine chamfer	

(b) Operation sequence

Fig. 13.8   Shaping a stepped component

Next, the slot would be machined using a parting or slotting tool as shown. The initial slot would be cut in the middle of the scribed lines and then the slot would be gradually opened up.

Finally, the chamfer would be machined. Neither the angle nor the position of the chamfer is toleranced, so the tool slide can be set over using the protractor scale on the ram head. The chamfer can then be machined down to the scribed line. The component would then be removed from the vice and any burrs would be removed with a file.

Profiles for such components as press tool punches can easily be cut out on the shaping machine as shown in Fig. 13.9. A very fine cross traverse is set and the machine is run more slowly than usual. As the job traverses under the tool, the operator winds the tool slide up and down to follow the scribed line. This requires very considerable skill and, since the operator is facing the cut, goggles **must** be worn.

It has already been shown in section 13.3 how large components can be mounted directly onto the shaping machine table. A typical example will now be considered.

The faces AA and BB of the bracket shown in Fig. 13.10(*a*) are to be machined at right angles to each other. The operation sequence could be as follows:

1. Clamp the bracket to the side of the shaping machine table so that the base AA is uppermost as shown in Fig. 13.10(*b*). Use feeler gauges between the tool point and the casting to get the surface being machined as nearly horizontal as possible.

    Some difficulty may be experienced in getting a rough, and possibly distorted, casting to lie flat against the machine table. A wad of paper between the casting and the table will help to prevent damage to the machined surface of the table and also help the casting to bed down. However, if it is too irregular or distorted, shims should be arranged so

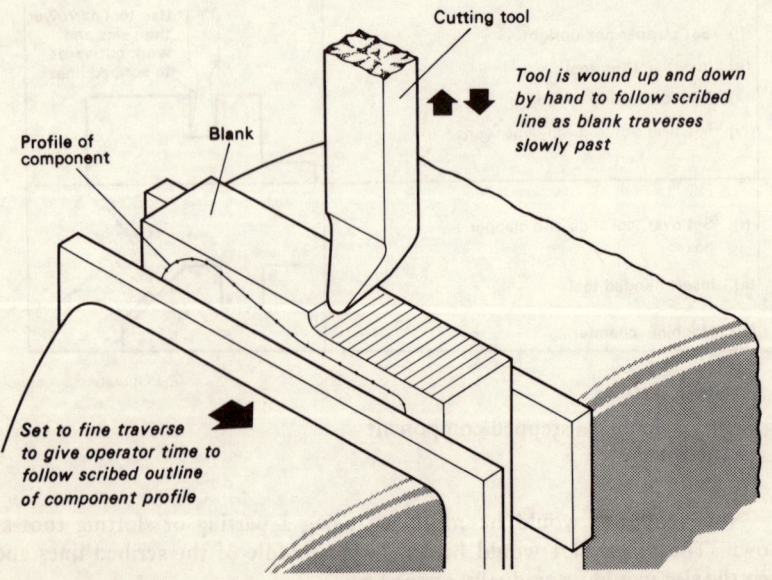

Fig. 13.9   Use of a shaping machine to profile a component

359

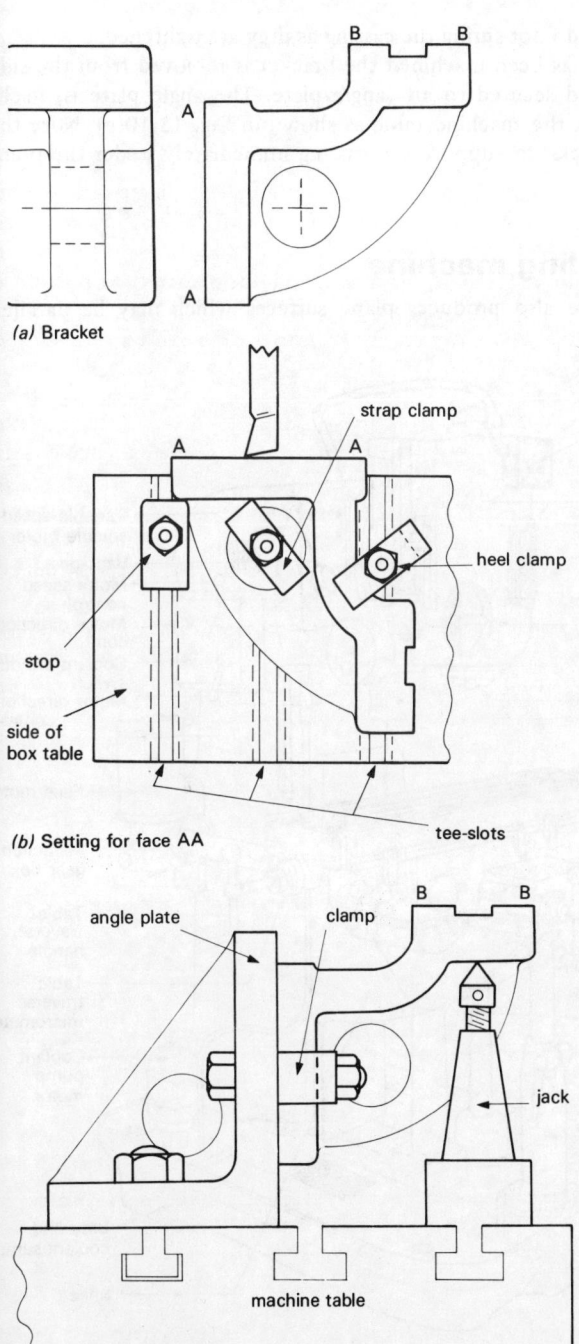

(a) Bracket

(b) Setting for face AA

strap clamp

heel clamp

stop

side of
box table

tee-slots

(c) Setting for face BB

angle plate

clamp

jack

machine table

Fig. 13.10 Bracket

that the clamps do not spring the casting as they are tightened.

2. After the base has been machined the bracket is removed from the side of the table and secured to an angle plate. The angle plate is, itself then secured to the machine table as shown in Fig. 13.10(c). Note the use of screw jacks to support the casting immediately under the point of cutting.

13.6 The milling machine

The milling machine also produces plane surfaces which may lie parallel

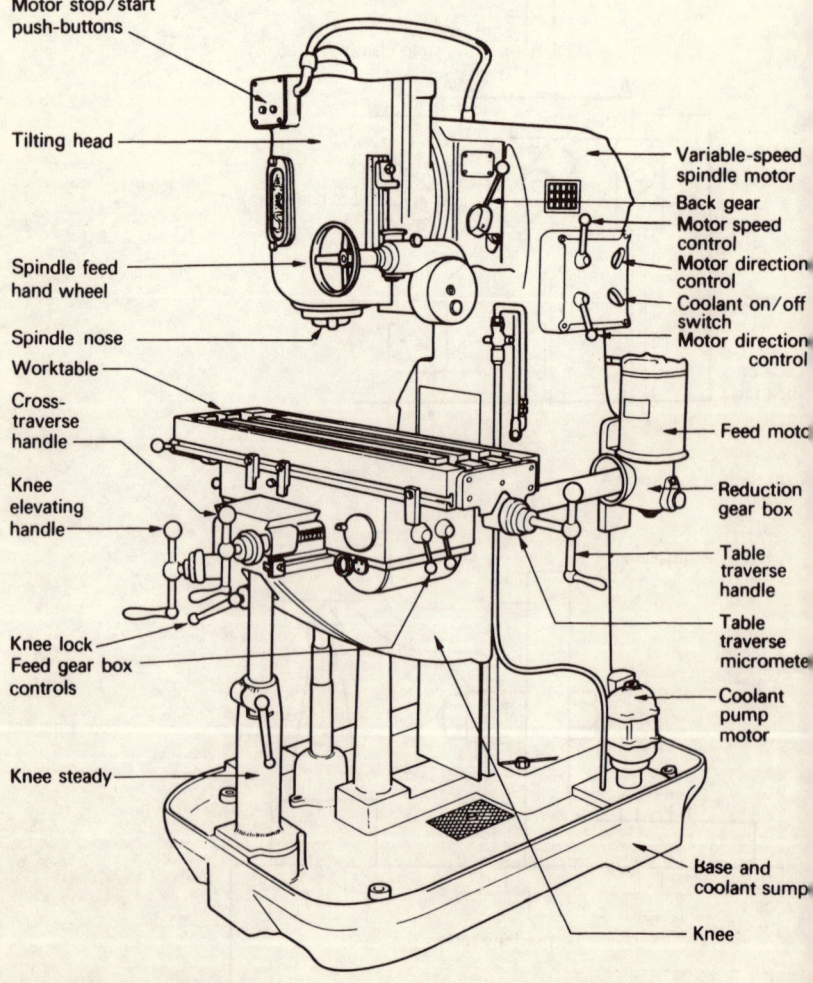

(a) Typical vertical machine

Fig. 13.11 Milling machines

perpendicular, or at an angle to the worktable. Unlike the shaping machine, previously described in this chapter, the milling machine uses a rotary, multi-tooth cutter. Since the milling cutter has a number of cutting edges, it can remove metal more rapidly than the single-point tools discussed so far.

Figure 13.11 shows two basic types of milling machine. There are other types for special applications. The vertical milling machine is shown in Fig. 13.11(*a*) and it can be seen that it gets its name from the fact that the axis of its spindle lies in a vertical plane. Figure 13.11(*b*) shows a horizontal milling machine. In this instance the machine derives its name from the fact that its spindle axis lies in a horizontal plane.

13.7 Milling cutters

Figure 13.12(*a*) compares the basic geometry of the single-point cutting tool used on the shaping machine with the geometry of a milling cutter tooth. It can be seen that the basic cutting angles of rake and clearance apply similarly to the cutting edges of both tools. In addition, the milling cutter tooth has a secondary clearance angle to prevent interference between the heel of the tooth and the workpiece.

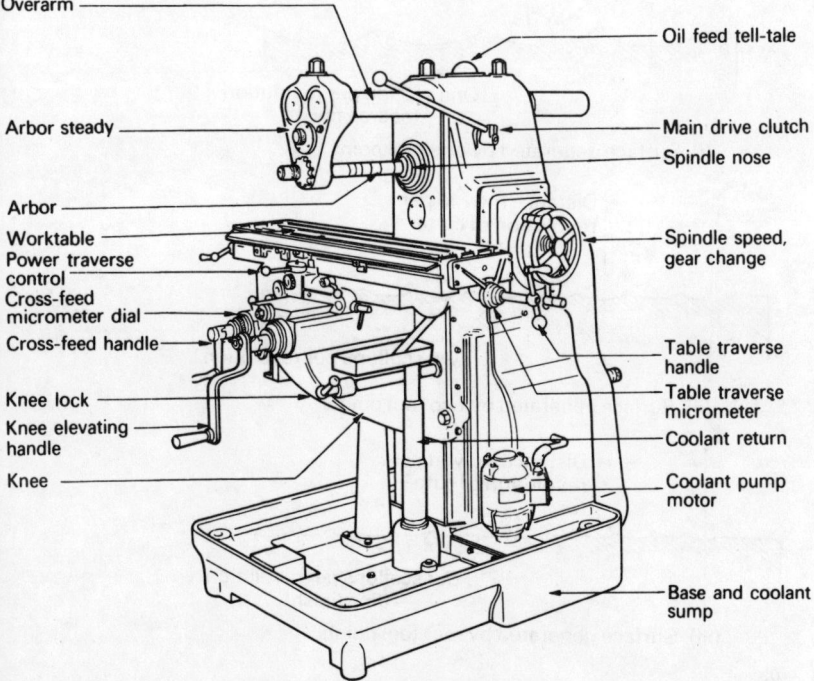

(b) **Typical horizontal machine**

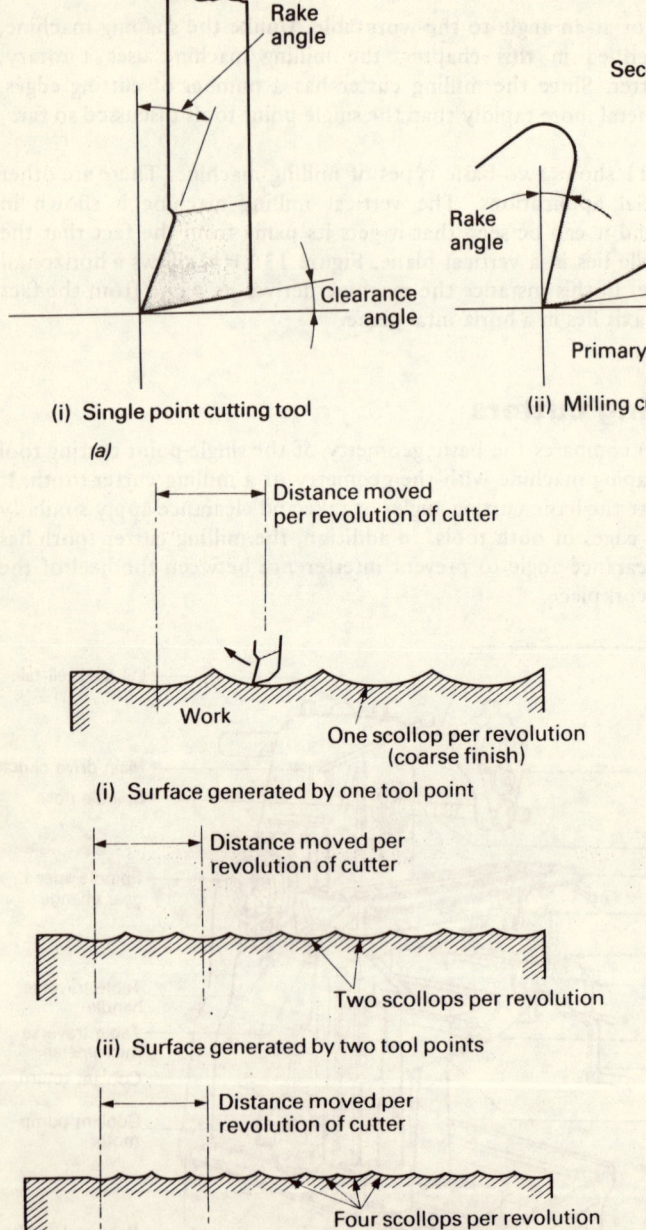

(i) Single point cutting tool

(a)

(ii) Milling cutter tooth

(i) Surface generated by one tool point

(ii) Surface generated by two tool points

(iii) Surface generated by four tool points

(b)

Fig. 13.12 Basic cutter geometry

Figure 13.12(*b*) shows how a plane surface is produced by a rotating cutter. By increasing the number of teeth the number of scollops per revolution is also increased and they become less apparent. However, increasing the number of teeth also reduces the swarf clearance between the teeth and the strength of the teeth. Thus the number of teeth round a cutter of any given diameter is a compromise between rate of metal removal, tooth strength and surface finish. The surface finish can also be improved by reducing the rate of feed, but this also reduces the rate of metal removal.

There are many different sorts of milling cutter and some of those used on the horizontal milling machine are shown in Fig. 13.13, together with typical applications.

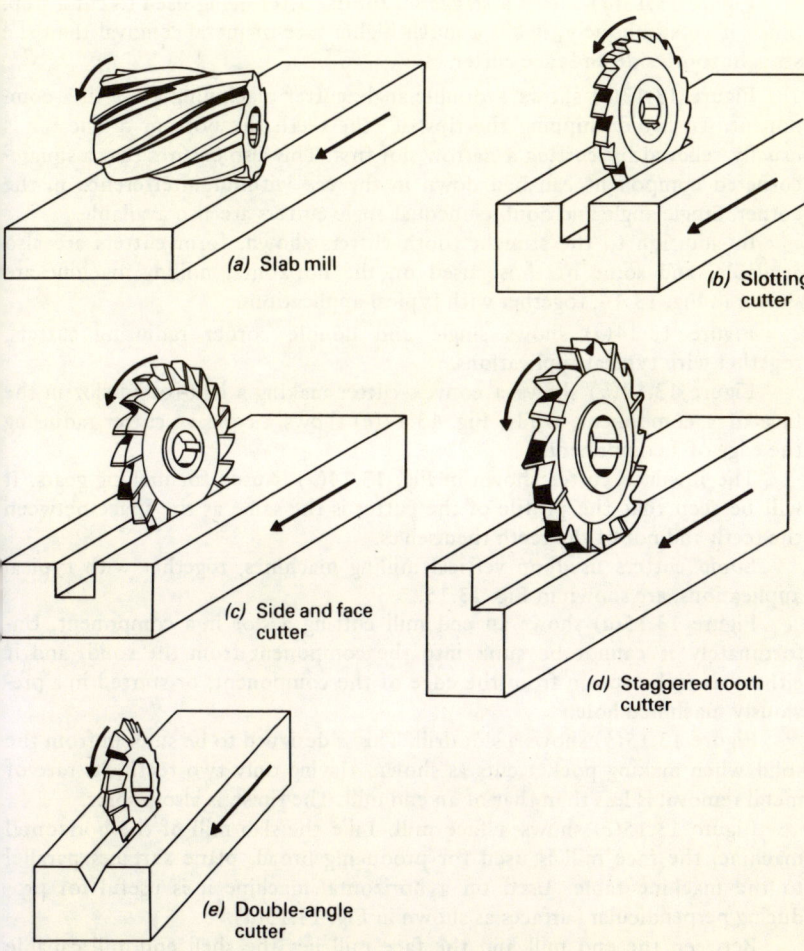

(a) Slab mill

(b) Slotting cutter

(c) Side and face cutter

(d) Staggered tooth cutter

(e) Double-angle cutter

Fig. 13.13 Milling cutter applications (straight tooth)

Figure 13.13(*a*) shows a slab or roller milling cutter. As can be seen, it is used for producing broad plane surfaces parallel to the machine table.

Figure 13.13(*b*) shows a slotting cutter. As its name implies, it is used for machining slots. The plain sides of the cutter help to guide it in the slot and keep it running straight. Care must be taken to limit the rate of metal removal so that the chip clearance between the teeth does not become choked with swarf so that the teeth are broken.

Figure 13.13(*c*) shows a side and face milling cutter. This can also be used for cutting slots and it has a higher rate of metal removal than the slotting cutter. Unfortunately, it is more prone to wander. It can also be used for cutting a step in a component; and since it can cut on both faces at the same time it is much more efficient for cutting out steps than was the single point tool of the shaping machine.

Figure 13.13(*d*) shows a staggered tooth cutter being used to cut a step. Since it cuts obliquely, it has a much higher rate of metal removal than the straight tooth side and face cutter.

Figure 13.13(*e*) shows a double-angle cutter machining a *vee* in a component. To avoid chipping the tips of the teeth the bottom of the *vee* is usually relieved by cutting a narrow slot first. This also ensures that a square-cornered component can bed down in the *vee* without interference in the corner. Single angle and double-unequal angle cutters are also available.

In addition to the straight tooth cutters shown, form cutters are also available, and some of those used on the horizontal milling machine are shown in Fig. 13.14, together with typical applications.

Figure 13.14(*a*) shows single and double corner radiusing cutters, together with typical applications.

Figure 13.14(*b*) shows a convex cutter making a half-round slot in the face of a component, whilst Fig. 13.14(*c*) shows a concave cutter radiusing the edge of a component.

The involute cutter shown in Fig. 13.14(*d*) is used for making gears. It will be seen that the profile of the cutter is the same as the space between the teeth and not of the teeth themselves.

Some cutters used on vertical milling machines, together with typical applications, are shown in Fig. 13.15.

Figure 13.15(*a*) shows an end mill cutting a slot in a component. Unfortunately it cannot be sunk into the component from the solid, and it either has to be run in from the edge of the component, or started in a previously machined hole.

Figure 13.15(*b*) shows a slot drill. This is designed to be sunk in from the solid when making pocket cuts as shown. Having only two teeth, its rate of metal removal is less than that of an end mill. The finish is also poorer.

Figure 13.15(*c*) shows a face mill. Like the slab mill of the horizontal machine, the face mill is used for producing broad, plane surfaces parallel to the machine table. Used on a horizontal machine it is useful for producing perpendicular surfaces as shown in Fig. 13.15(*d*).

Between the end mill and the face mill lies the shell end mill capable of large end-milling applications, or small facing applications. Its name

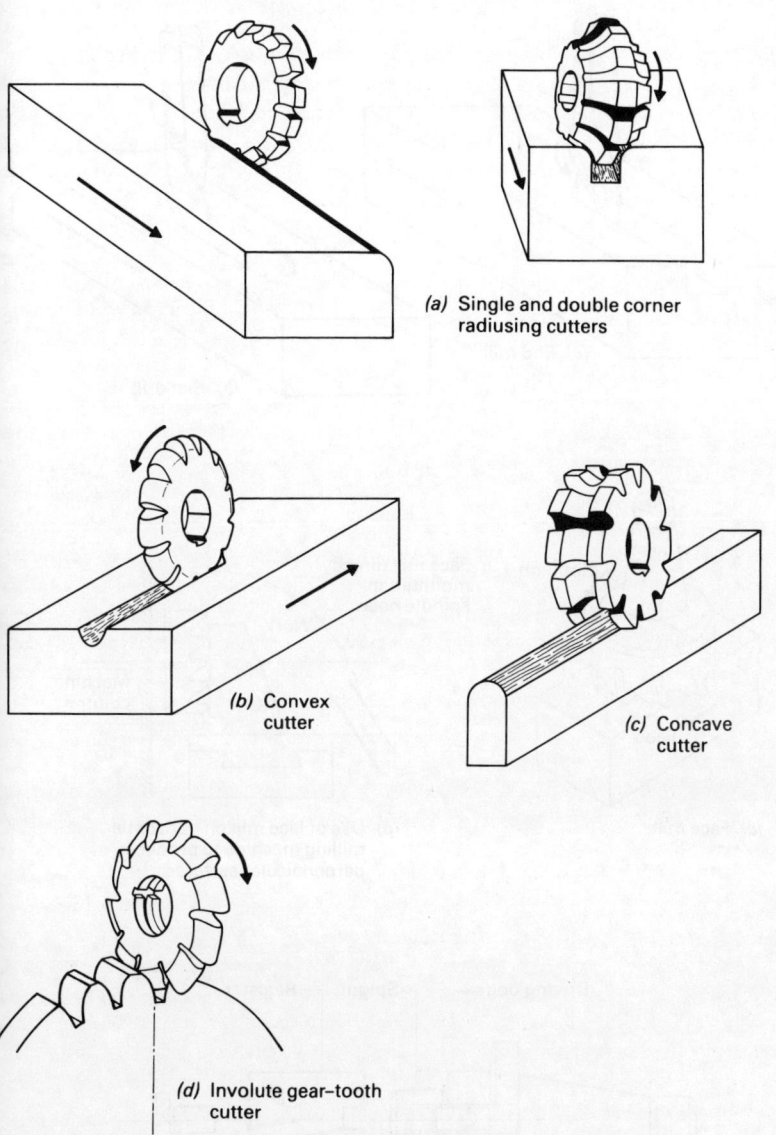

(a) Single and double corner radiusing cutters

(b) Convex cutter

(c) Concave cutter

(d) Involute gear–tooth cutter

Fig. 13.14 Form milling cutter applications

arises from the fact that it is hollow and fits onto a standard arbor as shown in Fig. 13.15(e).

Form cutters are also available for the vertical milling machine and some examples are shown in Fig. 13.16.

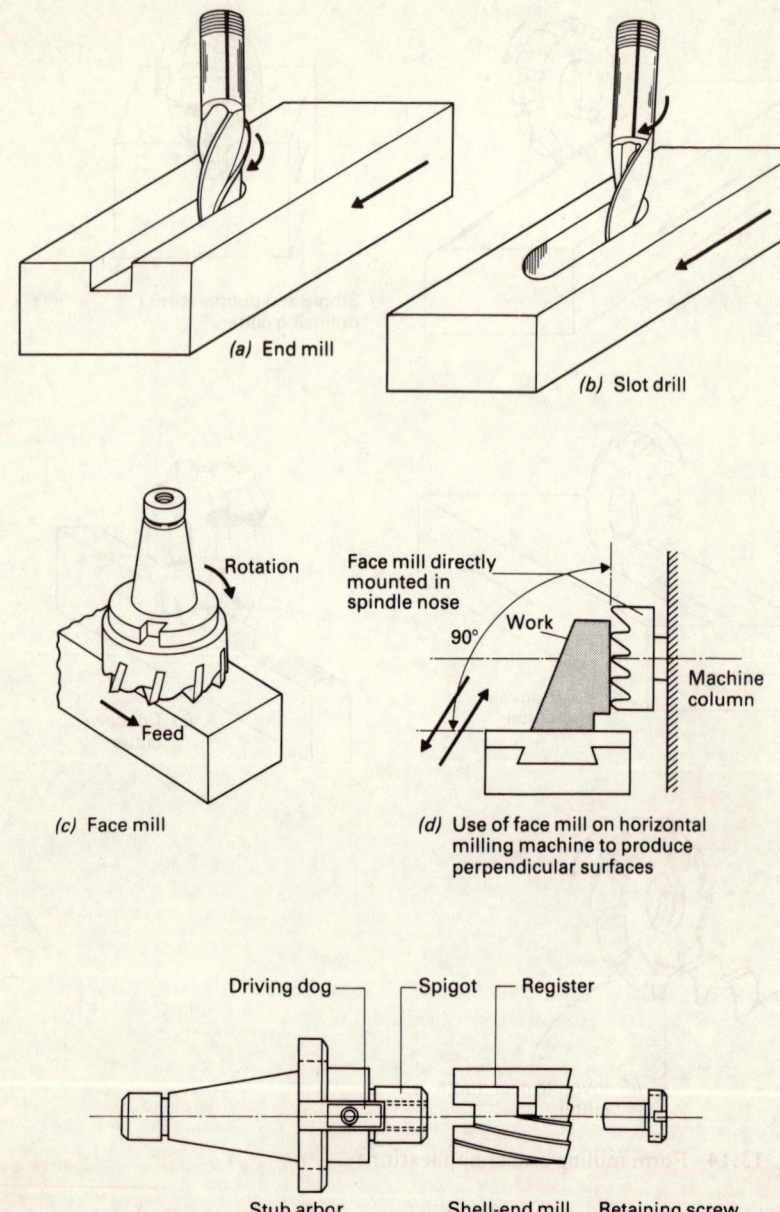

(a) End mill

(b) Slot drill

(c) Face mill

(d) Use of face mill on horizontal milling machine to produce perpendicular surfaces

(e) Shell-end mill

Fig. 13.15 Milling cutters for the vertical milling machine

366

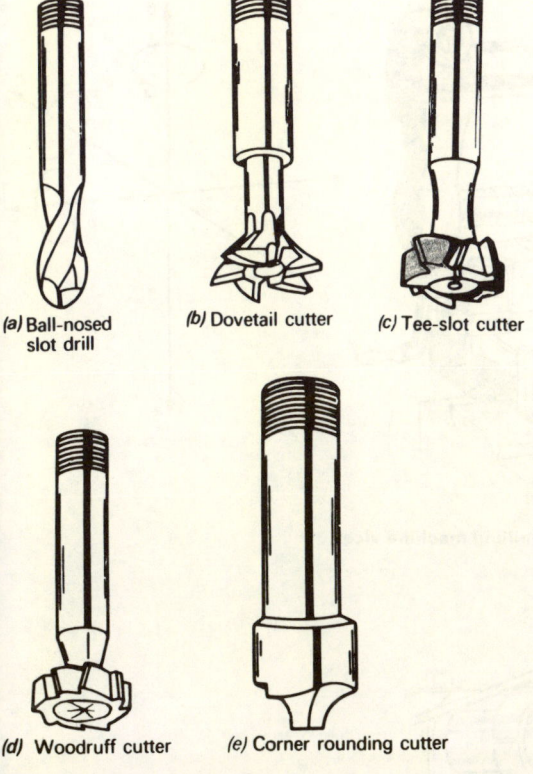

(a) Ball-nosed slot drill   (b) Dovetail cutter   (c) Tee-slot cutter

(d) Woodruff cutter   (e) Corner rounding cutter

Fig. 13.16 Form milling cutters for the vertical milling machine

The ball-nosed slot drill as shown in Fig. 13.16(*a*) is used for machining oil grooves. It is also used on die-sinking machines.

The dovetail cutter shown in Fig. 13.16(*b*) is used, as its name implies, for making the re-entrant cut in a dovetail slideway.

Figure 13.16(*c*) shows a tee-slot cutter and, as its name implies, is used for making the undercut in tee-slots.

Figure 13.16(*d*) shows the woodruff cutter used for making pocket cuts in shafts to take woodruff keys.

Finally, the corner rounding cutter shown in Fig. 13.16(*e*) is used for removing sharp corners from components.

13.8 Work holding

For jobbing applications, the component is usually held in a machine vice as shown in Fig. 13.17(*a*) if the work is small, or directly clamped to the

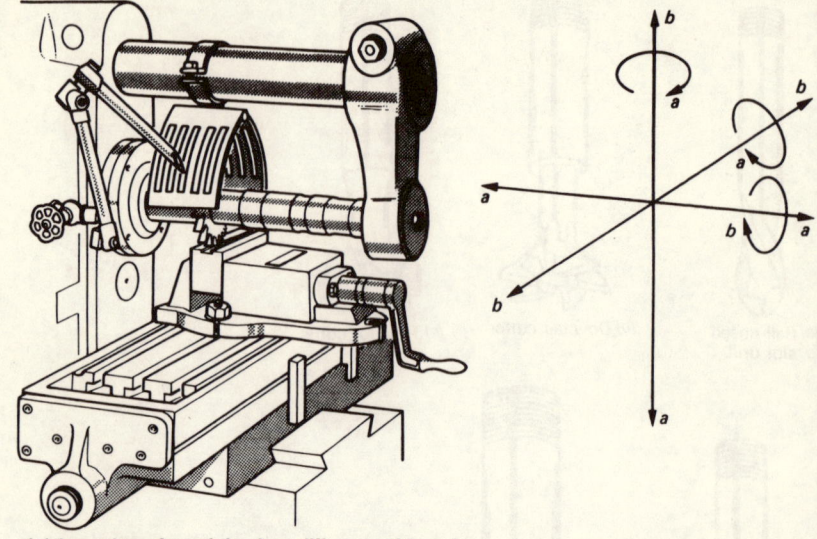

(a) Location of work in the milling machine vice

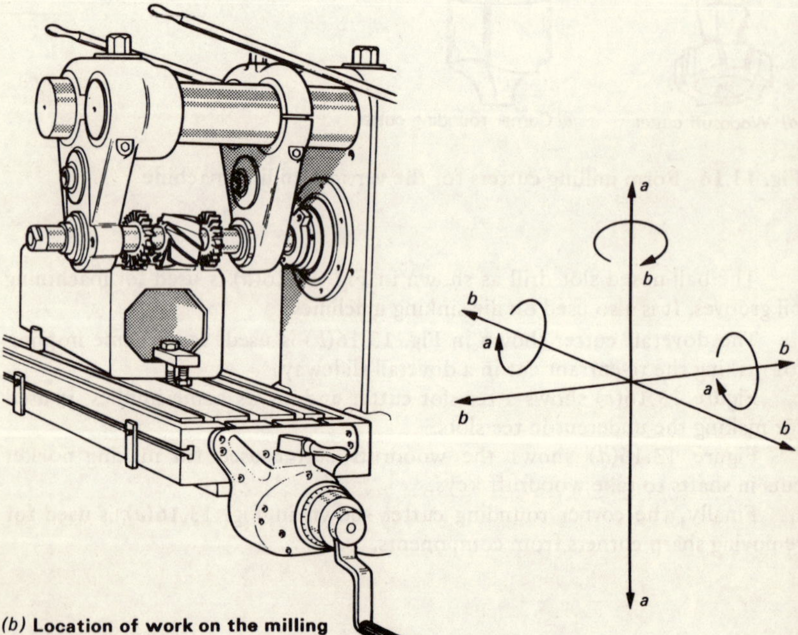

(b) Location of work on the milling
machine table

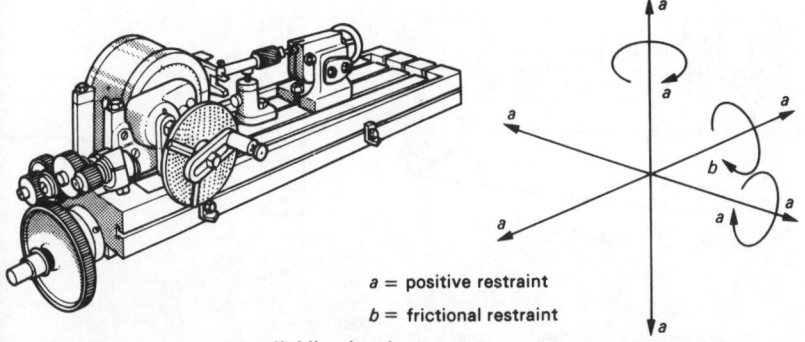

a = positive restraint
b = frictional restraint

(c) **Location of work on the dividing head**

Fig. 13.17 Work-holding on the milling machine

machine table as shown in Fig. 13.17(*b*) if the work is large. This example shows a number of cutters mounted on the machine arbor all being used at the same time. This is called *gang-milling*.

Where a number of cuts have to be made round a component. for example splines, or gear teeth, the work is held in a dividing head as shown in Fig. 13.17(*c*).

For production milling, work-holding fixtures are usually employed. These ensure that the components being machined can be loaded and unloaded quickly, and that they are located in the correct position relative to the cutter every time. Figure 13.18 shows a typical milling fixture. This example is fitted with a setting block. The gap between the setting block and the cutter teeth is set with a specified feeler gauge. The setting block is then removed before cutting commences, and all subsequent components will be located correctly relative to the cutter.

13.9 Some typical applications of milling machines

1. Squaring up a blank on a horizontal milling machine

Figure 13.19(*a*) shows a blank sawn from hot-rolled (black) mild steel bar. This blank is to be squared up all over to the dimensions shown in Fig. 13.19(*b*) on a horizontal milling machine. The cutters that will be used are shown in Fig. 13.19(*c*). The roller or slab mill will be used to square up the sides A, B, C and D, whilst a staggered tooth side and face mill will be used to square up the ends of the blank after the vice has been swung through 90°.

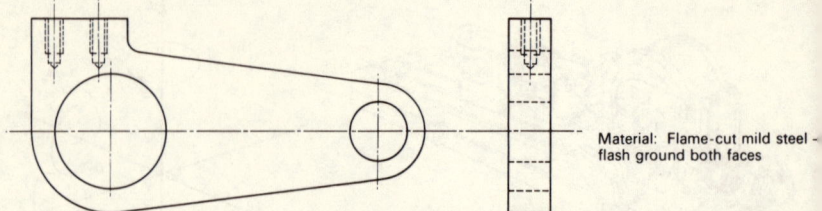

Material: Flame-cut mild steel –
flash ground both faces

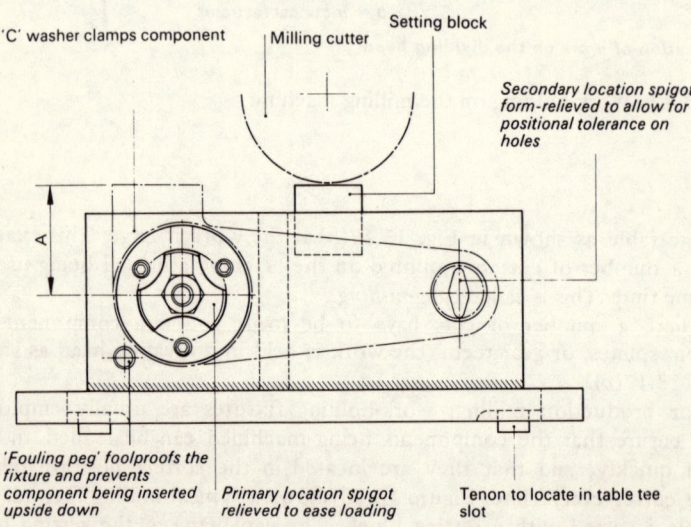

'C' washer clamps component

Milling cutter

Setting block

*Secondary location spigot
form-relieved to allow for
positional tolerance on
holes*

A

'Fouling peg' foolproofs the
fixture and prevents
component being inserted
upside down

Primary location spigot
relieved to ease loading

Tenon to locate in table tee
slot

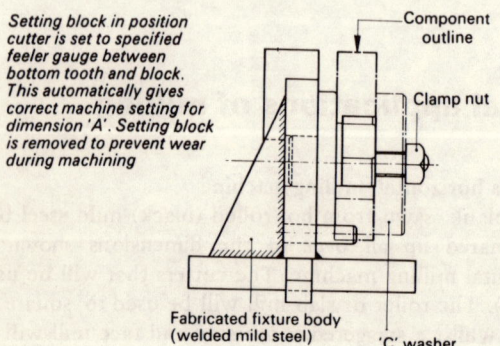

*Setting block in position
cutter is set to specified
feeler gauge between
bottom tooth and block.
This automatically gives
correct machine setting for
dimension 'A'. Setting block
is removed to prevent wear
during machining*

Component
outline

Clamp nut

Fabricated fixture body
(welded mild steel)

'C' washer

Fig. 13.18 Milling fixture

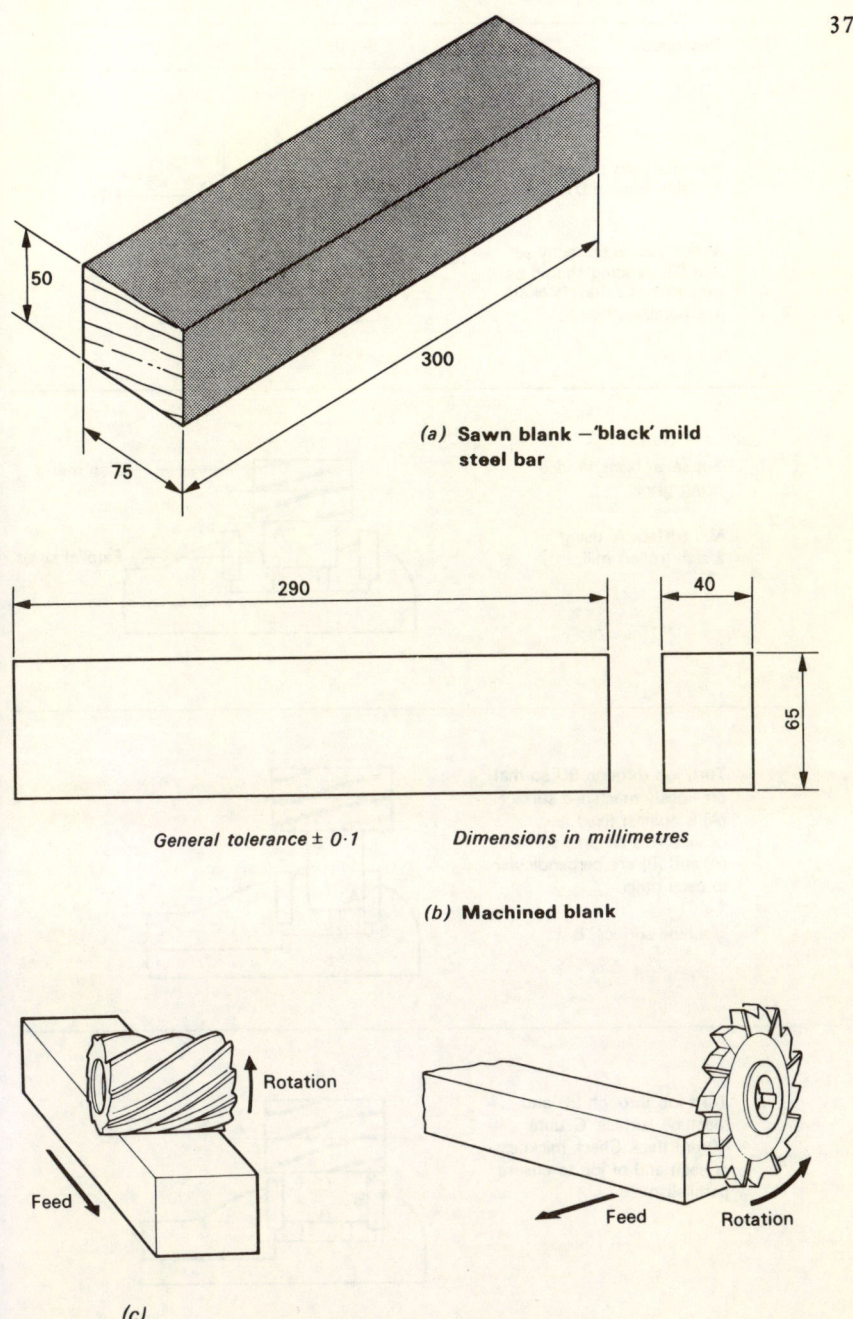

(a) Sawn blank —'black' mild steel bar

General tolerance ± 0·1   *Dimensions in millimetres*

(b) Machined blank

Rotation

Feed

Feed   Rotation

(c)

Fig. 13.19 Cutters used to square up a sawn blank

OP	Description	Set up

| 1 | Set vice jaws parallel to table using a DTI

When vice is correctly set the DTI reading should be constant as it travels along the parallel strip. | 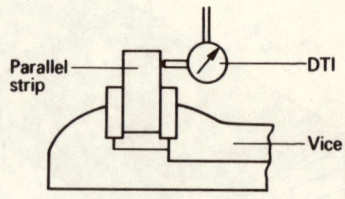 |

| 2 | Set sawn blank in vice using grips.

Mill surface'A' using a slab (roller) mill. | 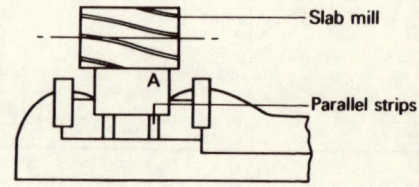 |

| 3 | Turn job through 90° so that previously machined surface (A) is against fixed jaw of vice. This ensures surface (A) and (B) are perpendicular to each other.

Machine surface 'B'. | 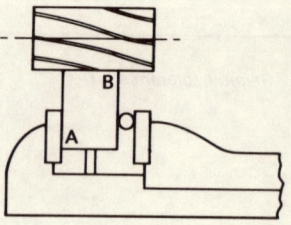 |

| 4 | Turn job through 90° and machine surface 'C' until 40mm thick. Check thickness at each end of job to ensure parallelism. | 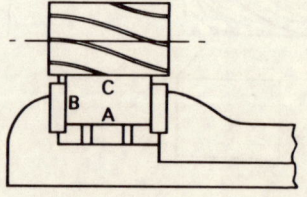 |

Fig. 13.20 Operation sequence to square up a blank on a horizontal milling machine

Op.	Description	Set-up

| 5 | Turn job through 90° again and machine surface 'D' until job is 65mm wide. Check width at each end to ensure parallelism. | 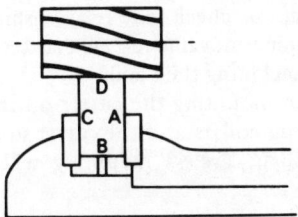 |

| 6 | Turn vice through 90° and check with DTI parallel to spindle axis. | 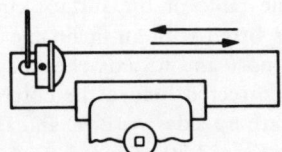 |

| 7 | Use side and face milling cutter to machine end square. | 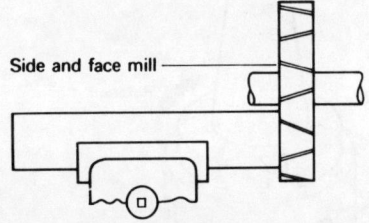 Side and face mill |

| 8 | Wind table across and machine to length. Check length with vernier caliper. | As 7 above |

374 The correct sequence of operations is set out in Fig. 13.20. It will be seen that the top surface of the moving jaw slide of the vice and the face of the fixed jaw provide the datum surfaces for this operation. It is essential, therefore, to check the relationship of these surfaces to the machine table and machine movements. A dial test indicator is used to make these checks before machining the blank.

When mounting the cutter on the arbor, it is essential to clean the arbor, the spacing collars and the cutter so that it runs true. A cutter that is running out through careless mounting will leave a poor finish as well as cutting oversize.

2. Squaring up a blank on a vertical milling machine

The blank shown in Fig. 13.19 can also be squared-up on a vertical milling machine using the cutters shown in Fig. 13.21. The face mill shown in Fig. 13.21(a) should be large enough to span the width of the work and care must be taken to check that the spindle of the machine is perpendicular to the machine table or the surface cut by the face mill will be hollow. If the machine is fitted with an inclinable head, a test arbor should be inserted in the spindle nose and its axis checked with a dial test indicator so that any error can be corrected before the cutter is mounted in place. The face mill is used to square up sides A, B, C and D in the same sequence of operations as described in Fig. 13.20.

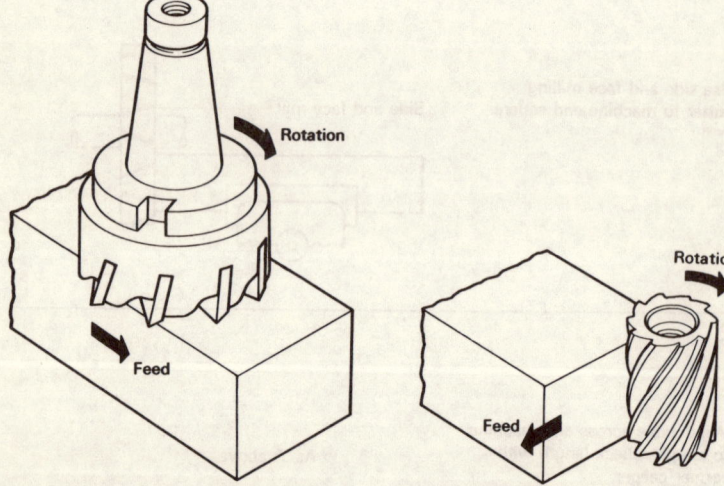

(a) A face mill large enough in diameter to span the width of the work would be used to square up sides A, B, C and D in the same sequences.

(b) A long reach shell end mill is used to square up the end of the blank. There is no need to re-set the vice, as in Fig. 9.36, as use of the cross-feed and slides will guarantee perpendicularity.

Fig. 13.21 Cutters used to square up a blank on a vertical milling machine

A long reach shell end mill is used to square up the ends of the blank, as shown in Fig. 13.21(b). There is no need to rotate the vice through $90°$ and re-set it when using a vertical milling machine. Use of the cross-feed motion and slides will ensure perpendicularity of the ends to the sides of the blank.

3. Milling a vee-jaw

Figure 13.22(a) shows a small vee-jaw from some work-holding device. This component is to be machined from a case-hardening quality mild steel and the dimensions given allow for grinding after case-hardening. Therefore the tolerances are quite generous as the grinding operation will impart the final accuracy. It is often difficult to determine when a vee is the correct size. In this example, the draughtsman has given a dimension over a standard roller. Having centred his double equal-angle milling cutter, the craftsman has merely to increase the depth of cut until the checking dimension is reached. The vee will then be the required size. The draughtsman is not always so obliging and the craftsman may often have to make the calculation himself using trigonometry.

One area of discussion is the slot at the bottom of the vee which provides corner clearance for the component being clamped. The alternative procedures are:

(a) Mill the slot first and then the vee.
(b) Mill the vee first and then the slot.

Method (a) has the advantage of removing the cutting load from the corner of the angle cutter teeth. This is good practice as the corners of the teeth are liable to chip under these conditions. However, it is difficult to cut a deep, narrow slot with a slitting saw and great care has to be taken to prevent the saw from jamming in the slot and breaking. It may also wander in a deep slot.

Method (b) puts more strain on the angle cutter and a number of shallow cuts have to be taken. On the other hand, there is less chance of breaking the slitting saw.

As will be seen from the operation schedule, Fig. 13.22(b), the author prefers method (a) where a double angle cutter is used to cut the vee. A good craftsman will have little trouble with a 6 mm thick slitting saw and, in any case, a broken slitting saw is cheaper to replace than a damaged angle cutter.

Method (b) would be satisfactory if the vee was produced using a more robust cutter, such as a shell end mill, as shown in Fig. 13.22(c). In this alternative set up it will be seen that a vertical spindle milling machine is used with the head inclined at $45°$.

4. Milling a cast iron angle plate

Figure 13.23(a) shows a cast iron angle-plate which is to be machined on the two faces as indicated. This example exploits the ability of a horizontal milling machine to generate a surface at right angles to the machine table when using a face mill on a stub arbor as shown in Fig. 13.23(b). When setting a

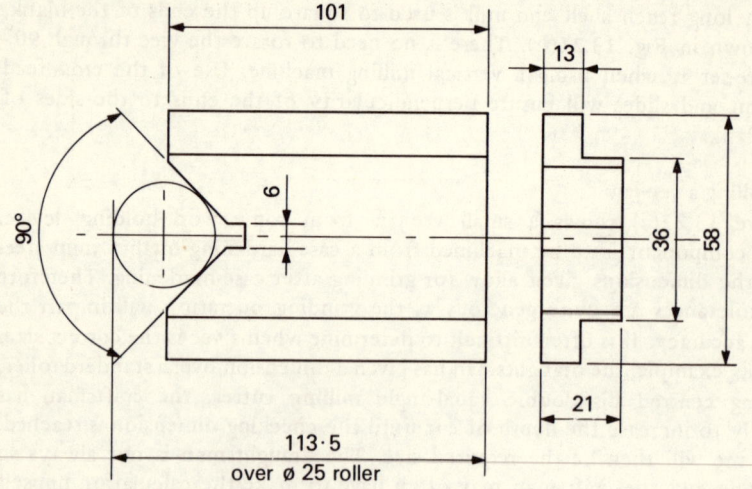

dimensions in millimetres

(a) Vee-jaw

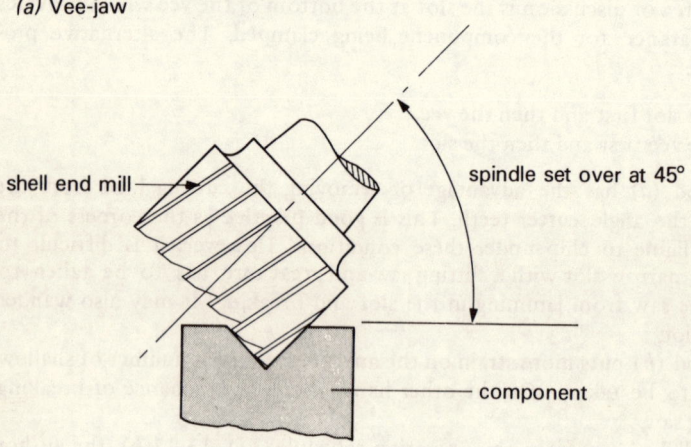

shell end mill

spindle set over at 45°

component

(c) Cutting 90° vee with shell end mill

Fig. 13.22 Milling a vee-jaw

rough casting on a machine table care has to be taken:

(*a*) to ensure that the rough casting does not damage the ground reference surface of the machine table;

(*b*) to arrange the clamping so that the casting is not distorted.

Providing that one surface of the casting is reasonably flat the casting can often be bedded down on several thicknesses of newspaper. This keeps the rough casting off the ground surfaces and provides a reasonable bed for

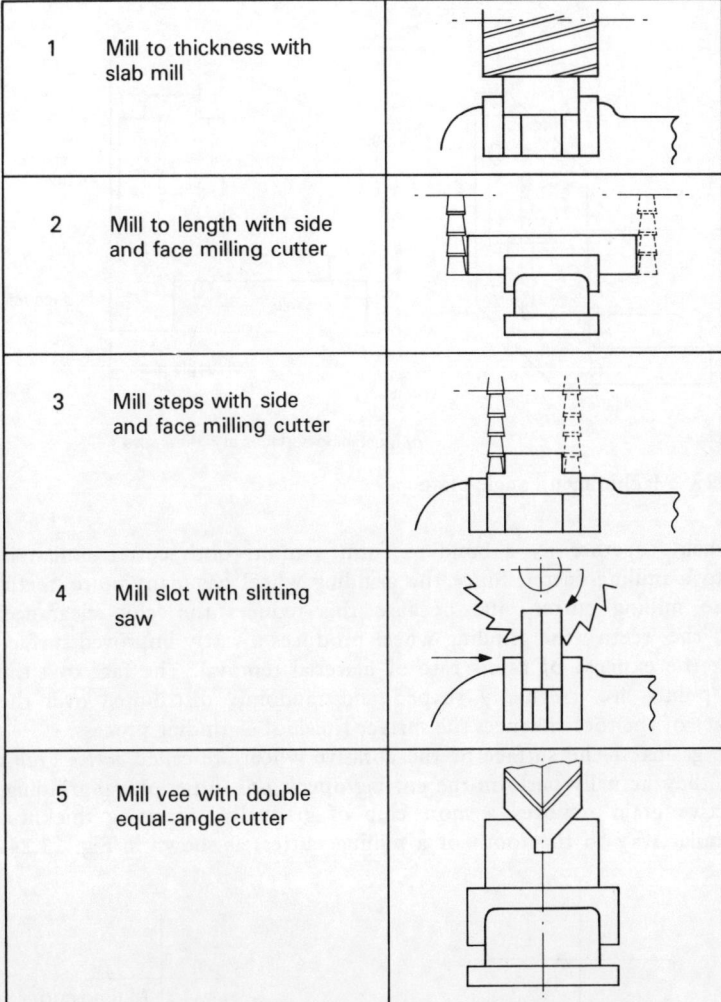

1	Mill to thickness with slab mill	
2	Mill to length with side and face milling cutter	
3	Mill steps with side and face milling cutter	
4	Mill slot with slitting saw	
5	Mill vee with double equal-angle cutter	

(b) Operation sequence for vee-jaw

the casting. However, if the casting is severely distorted, packing may be necessary.

Care must be taken to ensure that, on the first cut, the tips of the cutter teeth are operating below the hard and abrasive skin of the casting. Preferably an inserted, carbide-tipped face mill should be used with cast iron.

13.10 Grinding

Grinding wheels consist of a large number of hard, abrasive particles called

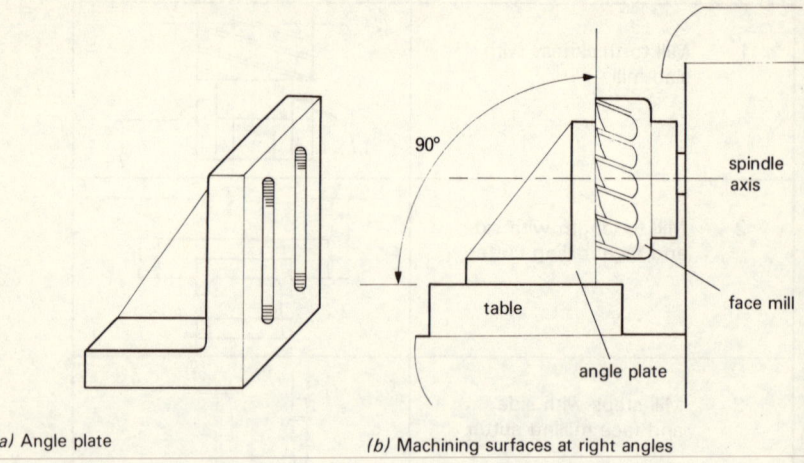

90°

spindle
axis

table

face mill

angle plate

(a) Angle plate

(b) Machining surfaces at right angles

Fig. 13.23 Machining an angle-plate

grains, held together by a bond to form a multi-tooth cutter similar in action to a milling cutter. Since the grinding wheel has many more 'teeth' than the milling cutter, and because this reduces the 'chip clearance' between the 'teeth', the grinding wheel produces a vastly improved surface finish at the expense of a low rate of material removal. The fact that the cutting points are irregularly shaped and randomly distributed over the active face of the tool enhances the surface finish of a grinding process.

The grains at the surface of the abrasive wheel are called *active grains* because they actually perform the cutting operation. In peripheral grinding, each active grain removes a short chip of gradually increasing thickness (in a similar way to the tooth of a milling cutter) as shown in Fig. 13.24.

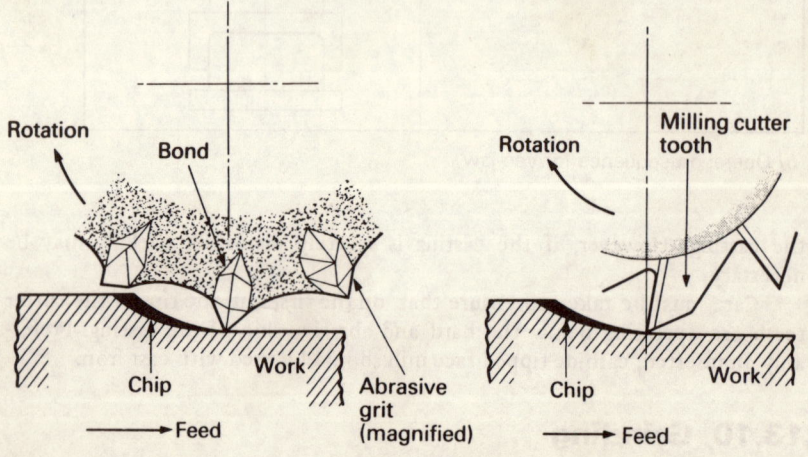

Rotation

Bond

Milling cutter
tooth

Rotation

Chip

Work

Abrasive
grit
(magnified)

Feed

Chip

Work

Feed

Fig. 13.24 Cutting action of abrasive wheel grains

As grinding proceeds, the cutting edges of the grains become dulled and the forces acting on the grains increase until either the dulled grains fracture and expose new cutting surfaces, or the whole of the dulled grains are ripped out from the wheel exposing new active grains. Therefore, grinding wheels have self-sharpening characteristics.

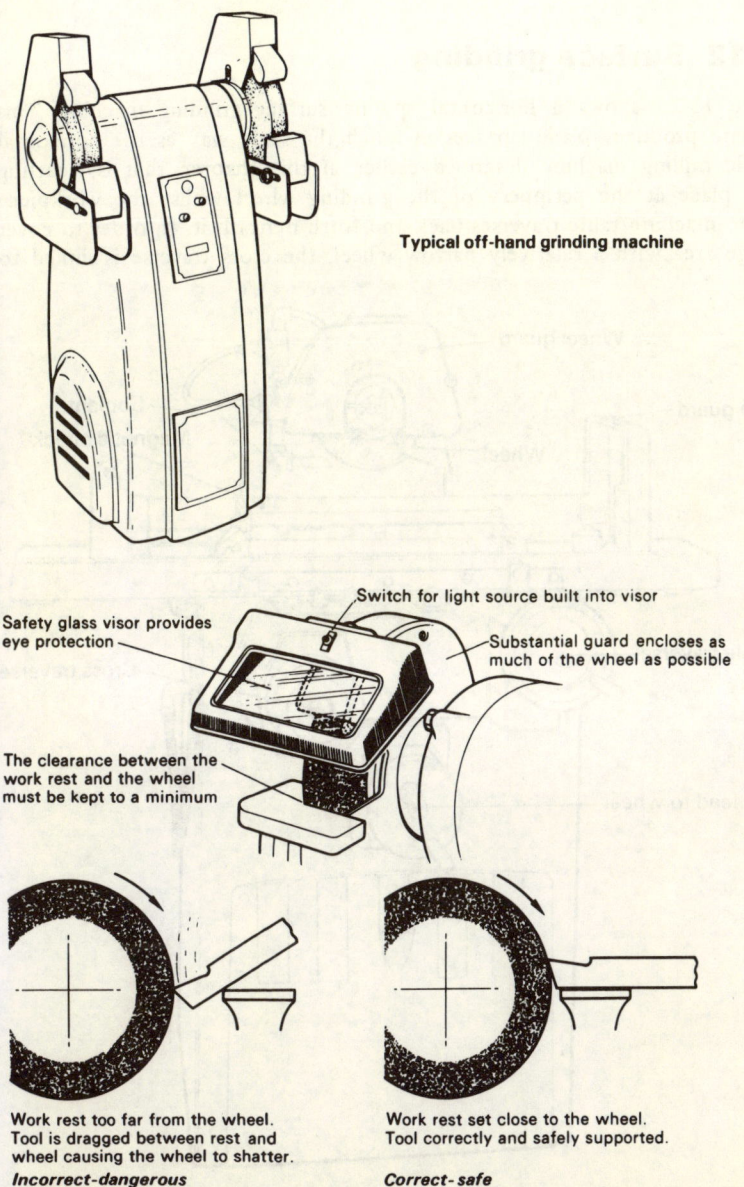

Typical off-hand grinding machine

Switch for light source built into visor

Safety glass visor provides eye protection

Substantial guard encloses as much of the wheel as possible

The clearance between the work rest and the wheel must be kept to a minimum

Work rest too far from the wheel. Tool is dragged between rest and wheel causing the wheel to shatter.
Incorrect-dangerous

Work rest set close to the wheel. Tool correctly and safely supported.
Correct-safe

Fig. 13.25 Double-ended off-hand grinding machine

13.11 The off-hand grinding machine

Figure 13.25 shows a typical double-ended off-hand grinding machine. The use of this type of machine for the re-sharpening of hand tools, single point lathe and shaping machine tools, and twist drills was described in Chapter 9.

13.12 Surface grinding

Figure 13.26 shows a horizontal spindle surface grinding machine. This machine produces plane surfaces in much the same way as the horizontal spindle milling machine described earlier in this chapter; that is, cutting takes place at the periphery of the grinding wheel whilst the workpiece on the machine table traverses back and forth beneath it. In order to cover a large area with a relatively narrow wheel, the cross traverse is linked to

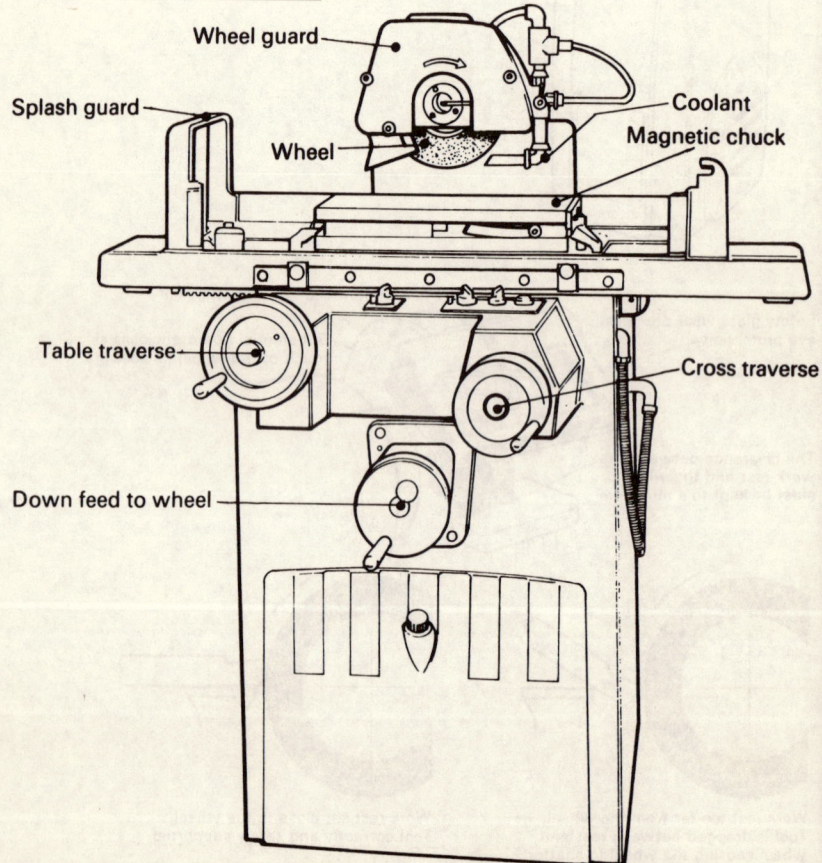

Fig. 13.26 A horizontal spindle surface grinding machine

the longitudinal table traverse so that the table moves across by one feed increment at the completion of each longitudinal traverse cycle.

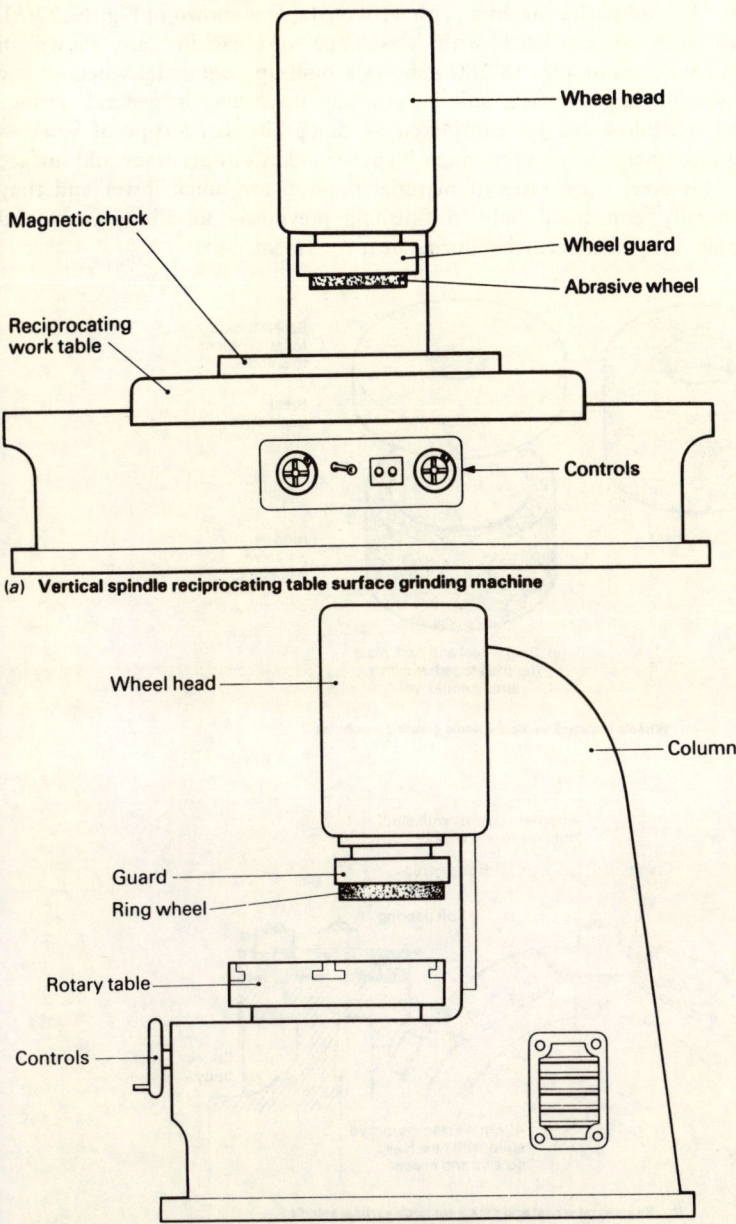

(a) **Vertical spindle reciprocating table surface grinding machine**

(b) **Vertical spindle rotary table surfaces grinding machine**

Fig. 13.27 Typical production surface grinding machines

Figure 13.27(*a*) shows a vertical spindle surface grinding machine with a reciprocating table. This machine produces plane surfaces in much the same way as the vertical spindle milling machine described earlier in this chapter. An alternative machine with a rotary table is shown in Fig. 13.27(*b*). The abrasive wheels used with this type of machine are shown in Fig. 13.28(*a*), whilst Fig. 13.28(*b*) shows a built-up, segmental wheel of the type used on large vertical spindle grinding machines. In general, surface grinding machines can be considered as doing the same type of work as milling machines, but to very much higher standards of accuracy and surface finish. However, their rates of material removal are much lower and they are generally concerned only in finishing previously machined work and correcting distortion occurring during heat treatment.

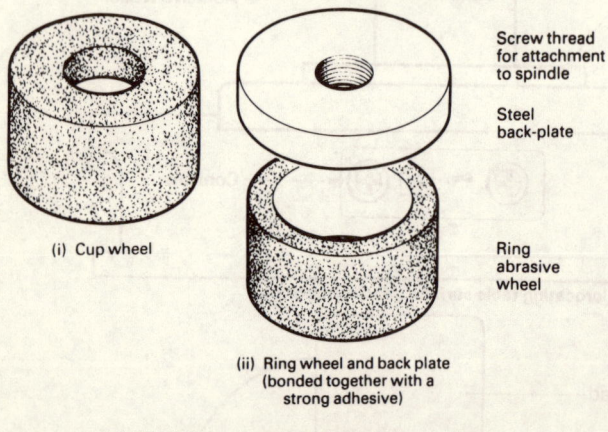

(i) Cup wheel

Screw thread for attachment to spindle

Steel back-plate

Ring abrasive wheel

(ii) Ring wheel and back plate (bonded together with a strong adhesive)

(*a*) **Wheels for small vertical spindle grinding machines**

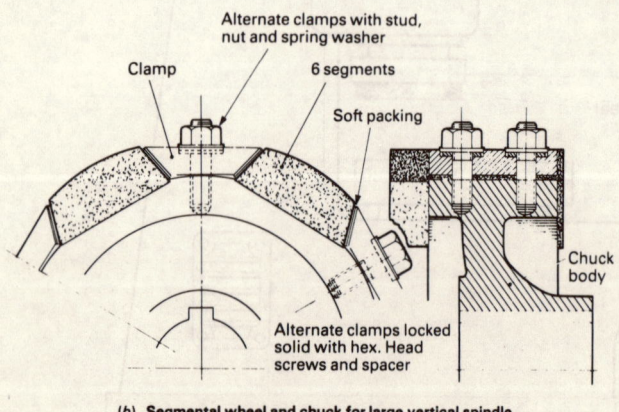

Alternate clamps with stud, nut and spring washer

Clamp

6 segments

Soft packing

Chuck body

Alternate clamps locked solid with hex. Head screws and spacer

(*b*) **Segmental wheel and chuck for large vertical spindle grinding machines**

Fig. 13.28 Grinding wheels for vertical spindle machines

Figure 13.29 shows some typical surface grinding operations.

(a) **Surface grinding a fixture**

(b) **Surface grinding small components** (Note the two strips used to give additional support).

(c) **A stepped component set ready for grinding**

(The vertical and horizontal faces of the step are to be ground at the same setting).

Fig. 13.29 Some typical surface grinding operations

384   Figure 13.30 shows some typical surface grinding operations.

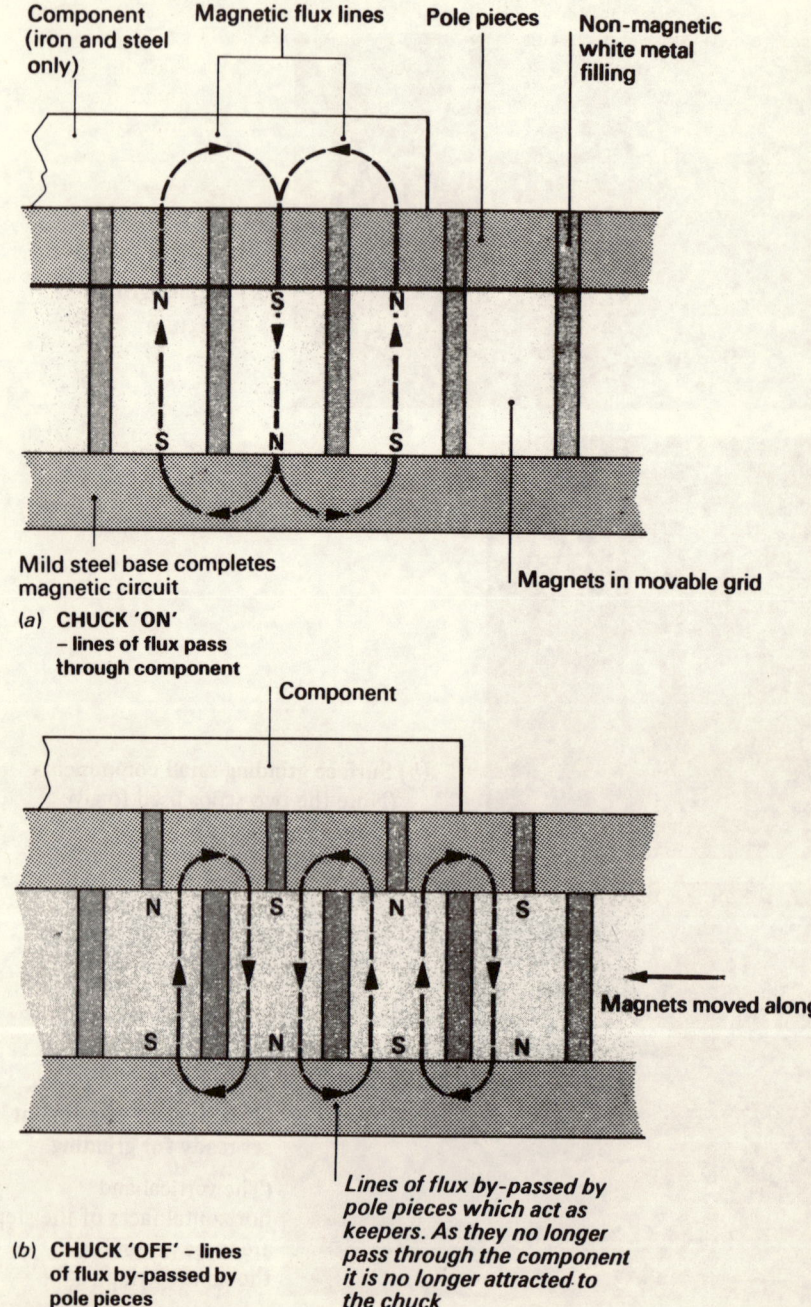

Component (iron and steel only)   Magnetic flux lines   Pole pieces   Non-magnetic white metal filling

N   S   N

S   N   S

Mild steel base completes magnetic circuit

Magnets in movable grid

(a) CHUCK 'ON'
– lines of flux pass through component

Component

N   S   N   S

S   N   S   N

Magnets moved along

Lines of flux by-passed by pole pieces which act as keepers. As they no longer pass through the component it is no longer attracted to the chuck

(b) CHUCK 'OFF' – lines of flux by-passed by pole pieces

Fig. 13.30   Permanent magnetic chuck

The main dissimilarity between grinding and milling lies in the method of work holding. Work holding on surface grinding machines is usually affected by means of a magnetic chuck. Figure 13.30(a) shows a section through such a chuck in the 'ON' position. It can be seen that the lines of flux pass through the workpiece which must be made of magnetic materials. The magnets are mounted in a grid which can be moved sideways by the operating handle. When this is moved to the 'OFF' position as shown in Fig. 13.30(b), the magnetic field no longer passes through the component. The flux field does not hold the component against the cutting forces directly, but provides a friction force between the component and the chuck. It is the friction that prevents the workpiece from moving. Mechanical clamping has to be used for non-magnetic materials.

13.13 Cylindrical grinding

Figure 13.31(a) shows a typical cylindrical grinding machine for grinding the external surfaces of cylindrical components. Such components may be mounted in chucks or between centres in the same way as on a centre lathe. Figure 13.31(b) shows the same machine set up for internal grinding. When

Fig. 13.31 Typical cylindrical grinding machine

(a) External cylindrical grinding between centres

(b) External cylindrical grinding work held in the chuck

(c) Internal cylindrical grinding

Fig. 13.32 Some typical cylindrical grinding operations

turning on the lathe, it is the speed at which the work rotates that controls the cutting speed. However, on the grinding machine it is the speed at which the grinding wheel rotates that controls the cutting speed, and the work speed can be chosen to give the finish and accuracy required.

Figure 13.32 shows some typical cylindrical grinding operations.

Problems

Section A

1. The shaping machine produces plane surfaces using a: (*a*) reciprocating multi-tooth cutter; (*b*) rotating multi-tooth cutter; (*c*) reciprocating single point tool; (*d*) rotating single point tool.
2. The milling machine produces plane surfaces using a: (*a*) reciprocating multi-tooth cutter; (*b*) rotating multi-tooth cutter; (*c*) reciprocating single point tool; (*d*) rotating single point tool.
3. The abrasive wheel used on a grinding machine removes metal by: (*a*) rubbing it off; (*b*) burning it off by friction; (*c*) cutting it off; (*d*) chemical attack.
4. Surface grinding is a finishing operation producing surfaces in a similar manner to a: (*a*) horizontal milling machine; (*b*) vertical milling machine; (*c*) shaping machine; (*d*) centre lathe.
5. The most suitable machine for roughing out a slot in a mild steel blank is a: (*a*) milling machine; (*b*) shaping machine; (*c*) surface grinding machine; (*d*) cylindrical grinding machine.

Section B

6. Compare the advantages and limitations of the *shaping machine* and the *horizontal milling machine* for producing a stepped component from a cast iron blank.
7. Compare and contrast the grinding wheel and the side and face milling cutter as means of removing metal.
8. Sketch suitable components for production on: (*a*) a horizontal milling machine; (*b*) a vertical milling machine; and in each instance explain how the component chosen exploits the particular advantages of each machine.
9. Choose a suitable process for each of the following operations giving reasons for your choice: (*a*) Sharpening a single-point shaping machine tool; (*b*) machining a deep, narrow slot; (*c*) sharpening a press tool die; (*d*) cutting a gear; (*e*) squaring up a die steel blank; (*f*) finish machining a hardened steel shaft; (*g*) making a pocket out for a woodruff key; (*h*) profiling round a press-tool punch to a scribed line.
10. Compare and contrast the cylindrical grinding machine and the centre lathe as means of producing cylindrical components.

Index